Dynamic Characteristics of Layered Structures

This book covers topics such as seismic and surface wave propagation, nonlinear dynamics, heat and mass transfer, convection phenomena, complex fluid mixing, numerical integration techniques, and crack behaviour in piezoelectric materials. Bringing together analytical, numerical, and theoretical and experimental perspectives, this book serves as a comprehensive reference for understanding cutting-edge developments that shape modern scientific and engineering applications. It is an essential resource for researchers, academicians, and professionals engaged in modelling, simulation, and analysis of sophisticated physical systems.

The book features:

- Comprehensive coverage of advanced wave propagation, including Love-type, Rayleigh-type, and seismic wave scattering in layered, anisotropic, and heterogeneous media.
- In-depth studies on heat and mass transfer phenomena, featuring thermoconvective behaviour, double-diffusive convection, and fluid–thermal interactions.
- Robust analytical and numerical methodologies, such as stability analysis, mixed quadrature rules, nonlinear dynamic modelling, and techniques addressing dissipative processes.
- Focused exploration of modern and complex materials, including piezoelectric composites, orthotropic structures, viscoelastic media, and micro-mechanical systems.

It demonstrates interdisciplinary applicability and provides valuable insights for researchers and professionals in applied mathematics, engineering, geophysics, materials science, and computational mechanics. It is primarily written for graduate students and academic researchers in the fields of mechanical engineering, engineering design, civil engineering, materials science, aerospace engineering, and engineering mathematics.

Mathematical Engineering, Manufacturing, and Management Sciences

Series Editor: Mangey Ram, *Professor, Assistant Dean (International Affairs), Department of Mathematics, Graphic Era University, Dehradun, India*

This series publishes research studies and articles that highlight the latest developments and the latest research applied to mathematics and the applications of mathematics in the manufacturing and management sciences areas.

Applied Mathematical Modeling and Analysis in Renewable Energy
Edited by Manoj Sahni and Ritu Sahni

Swarm Intelligence
Foundation, Principles, and Engineering Applications
Abhishek Sharma, Abhinav Sharma, Jitendra Kumar Pandey, and Mangey Ram

Advances in Sustainable Machining and Manufacturing Processes
Edited by Kishor Kumar Gajrani, Arbind Prasad and Ashwani Kumar

Advanced Materials for Biomechanical Applications
Edited by Ashwani Kumar, Mangey Ram and Yogesh Kumar Singla

Biodegradable Composites for Packaging Applications
Edited by Arbind Prasad, Ashwani Kumar and Kishor Kumar Gajrani

Computing and Stimulation for Engineers
Edited by Ziya Uddin, Mukesh Kumar Awasthi, Rishi Asthana and Mangey Ram

Advanced Manufacturing Processes
Edited by Yashvir Singh, Nishant K. Singh and Mangey Ram

Additive Manufacturing
Advanced Materials and Design Techniques
Pulak M. Pandey, Nishant K. Singh and Yashvir Singh

Advances in Mathematical and Computational Modeling of Engineering Systems
Mukesh Kumar Awasthi, Maitri Verma and Mangey Ram

Biowaste and Biomass in Biofuel Applications
Edited by Yashvir Singh, Vladimir Strezov, and Prateek Negi

Lean Manufacturing and Service
Fundamentals, Applications, and Case Studies
Kanchan Das and Miranda Dixon

Multi-Attribute Decision-Making
Theory, Emerging Methods, and Applications
Edited by Rajesh S. Prabhu Gaonkar

Dynamic Characteristics of Layered Structures
Mathematical Modelling, and Numerical Simulation
Edited by Saurabh Kapoor, Anil Negi, and Ahmed Kadhim Hussein

For more information about this series, please visit: www.routledge.com/Mathematical-Engineering-Manufacturing-and-Management-Sciences/book-series/CRCMEMMS

Dynamic Characteristics of Layered Structures

Mathematical Modelling and Numerical Simulations

Edited by
Saurabh Kapoor, Anil Negi,
and Ahmed Kadhim Hussein

CRC Press
Taylor & Francis Group
Boca Raton London New York

CRC Press is an imprint of the
Taylor & Francis Group, an **informa** business

First edition published 2026
by CRC Press
2385 NW Executive Center Drive, Suite 320, Boca Raton FL 33431

and by CRC Press
4 Park Square, Milton Park, Abingdon, Oxon, OX14 4RN

CRC Press is an imprint of Taylor & Francis Group, LLC

ISBN: 978-1-032-74785-9 (hbk)
ISBN: 978-1-032-74784-2 (pbk)
ISBN: 978-1-003-47088-5 (ebk)

DOI: 10.1201/9781003470885

Typeset in Sabon LT Std
by KnowledgeWorks Global Ltd.

Contents

About the editors

Saurabh Kapoor works as an Assistant Professor in the Department of Mathematics at the Regional Institute of Education, Bhubaneswar, India. His research interests include computational fluid dynamics (CFD), hydrodynamic stability, and finite element method. He has published more than 40 research articles in national and international peer-reviewed journals. His taught courses have included real analysis, complex analysis, engineering mathematics, numerical methods, probability and statistics, discrete mathematics, operations research, ordinary and partial differential equations, and advanced calculus. He is a member of international societies such as the International Association of Computer Science and Information Technology (IACSIT), the International Association of Engineers (IAEG), and the International Innovative Scientific and Research Organization.

Anil Negi is currently serving as an Assistant Professor in the Department of Mathematics, Applied Sciences Cluster, School of Advanced Engineering at University of Petroleum and Energy Studies (UPES), Dehradun, Uttarakhand, India. Prior to this, he gained over five years of teaching and research experience as an Assistant Professor at VIT-AP University. His fields of research are computational solid mechanics, geomechanics, theoretical seismology, electrodynamics, piezoelectricity, and fracture mechanics. He has published research papers in international SCI-indexed journals. He has taught courses, including calculus for engineers, applied linear algebra, and applications of differential equations. He is a life member of the International Association of Engineers (IAENG).

Ahmed Kadhim Hussein is currently working as a Professor in the Department of Mechanical Engineering at the University of Babylon, Iraq. His research work includes numerical modelling of various areas of heat transfer, computational fluid dynamics, aerodynamics, nanotechnology, clean energy, and extended irreversible thermodynamics. He has published about 120 papers, most of them in Scopus-indexed papers in many international journals and conferences.

List of contributors

Parvez Alam
Department of Mathematics, School of Advanced Sciences, Vellore Institute of Technology
Vellore, Tamil Nadu, India

Shahad Z. Al-Najjar
Chemical Engineering Department, College of Engineering, Al-Nahrain University
Baghdad, Iraq

Zainab T. Al-Sharify
Department of Oil and Gas Refining Engineering, Al Hikma University College
Baghdad, Iraq

Nidhi Asthana
Department of Mathematics, Faculty of Engineering & Technology, V.B.S. Purvanchal University
Jaunpur, Uttar Pradesh, India

Dwiti Krushna Behera
Department of Mathematics, Ravenshaw University
Cuttack, Odisha, India

V. Dabral
Navajyoti +3 Science College
Sisupalgarh, Bhubaneswar, Odisha, India

Pradeep Gautam
Sr Research Associate Mewbourne School of Petroleum & Geological Engineering, The University of Oklahoma
Norman, Oklahoma, United States

Sayantan Guha
Centre for Data Science, Institute of Technical Education and Research, Siksha 'O' Anusandhan (Deemed to be University)
Bhubaneswar, Odisha, India

Ahmed Kadhim Hussein
College of Engineering, Mechanical Engineering Department, University of Babylon
Babylon City, Hilla, Iraq

Manish Kumar Jha
Mechanical Engineering, UCET-Vinoba Bhave University
Hazaribag, Jharkhand, India

Saurabh Kapoor
Department of Education in Science and Mathematics, National Council of Educational Research and Training, Regional Institute of Education
Bhubaneswar, Odisha, India

Siddhartha Koley
Department of Mathematics & Computing, IIT(ISM)
Dhanbad, Jharkhand, India

Pulkit Kumar
Center for Nondestructive Evaluation (CNDE), Iowa State University (ISU)
Ames, Iowa, USA

Raj Kumar
Department of Mathematics, Faculty of Engineering & Technology, V.B.S. Purvanchal University
Jaunpur, Uttar Pradesh, India

Moumita Mahanty
Centre for Data Science, Institute of Technical Education and Research, Siksha 'O' Anusandhan (Deemed to be University)
Bhubaneswar, Odisha, India

Kumari Manju
Faculty of Engineering and Technology, Siksha 'O' Anusandhan (Deemed to be University)
Bhubaneswar, Odisha, India

Tabinda Nahid
Department of Mathematics, School of Advanced Sciences, Vellore Institute of Technology
Vellore, Tamil Nadu, India

Suman Nain
Department of Mathematics, Indraprastha College for Women, University of Delhi
New Delhi, India

Durgaprasad Nayak
Department of Mathematics, Utkal University, Vani Vihar
Bhubaneswar, Odisha, India

Anil Negi
Department of Mathematics, Applied Sciences Cluster, University of Petroleum and Energy Studies (UPES)
Dehradun, Uttarakhand, India

Helen Onyeaka
School of Chemical Engineering, University of Birmingham
Edgbaston, Birmingham, United Kingdom

Pranav P
Department of Mathematics, School of Advanced Sciences (SAS), Vellore Institute of Technology
Chennai, Tamil Nadu, India

Venkatesan P
Department of Mathematics, School of Advanced Sciences, Vellore Institute of Technology
Vellore, Tamil Nadu, India

Kamlesh Kumar Pankaj
Government Polytechnic Kaimur, DSTTE Government of Bihar, Mohania
Bihar, India

Sarita Pippal
Department of Mathematics, Punjab University
Chandigarh, India

Amit Kumar Rahul
Department of Mathematics, School of Advanced Sciences (SAS), Vellore Institute of Technology
Chennai, Tamil Nadu, India

Sachindra Kumar Rout
Department of Mechanical Engineering, C.V. Raman Global University
Bhubaneswar, Odisha, India

Rostam Kareem Saeed
Department of Mathematics, College of Science, Salahaddin University-Erbil
Erbil, Iraq

A. K. Sahoo
Regional Institute of Education, Utkal University, Vani Vihar
Bhubaneswar, Odisha, India

Siprarani Sahoo
Department of Mathematics, Utkal University
Bhubaneswar, Odisha, India

Abhishek Kumar Singh
Department of Mathematics & Computing, IIT(ISM)
Dhanbad, Jharkhand, India

Ajeet Kumar Singh
College of Post Harvest Technology and Food Processing, Sardar Vallabhbhai Patel University of Agriculture and Technology
Meerut, Uttar Pradesh, India

Manoj Kumar Singh
P.G. Department of Mathematics, Magadh University
Bodh Gaya, Bihar, India

Sonam Singh
Department of Applied Mechanics, IIT Delhi
Delhi, India

Tusar Singh
Department of Mathematics, Institute of Technical Education and Research, Siksha 'O' Anusandhan (Deemed to be University)
Bhubaneswar, Odisha, India

Dig Vijay Tanwar
Department of Mathematics, Graphic Era (Deemed to be University)
Dehradun, Uttarakhand, India

Pooja Raj Verma
Department of Mathematics, Shaheed Bhagat Singh College, University of Delhi
Delhi, India

Sumit Kumar Vishwakarma
Department of Mathematics, BITS-Pilani, Hyderabad Campus
Hyderabad, Telangana, India

Ram Prasad Yadav
Department of Mathematics, SRM University
Sonepat, Haryana, India

Chapter 1

Love-type waves in heterogeneous Kelvin–Voigt-type viscoelastic media with coulomb frictional interface

Kamlesh Kumar Pankaj, Manoj Kumar Singh, and Manish Kumar Jha

1.1 INTRODUCTION

The Earth comprises different layers that generally display the viscoelastic behavior. Viscoelastic property of the Earth is due to the geophysical activity which involves the time interval ranging from a few years to millions of years to operate. Movements of tectonic plates or volcanic eruptions give rise to low-frequency earthquakes but the effect of these phenomena is minimized to some extent by the Earth's viscoelastic layers. This process is influenced by the varying temperature conditions as well as the gravity of the Earth. The response of viscoelastic property of Earth is an area of comprehensive interest for many researchers as the viscoelasticity results from the long-term arrangement of the Earth's rheology. The study of Love-type waves in different viscoelastic media give insight into the damping (or attenuation) behavior along with the dispersion behavior of the seismic waves [1]. Viscoelastic medium diminishes the destructive effect of vibrational energy by absorbing the shock created by seismic waves. Attenuation corresponds to the dissipation of mechanical energy in the mechanical system. Therefore, problems concerned with vibration damping have been a subject of comprehensive interest of scientists and researchers from several decades [2]. Carcione [3] investigated the wave propagation characteristic in viscoelastic model by considering Boltzmann's superposition principle in the equation of motion. Mazzotti et al. [4] analyzed the guided wave propagation in pre-stressed viscoelastic layer. Effect of harmonic oscillation on viscoelastic layer of Kelvin–Voigt-type by using the Fourier integral transform is studied by Sumbatyan et al. [5]. Torsional wave and Love-type wave propagation studies in interfaced orthotropic and viscoelastic composite structures were investigated by Singh and Alam [6, 7] and Singh et al. [8]. Propagation behavior of SH-waves in layered structure considering Kelvin–Voigt-type viscoelastic layer over inhomogeneous isotropic half-space has been analyzed by Majhi et al. [9]. Recently, Ren et al. [10] investigated dispersion and attenuation of the Rayleigh wave in functionally graded Kelvin–Voigt viscoelastic model by combining fractional

DOI: 10.1201/9781003470885-1

derivatives with the power series technique. In order to properly analyze the internal structure of the Earth, the consideration of heterogeneity in its layers is crucial for seismologists and geoscientists [11, 12]. In the practical scenario, the mechanical properties of the geo-medium (Earth) alter gradually in the vertical direction which causes heterogeneity in the geo-medium. To better comprehend the practical aspects of the problem, the present study deals with the exponential heterogeneity inside the medium. Due to the involvement of various geo-mechanical processes viz. process of creep and gravitational force, initial stress already exist in the Earth. Biot [13] discussed the effect of initial stress on elastic waves. He enlightened the theory that the initial stress is the component of internal equilibrium. Thus, introducing initial stress in any model provides the more general result as compared to results observed in a model without initial stress. Several literatures are available regarding the study of wave propagation in pre-stressed medium [14–16]. Singh et al. [17] have discussed the moving punch induced Love-type wave propagation in a pre-stressed viscoelastic medium. Kumari et al. [18] analyzed dissipation and attenuation curves of point source–induced shear waves in magneto-viscoelastic layered media. Propagation of Love waves in a fiber-reinforced medium sandwiched between an isotropic layer and elastic half-space under the influence of gravity was discussed by Kundu et al. [19]. Singh et al. [20] have investigated generalized Rayleigh waves in a pre-stressed piezoelectric medium.

Fiber-reinforced materials are stronger and stiffer anisotropic materials in which reinforced material is merged together with matrix materials to act as single unit. The Earth's crust contains both soft and strong rocks that can be modeled as fiber-reinforced media. Human-made reinforced materials include reinforced cement concrete (RCC), which is extensively utilized in the construction of buildings, bridges, dams, and other infrastructure. These materials are characterized by their high energy absorption capacity, structural stiffness, relatively low weight, and enhanced resistance to corrosion, making them suitable for durable and load-bearing applications. Construction and sustainability of reinforced structures like embankments and bridges are the major areas of thrust regarding its interaction with the seismic wave propagation. A layer of geotextiles or geogrids are also applied to provide reinforcement and separate this structure from the soft and native soil. These materials allow the disturbance to transmit through them with ease, so these materials are used in construction to minimize the effect of shaking and vibrations caused by seismic waves and human-made blasting. Surface wave propagation in layered self-reinforced media is studied by Chattopadhyay et al. [21]. The frequency relation for various surface waves in electrically and magnetically conducting fiber-reinforced medium is investigated by Acharya and Roy [22]. Belfield et al. [23] discussed the stress analysis for the fiber-reinforced material. Effective elastic modulus for fiber-reinforced material is given by Hashin and Rosen [24]. Plane wave in generalized fiber-reinforced thermo-elastic media is investigated by Singh [25]. By considering the reinforced concrete building specimen, Ebrahimian et al. [26] studied the shear wave propagation in viscoelastic medium for structural health monitoring. Recently Srivastava et al. [27] investigated the influence of frictional boundary on the dispersion of Rayleigh-type wave in pre-stressed fiber-reinforced layer overlying a homogeneous semi-infinite medium.

An elastic medium shows different types of anisotropic nature depending upon the elastic properties of the medium. Orthotropic materials are the special class of anisotropic materials which exhibit unique characteristic based on three principle axes and nine independent elastic constants. Several literatures are available which emphasize the wave propagation aspects in an orthotropic medium with numerous applications. Singh et al. [28] examined the elastic wave propagation in orthotropic half-space with rotation. Vinh et al. [29] gave the exact secular equation for the seismic waves in the orthotropic medium. Effect of pre-stressed functionally graded orthotropic medium on the guided wave velocity is given by Yu and Zhang [30]. The assumption of a perfectly bonded interface of two media may not be always true. The creep process in viscoelastic materials may debilitate the interface of two media and cause to be frictional interface. The aging of adhesive bonds, the presence of impurities, material diffusion, or other types of degradation can all lead to the development of frictional surfaces between two materials in human-made structures. The distribution of stress at the contacting fractional surface has a significant impact on the wave propagation [31]. Miller [32] examined the Love-type wave propagation in elastic media with frictional interface. Li et al. [33] addressed the seismic wave propagation in viscoelastic rock mass that are joined but not continuous. Effect of sliding interface in between two isotropic elastic medium is studied by Bartolomeo et al. [34]. Singh et al. [35, 36] used the WKB (Wentzel–Kramers–Brillouin) approach to provide an asymptotic estimation for the frequency of Love waves in piezo-composite structures. Kaur et al. [37] have examined the effect of imperfect bonding on the dispersion and damping behavior of shear waves in heterogeneous viscoelastic layer bonded imperfectly with micropolar elastic half-space.

The present problem has been framed out to study the Love-type wave propagation in initially stressed heterogeneous viscoelastic layered structure. The layered structure consists of self-reinforced layer bedded over the orthotropic half-space. The interface between two media is assumed as Coulomb frictional interface. Viscoelasticity model of Kelvin–Voigt-type is taken into consideration. The frequency relation is obtained in a complex form. Prominent effects of various parameters are plotted to study the dispersive and damping nature of Love-type waves in considered media.

1.2 MATHEMATICAL FORMULATION AND SOLUTION

Examining a reference Cartesian coordinate system, we can see that Figure 1.1 defines the origin. Fiber-reinforced heterogeneous viscoelastic layer occupying

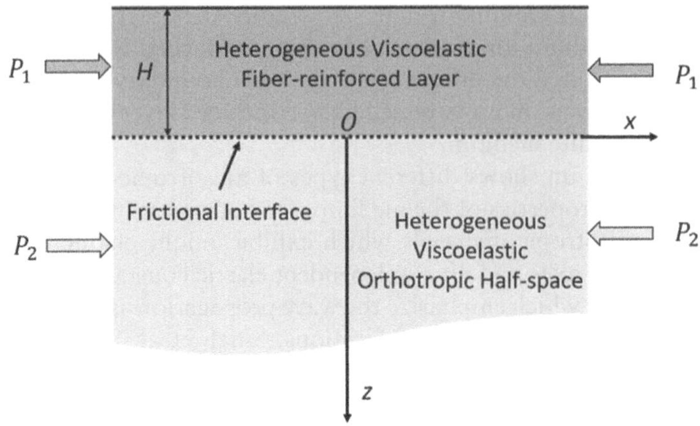

Figure 1.1 Geometric configuration of the problem.

the domain $\Omega^F = \{(x,z);0 \geq z \geq -H\}$ is embedded over orthotropic heterogeneous viscoelastic half-space with domain $\Omega^O = \{(x,z);0 \leq x < \infty\}$. Both media are under the influence of horizontal component of initial stresses P_1 and P_2, respectively. The two media are bonded by Coulomb friction. The Kelvin–Voigt model of the viscoelasticity is considered. In both media, heterogeneity is thought to change exponentially with depth (along z-direction). Love-type surface waves travel in the x-direction and exhibit shear motion polarized along the y-direction.

For the propagation of Love-type waves, following assumptions are made

$$u = w = 0, \quad v = v(x,z,t), \quad \frac{\partial}{\partial y} \equiv 0, \tag{1.1}$$

where the components of displacement u,v,w are measured along the x,y, and z directions, respectively.

1.2.1 For fiber-reinforced heterogeneous viscoelastic layer

The elastodynamic equation of motion in the medium under the horizontal component of initial stress P_1 is given by Biot [38] as following:

$$\frac{\partial \tau_{xy}}{\partial x} + \frac{\partial \tau_{yz}}{\partial z} - \frac{P_1}{2}\frac{\partial^2 v_1}{\partial x^2} = \rho_1 \frac{\partial^2 v_1}{\partial t^2}, \tag{1.2}$$

where v_1 is the displacement component along y-direction, $\tau_{ij}(i,j = x,y,z)$ represents incremental stress components, and ρ_1 is the density of fiber-reinforced layer.

The constitutive relations with direction of reinforcement along unit vector $\vec{a} = (a_x, a_y, a_z)$ are [39]

$$\tau_{xy} = \mu_T^* \frac{\partial v_1}{\partial x} + \left(\mu_L^* - \mu_T^*\right)\left(\frac{\partial v_1}{\partial x} a_x^2 + a_x a_z \frac{\partial v_1}{\partial z}\right),$$
$$\tau_{yz} = \mu_T^* \frac{\partial v_1}{\partial z} + \left(\mu_L^* - \mu_T^*\right)\left(a_z^2 \frac{\partial v_1}{\partial z} + a_x a_z \frac{\partial v_1}{\partial x}\right). \tag{1.3}$$

where μ_L^* and μ_T^* are elastic parameters of the medium.

Since the medium exhibits Voigt-type viscoelasticity, elastic parameters are regarded as

$$\mu_L^* = \mu_L + \bar{\mu}_L \frac{\partial}{\partial t}, \quad \mu_T^* = \mu_T + \bar{\mu}_T \frac{\partial}{\partial t}, \tag{1.4}$$

where μ_L, μ_T are shear modulus and $\bar{\mu}_L, \bar{\mu}_T$ are internal frictions of the layer in longitudinal and transverse direction of reinforcement, respectively.

Consideration of the layer's heterogeneity has taken in the following shape:

$$\mu_L = \mu_L^{(1)} e^{b_1 z}, \ \bar{\mu}_L = \bar{\mu}_L^{(1)} e^{b_1 z}, \ \mu_T = \mu_T^{(1)} e^{b_1 z}, \ \bar{\mu}_T = \bar{\mu}_T^{(1)} e^{b_1 z}, \ P_1 = P_1^{(1)} e^{b_1 z} \ \ \rho_1 = \rho_1^{(1)} e^{b_1 z}. \tag{1.5}$$

where $\mu_L^{(1)}, \bar{\mu}_L^{(1)}, \mu_T^{(1)}, \bar{\mu}_T^{(1)}, P_1^{(1)}, \rho_1^{(1)}$ are material properties at $z = -H$ and b_1 is the heterogeneity parameter having dimension inverse of length.

We assume the solutions for Eq. (1.2) in harmonic form as

$$v_1 = V_1(z) e^{i(pt - kx)}, \tag{1.6}$$

where V_1 is unknown functions, k is wave number, p is frequency parameter.

Using Eqs. (1.3)–(1.6) in Eq. (1.2), the following differential equation is obtained

$$A \frac{d^2 V_1}{dz^2} + B \frac{dV_1}{dz} + C\, V_1(z) = 0, \tag{1.7}$$

where A, B, C are given in Appendix.

The solution of Eq. (1.7) is given by

$$V_1(z) = e^{Q_1 z}\left(C_1 \cos Q_2 z + C_2 \sin Q_2 z\right) \tag{1.8}$$

where $Q_1 = \frac{-B}{2A}$, $Q_2 = \frac{\sqrt{4AC - B^2}}{2A}$ and C_1, C_2 are arbitrary constants.

Using Eq. (1.8) in Eq. (1.6), the displacement components for the fiber-reinforced layer can be expressed as

$$v_1 = e^{Q_1 z}\left(C_1 \cos Q_2 z + C_2 \sin Q_2 z\right)e^{i(pt-kx)}. \tag{1.9}$$

1.2.2 For orthotropic heterogeneous viscoelastic half-space

The equation of motion in the medium, considering the horizontal component of the initial stress P_2, is given by

$$\frac{\partial \sigma_{xy}}{\partial x} + \frac{\partial \sigma_{yz}}{\partial z} - \frac{P_2}{2}\frac{\partial^2 v_2}{\partial x^2} = \rho_2 \frac{\partial^2 v_2}{\partial t^2}, \tag{1.10}$$

where v_2 is the displacement component along y-direction, σ_{ij} $(i, j = x, y, z)$ represents incremental stress components, and ρ_2 is the density of orthotropic half-space.

The stress–displacement relationships for the half-pace is given by [40]

$$\sigma_{xy} = C_{66}^{*}\frac{\partial v_2}{\partial x}, \quad \sigma_{yz} = C_{44}^{*}\frac{\partial v_2}{\partial z}, \tag{1.11}$$

where $C_{66}^{*} = C_{66} + \bar{C}_{66}\frac{\partial}{\partial t}, C_{44}^{*} = C_{44} + \bar{C}_{44}\frac{\partial}{\partial t}; C_{44}, C_{66}$ are elastic coefficient and $\bar{C}_{44}, \bar{C}_{66}$ are internal frictions due to viscoelasticity of the half-space.

The heterogeneity in the half-space has been considered as follows

$$C_{44} = C_{44}^{(1)}e^{b_2 z},\ \bar{C}_{44} = \bar{C}_{44}^{(1)}e^{b_2 z},\ C_{66} = C_{66}^{(1)}e^{b_2 z},\ \bar{C}_{66} = \bar{C}_{66}^{(1)}e^{b_2 z},\ \rho_2 = \rho_2^{(1)}e^{b_2 z},\ P_2 = P_2^{(1)}e^{b_2 z}, \tag{1.12}$$

where $C_{44}^{(1)}, \bar{C}_{44}^{(1)}, C_{66}^{(1)}, \overline{C_{66}}^{(1)}, P_2^{(1)}, \rho_2^{(1)}$ are material properties at $z = 0$ and b_2 is the heterogeneity parameter.

The solution of Eq. (1.10) can be considered in the following form

$$v_2 = V_2(z)e^{i(pt-kx)}, \tag{1.13}$$

where V_2 is the unknown function.

Using Eqs. (1.11)–(1.13) in Eq. (1.10), the following differential equation is obtained as

$$\frac{d^2 V_2}{dz^2} + b_2\frac{dV_2}{dz} + D V_2 = 0, \tag{1.14}$$

where D is given in Appendix.

The solution of Eq. (1.14) is given by

$$V_2(z) = e^{\frac{-b_2}{2}z}\left(D_1 e^{\frac{\sqrt{Q_3}}{2}z} + D_2 e^{\frac{-\sqrt{Q_3}}{2}z}\right), \tag{1.15}$$

where $Q_3 = b_2^2 - 4D$, D_1 and D_2 are arbitrary constants.

In view of the condition $\lim_{z\to\infty} v_2 \to 0$, the appropriate solution of Eq. (1.14) is given by

$$V_2(z) = D_2 e^{\frac{-(b_2+\sqrt{Q_3})}{2}z}. \tag{1.16}$$

From Eq. (1.13) the displacement component for the orthotropic half-space is obtained as

$$v_2 = D_2 e^{\frac{-(b_2+\sqrt{Q_3})}{2}z} e^{i(pt-kx)}. \tag{1.17}$$

1.3 BOUNDARY CONDITIONS

To model the friction interface, an approximate linearization technique as suggested by Miller [27, 28] and Iwan [41] is adopted.

The relative slip at the frictional boundary is expressed as

$$s(x,t) = v_2(x,0,t) - v_1(x,0,t). \tag{1.18}$$

Until frictional stress (τ_f) is less than the slip stress (τ_s), it is expected that there won't be any slip between the two media. The frictional stress during slippage is provided by

$$\tau_f = \tau_s \operatorname{sgn}\left(\frac{\partial s}{\partial t}\right), \tag{1.19}$$

where sgn denotes the Signum function.

The frictional stress (τ_f) as a linear function of relative slip $s(x,t)$ is given by

$$\tau_f = L_1 s(x,t) + L_2 \frac{\partial s(x,t)}{\partial t}, \tag{1.20}$$

where L_1 and L_2 are constants given by

$$L_1 = \frac{1}{\pi E}\int_0^{2\pi} \tau_f (E\cos\theta)\cos\theta \, d\theta, \tag{1.21}$$

$$L_2 = -\frac{1}{\pi \omega E}\int_0^{2\pi} \tau_f (E\cos\theta)\sin\theta \, d\theta, \tag{1.22}$$

$\theta = \omega t - kx$ is the phase variable, ω represents frequency.

The relative slip may be represented as [42]

$$s(x,t) = E cos\theta, \tag{1.23}$$

where $E = E_0\, e^{-\omega\kappa t}$ is the effective amplitude of motion. κ represents the attenuation coefficient. The exponential decay of the amplitude is due to the energy dissipation.

The following conditions will serve as boundary conditions:

- The traction free condition at the free surface that is at $z = -H$

$$\tau_{yz} = 0. \tag{1.24}$$

- The continuity of shear and frictional stresses at the frictionally bonded interface of layer and half-space that is at $z = 0$

$$\tau_{yz} = \tau_f\left\{s(x,t)\right\}, \tag{1.25}$$

$$\sigma_{yz} = \tau_f\left\{s(x,t)\right\}. \tag{1.26}$$

The stress components needed for the boundary is obtained from Eqs. (1.3), (1.9), (1.11), and (1.17) as

$$\begin{aligned}\tau_{yz} &= \left(Q_1 v_1 + Q_2 e^{Q_1 z}\left(-C_1 \sin Q_2 z + C_2 \cos Q_2 z\right)e^{i(pt-kx)}\right)\left(\mu_T^{(1)} + \left(\mu_L^{(1)} - \mu_T^{(1)}\right)a_z^2\right)\\ &\quad + ip\left(Q_1 v_1 + Q_2 e^{Q_1 z}\left(-C_1 \sin Q_2 z + C_2 \cos Q_2 z\right)e^{i(pt-kx)}\right)\left(\bar{\mu}_T^{(1)} + \left(\bar{\mu}_L^{(1)} - \bar{\mu}_T^{(1)}\right)a_z^2\right)\\ &\quad + \left(\mu_L^{(1)} - \mu_T^{(1)}\right)\left(-ia_x a_z k v_1\right) + \left(\bar{\mu}_L^{(1)} - \bar{\mu}_T^{(1)}\right)\left(a_x a_z kp\, v_1\right),\end{aligned} \tag{1.27}$$

$$\sigma_{yz} = -\left(C_{44}^{(1)} + i\bar{C}_{44}^{(1)}\right)\left(\frac{b_2 + \sqrt{Q_3}}{2}\right)v_2. \tag{1.28}$$

1.4 FREQUENCY RELATION

Using Eqs. (1.9), (1.17), (1.18), (1.20), (1.27), and (1.28) into Eqs. (1.24)–(1.26) results in system of equations involving three arbitrary constants C_1, C_2, D_2. For nontrivial solutions the determinant of coefficient matrix of system of equations should vanish, which leads to the frequency equation for Love-type waves propagation in the considered structure as

$$\cot Q_2 H - \frac{G_3}{G_4} - \frac{G_3^2 + G_4^2}{G_4 \left(L_1 + ipL_2\right)} - \frac{2\left(G_3^2 + G_4^2\right)}{T_4\left(b_2 + \sqrt{Q_3}\right)\left(C_{44}^{(1)} + ip\bar{C}_{44}^{(1)}\right)} = 0, \tag{1.29}$$

where G_1, G_2, G_3 and G_4 are given in Appendix.

Due to the energy dissipation in the system, the frequency parameter is considered in complex form as

$$p = \omega(1 + i\kappa). \tag{1.30}$$

Considering c as the phase velocity, following non-dimensional parameters and variables are introduced

$$c_d = \frac{c}{\beta_1},\ n_d = kH,\ \beta_1 = \sqrt{\frac{\mu_T^{(1)}}{\rho_1^{(1)}}},\ \beta_2 = \sqrt{\frac{C_{66}^{(1)}}{\rho_2^{(1)}}}, v_1 = \omega \frac{\bar{\mu}_L^{(1)}}{\mu_L^{(1)}}, v_2 = \omega \frac{\bar{\mu}_T^{(1)}}{\mu_T^{(1)}}, v_3 = \omega \frac{\bar{C}_{66}^{(1)}}{C_{66}^{(1)}},$$
$$v_4 = \omega \frac{\bar{C}_{44}^{(1)}}{C_{44}^{(1)}},\ \xi_1 = \frac{P_1^{(1)}}{2\mu_T^{(1)}},\ \xi_2 = \frac{P_2^{(1)}}{2C_{66}^{(1)}},\ \bar{L}_1 = \frac{L_1\beta_1}{\mu_T^{(1)}\omega},\ \bar{L}_2 = \frac{L_2\beta_1}{\mu_T^{(1)}}. \tag{1.31}$$

Using Eqs. (1.30) and (1.31) in Eq. (1.29), the non-dimensional frequency relation is obtained as

$$\cot Q_2 H - \frac{T_3}{T_4} - \frac{T_3^2 + T_4^2}{T_4 T_5} - \frac{T_3^2 + T_4^2}{T_4 T_6} = 0, \tag{1.32}$$

where T_3, T_4, T_5, T_6 are given in Appendix.

In view of Eqs. (1.19)–(1.22) and Eq. (1.31), we instate

$$\bar{L}_1 = 0, \bar{L}_2 = \frac{4\varsigma}{\pi c_d n_d}, \tag{1.33}$$

where $\varsigma = \frac{\tau_s}{\mu_T^{(1)}} \frac{H}{E}$ denotes frictional bonding parameter.

1.5 PARTICULAR CASES

1.5.1 Case I

When the considered layered structure is reduced to homogeneous elastic layer bonded by Coulomb friction to homogeneous elastic half-space. We have,

$$a_x = a_z = b_1 = b_2 = \bar{C}_{66}^{(1)} = \bar{C}_{44}^{(1)} = \bar{\mu}_L^{(1)} = \bar{\mu}_T^{(1)} = 0, C_{66}^{(1)} = C_{44}^{(1)} = \mu_2, \mu_L^{(1)} = \mu_T^{(1)} = \mu_1. \tag{1.34}$$

Using Eq. (1.34) in Eq. (1.29), we instate

$$\cot\left(\sqrt{\frac{p^2}{\beta_{10}^2}-k^2}\right)H-\frac{\left(\sqrt{\frac{p^2}{\beta_{10}^2}-k^2}\right)\mu_1}{\left(\sqrt{k^2-\frac{p^2}{\beta_{20}^2}}\right)\mu_2}-\frac{\left(\sqrt{\frac{p^2}{\beta_{10}^2}-k^2}\right)\mu_1}{L_1+ipL_2}=0, \qquad (1.35)$$

where $\beta_{10}^2=\frac{\mu_1}{\rho_1^{(1)}}$, $\beta_{20}^2=\frac{\mu_2}{\rho_2^{(1)}}$.

The obtained frequency Eq. (1.34) is in well agreement with the result obtained by Miller [28]

1.5.2 Case 2

Assuming dissipation of energy in the system is negligible and so dropping the imaginary part of the frequency parameter p. Also, for the welded contact between the two media $L_1+ipL_2\to\infty$. Then from Eq. (1.34) we have

$$\tan\left(kH\sqrt{\frac{c^2}{\beta_{10}^2}-1}\right)=\frac{\left(\sqrt{1-\frac{c^2}{\beta_{20}^2}}\right)\mu_2}{\left(\sqrt{\frac{c^2}{\beta_{10}^2}-1}\right)\mu_1}. \qquad (1.36)$$

Eq. (1.35) determines the Love-type waves velocity in homogeneous elastic layered structure having a layer over half-space, which is resembling with the classical Love wave equation [37].

1.6 NUMERICAL EXAMPLE AND DISCUSSIONS

The dispersion equation and the damping equation related to the propagation of Love-type waves in the media under consideration are provided by the real and imaginary parts of Eq. (1.32). Mathematica software has been used for computation and graphical representation. The main aspect of the present study is to highlight the influence of frictional bonding, reinforcement, heterogeneity, viscoelasticity, and initial stress parameters on the dispersive and damping behavior of Love-type surface waves in considered layered structure. The following values of elastic constants have been considered:

For fiber-reinforced layer [43]

$$\mu_L^{(1)}=4.4\times10^9\ \text{N/m}^2,\ \mu_T^{(1)}=1.89\times10^9\ \text{N/m}^2,\ \rho_1^{(1)}=5.6\times10^3\ \text{kg/m}^3.$$

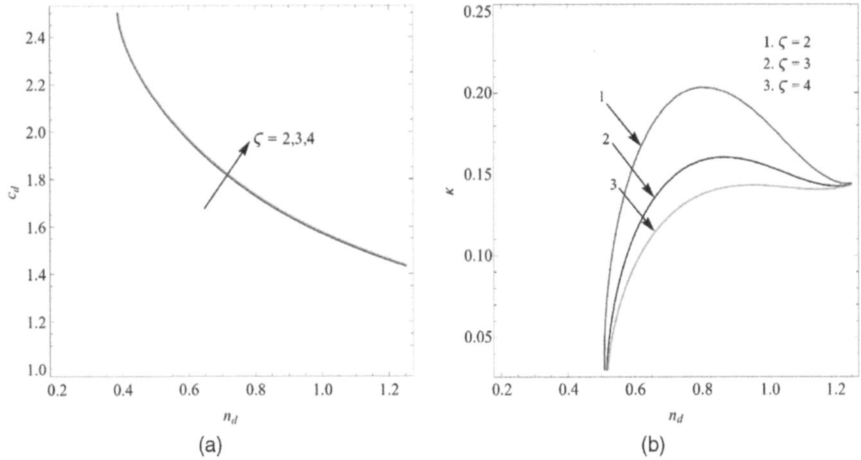

Figure 1.2 Variation of (a) dimensionless phase velocity (c_d) and (b) attenuation coefficient (κ) against dimensionless wave number (n_d) for different values of the bonding parameter (ς).

For orthotropic half-space [44]

$$C_{44}^{(1)} = 6.15\times10^{9}\ \text{N/m}^2,\ C_{66}^{(1)} = 3.32\times10^{9}\ \text{N/m}^2,\ \rho_2^{(1)} = 1.5\times10^{3}\ \text{kg/m}^3.$$

In each Figures 1.2–1.9, sub-figure (a) exhibits the variation in dimensionless phase velocity $\left(c_d = c/\beta_1\right)$ against the dimensionless wave number $\left(n_d = kH\right)$ while sub-figure (b) manifests the variation in attenuation

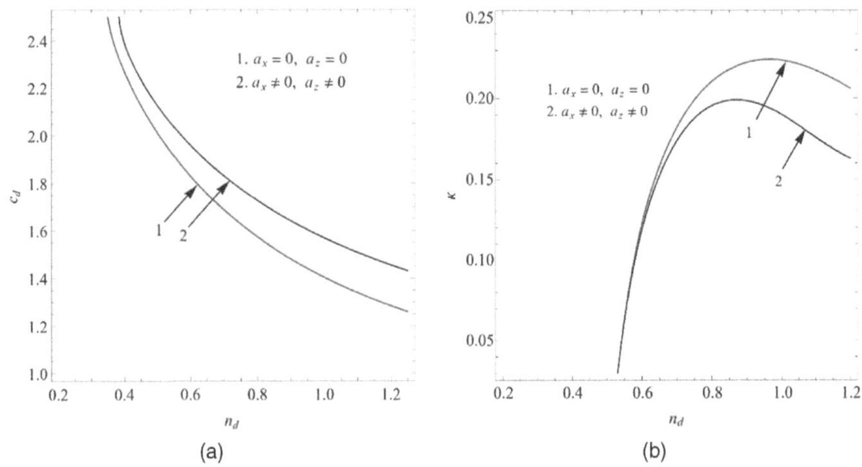

Figure 1.3 Variation of (a) dimensionless phase velocity (c_d) and (b) attenuation coefficient (κ) against dimensionless wave number (n_d) for presence and absence of reinforcement parameter.

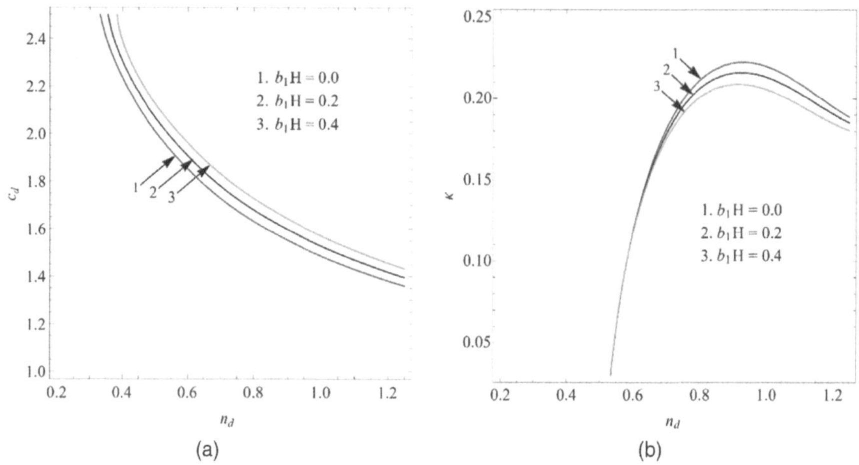

Figure 1.4 Variation of (a) dimensionless phase velocity (c_d) and (b) attenuation coefficient (κ) against dimensionless wave number (n_d) for different values of heterogeneity parameter (b_1H) associated with fiber-reinforced layer.

coefficient (κ) against the dimensionless wave number $(n_d = kH)$ for various affecting parameters.

Figure 1.2(a) depicts the effect of frictional bonding parameter on the dimensionless phase velocity. It is observed that all the phase velocity curves for different values of bonding parameter are overlapped, which indicates that the bonding parameter has no influence on the phase velocity of Love-type

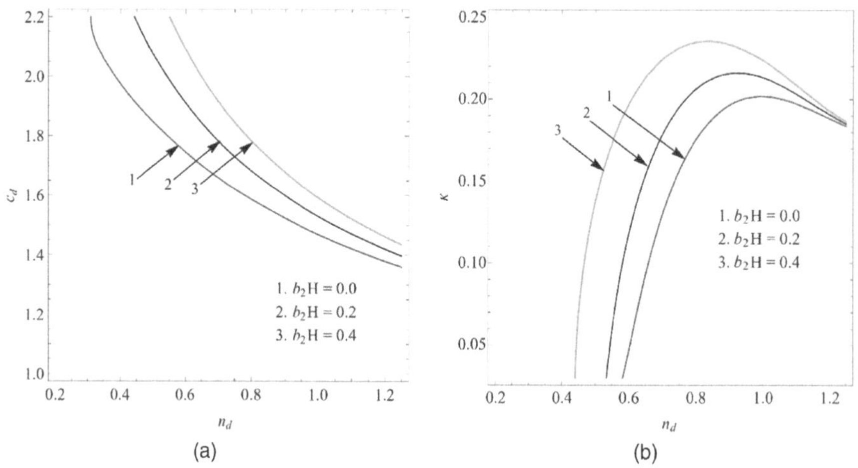

Figure 1.5 Variation of (a) dimensionless phase velocity (c_d) and (b) attenuation coefficient (κ) against dimensionless wave number (n_d) for different values of heterogeneity parameter (b_2H) associated with orthotropic half-space.

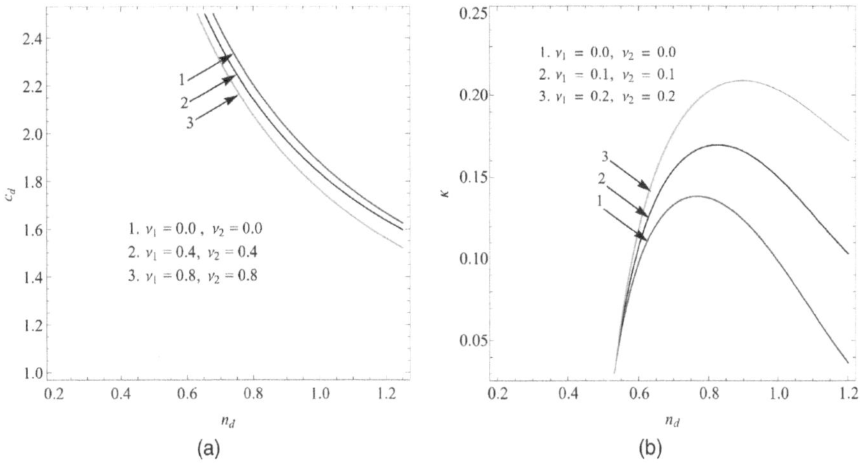

Figure 1.6 Variation of (a) dimensionless phase velocity (c_d) and (b) attenuation coefficient (κ) against dimensionless wave number (n_d) for different values of viscoelastic parameters (ν_1, ν_2) associated with fiber-reinforced layer.

waves in considered media. Figure 1.2(b) exhibits the effect of frictional bonding parameter on the attenuation coefficient. For different values of the frictional bonding parameter, the attenuation curves outset from a common value of wave number and finally accumulate at the particular value of wave number. By inspecting the curves in between the starting and accumulating points, it is noticed that the attenuation curve shifts downward as the frictional bonding parameter increases.

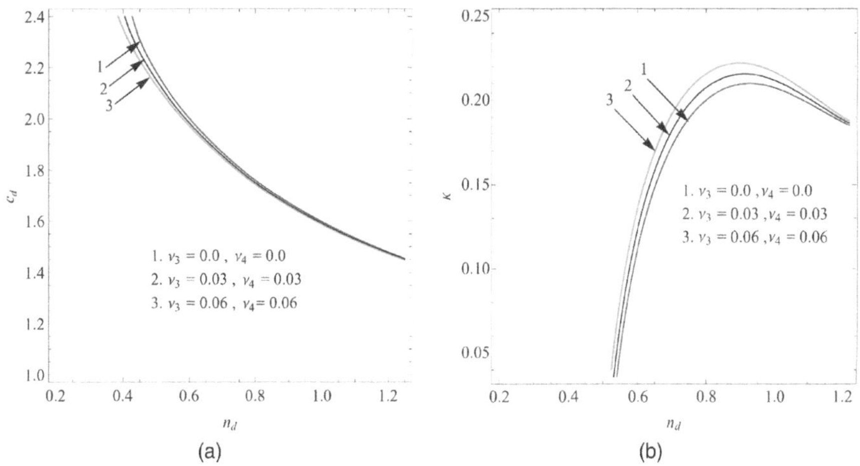

Figure 1.7 Variation of (a) dimensionless phase velocity (c_d) and (b) attenuation coefficient (κ) against dimensionless wave number (n_d) for different values of viscoelastic parameters (ν_3, ν_4) associated with orthotropic half-space.

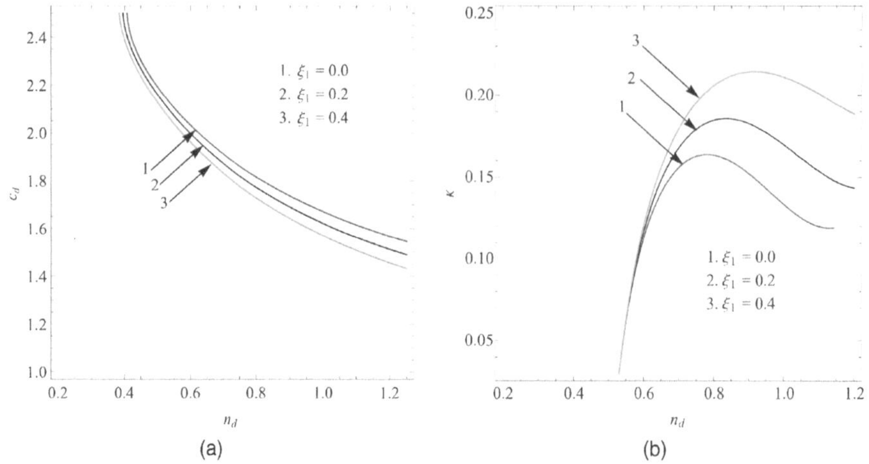

Figure 1.8 Variation of (a) dimensionless phase velocity (c_d) and (b) attenuation coefficient (κ) against dimensionless wave number (n_d) for different values of initial stress parameter (ξ_1) associated with fiber-reinforced layer.

The phase velocity and attenuation coefficient have been plotted with respect to the wave number for the presence and absence of reinforcement parameters associated with the fiber-reinforced layer in Figure 1.3(a) and (b), respectively. In the situated figure, curve 1 shows the case of absence of reinforcement and curve 2 shows the case of presence of reinforcement in the layer. It is clearly visible that the reinforcement increases the phase velocity, while the reinforcement has adverse influence on the attenuation coefficient.

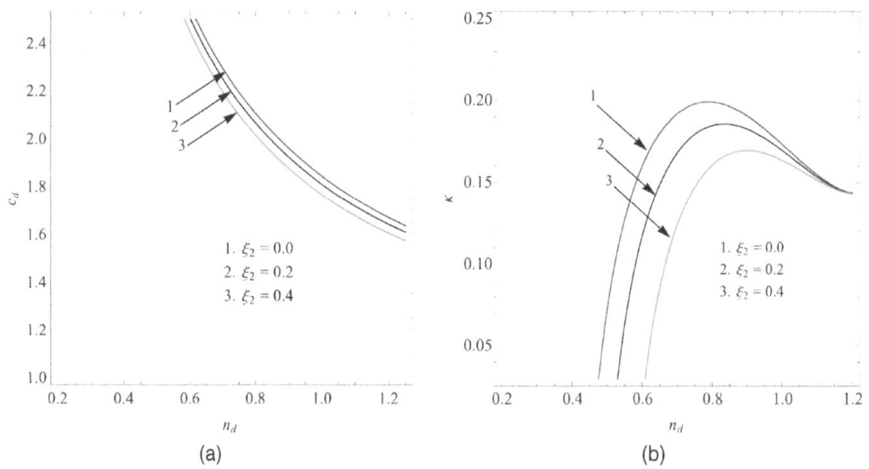

Figure 1.9 Variation of (a) dimensionless phase velocity (c_d) and (b) attenuation coefficient (κ) against dimensionless wave number (n_d) for different values of initial stress parameter (ξ_2) associated with orthotropic half-space.

Moreover, the effect of reinforcement is more prominent for the higher value of wave number.

The variations in phase velocity and attenuation coefficient have been carved out against wave number for different values of heterogeneity parameter allied with fiber-reinforced layer in Figure 1.4(a) and (b), respectively. It is evident from the figure that an increment in the heterogeneity parameter steps up the phase velocity of the Love-type waves while the increment in heterogeneity parameter steps down the attenuation coefficient. It is worthwhile to note that the attenuation curves are assembled for smaller values of wave number which spread out as the wave number increases.

The pronounced influence of heterogeneity parameter allied with orthotropic half-space on the phase velocity and attenuation coefficient have been depicted in Figure 1.5(a) and (b), respectively. Both the phase velocity and attenuation curve rise up considerably with an increment in the heterogeneity parameter. The influence of heterogeneity is more prominent for lower values of wave number rather than for higher values. Moreover, the attenuation curves outset from different values of wave number which increase rapidly with increasing value of wave number then start to decline and finally accumulate.

Figure 1.6(a) and (b) depicts the effect of viscoelastic parameters allied with the fiber-reinforced layer on the phase velocity and attenuation coefficient, respectively. With the increment in viscoelastic parameters, the phase velocity decreases while the damping parameter increases. The attenuation curves outset from a common value of wave number which spread out as the wave number increases. Moreover, even for a slight increment in the values of viscoelastic parameters associated with the fiber-reinforced layer, attenuation curve rises up remarkably. Thus, the influence of viscoelastic parameters is more significant in attenuation coefficient as compared to phase velocity.

Figure 1.7(a) and (b) shows the variation in dimensionless phase velocity and attenuation coefficient versus wave number, respectively, for different values of viscoelastic parameters allied with the orthotropic half-space. It can be observed from Figure 1.7(a) that the viscoelastic parameters have negligible influence on the phase velocity for higher values of wave number, while for smaller values of wave number, increasing values of viscoelastic parameters diminish the phase velocity. Moreover, Figure 1.7(b) indicates that the increasing values of viscoelastic parameters endorse the attenuation coefficient.

The impact of the fiber-reinforced layer's initial stress parameter on the phase velocity and attenuation coefficient is shown in Figure 1.8(a) and (b), respectively. Observation of the figure signifies that the increasing value of initial stress parameter diminishes the phase velocity, while it endorses the attenuation coefficient. It is worthwhile to note that the attenuation curves outset from a common value of wave number and are assembled for smaller values of wave number which spread out as the wave number increases. Thus, the effect of aforesaid parameter is more prominent for higher values of wave numbers.

Figure 1.9(a) and (b) depicts the effect of initial stress parameter associated with the orthotropic half-space on the phase velocity and attenuation coefficient, respectively. The phase velocity and the attenuation coefficient are found to decrease as the initial stress parameter value increases. The attenuation curves outset from different values of wave number and finally accumulate at a point.

Each of Figures 1.2–1.9 makes it clear that the attenuation curve rises quickly with increasing wave number until it reaches the maximum value of the attenuation parameter, at which point it begins to decline, while the phase velocity curve starts at a specific value and falls off continuously as the wave number increases.

1.7 CONCLUDING REMARKS

Analysis of the propagation of Love-type surface waves in initially stressed heterogeneous viscoelastic layered media has been done. The considered layered media contain self-reinforced layer bedded over orthotropic half-space and these two media are bonded by Coulomb friction. Prominent effect of frictional bonding, reinforcement, heterogeneity, viscoelasticity, and initial stress on the dispersion and attenuation behavior of Love-type waves in the considered structure has been observed. The following consequences are the crux of meticulous examination of the present study:

- The increasing value of wave number has an adverse effect on the phase velocity, while the attenuation coefficient keeps on increasing upto certain value and then starts to decline as the wave number increases.
- The presence of the frictional bonding between the two media has no effective impact on the phase velocity of Love-type waves while increasing the value of frictional bonding parameter the value of attenuation coefficient decreases. This signifies that the dissipation of energy in the considered layered media will be less in case of the frictional bonding interface.
- The presence of reinforcement tends to increase the phase velocity but shows the reverse effect on the attenuation coefficient. The effect of reinforcement is more prominent for the higher values of wave number.
- The heterogeneity of both the layer and the half-space favors the phase velocity of the Love-type waves. The influence of heterogeneity of layer on the attenuation coefficient is opposite to the heterogeneity of half-space. Heterogeneity of the layer disfavors the attenuation, while the heterogeneity of half-space supports the attenuation. Moreover, influence of heterogeneity of the layer is significant for higher values of wave number, whereas influence of heterogeneity of the half-space is prominent for lower wave number.

- Viscoelasticity of both the media (layer and half-space) declines the phase velocity while favors the attenuation. Moreover, the viscoelastic parameters affect attenuation parameter more significantly as compared to the phase velocity.
- The initial stress of the layer enhances the attenuation, while it deteriorates the phase velocity. The presence of initial stress in half-space lessens both the phase velocity and attenuation. The impact of initial stress allied with the layer on the attenuation is prominent for higher wave number while contrary to this, initial stress allied with the half-space has prominent effect for lower wave number.

The outcomes of the study of Love-type waves in considered frictionally bonded heterogeneous viscoelastic layered media have possible application in the field of earthquake engineering especially to interpret the seismic activities near the crust-mantle boundary. The analysis may be beneficial in the construction sectors for designing disaster-resistant building, bridges, and towers which generally involve using reinforced materials such as reinforcement cement concrete (RCC).

1.8 APPENDIX

$$A=\mu_T^{(1)}+\bar{\mu}_T^{(1)}ip+\left(\mu_L^{(1)}-\mu_T^{(1)}\right)a_z^2+\left(\bar{\mu}_L^{(1)}-\bar{\mu}_T^{(1)}\right)a_z^2 ip,$$

$$B=\left(-2ika_xa_z+b_1a_z^2\right)\left(\mu_L^{(1)}-\mu_T^{(1)}\right)+\left(2a_xa_zpk+ib_1a_z^2p\right)\left(\bar{\mu}_L^{(1)}-\bar{\mu}_T^{(1)}\right)$$
$$+b_1\mu_T^{(1)}+b_1\bar{\mu}_T^{(1)}ip,$$

$$C=-k^2\mu_T^{(1)}-ik^2\bar{\mu}_T^{(1)}p+\frac{P_1^{(1)}}{2}k^2+\rho_1^{(1)}p^2-a_xk\left(a_xk+ia_zb_1\right)\left(\mu_L^{(1)}-\mu_T^{(1)}\right)$$
$$+a_xk\left(-ia_xkp+a_zb_1p\right)\left(\bar{\mu}_L^{(1)}-\bar{\mu}_T^{(1)}\right),$$

$$D=\frac{-k^2C_{66}^{(1)}-ik^2p\bar{C}_{66}^{(1)}+\frac{P_2^{(1)}}{2}k^2+\rho_2^{(1)}p^2}{C_{44}^{(1)}+ip\bar{C}_{44}^{(1)}},G_1=\mu_T^{(1)}+\left(\mu_L^{(1)}-\mu_T^{(1)}\right)a_z^2,$$
$$G_2=\bar{\mu}_T^{(1)}-\mu_T^{(1)}+\left(\bar{\mu}_L^{(1)}-\bar{\mu}_T^{(1)}\right)a_z^2,$$

$$G_3=Q_1G_1+ipQ_1G_2-ika_xa_z\left(\mu_L^{(1)}-\mu_T^{(1)}\right)+a_xa_zkp\left(\bar{\mu}_L^{(1)}-\bar{\mu}_T^{(1)}\right),$$
$$G_4=Q_2G_1+ipQ_2G_2,$$

$$T_3 = \frac{-b_1 H \mu_T^{(1)} R_1}{2}, T_4 = \mu_T^{(1)} \frac{\sqrt{-R_2 + 4R_3}}{2},$$

$$T_5 = \frac{C_{44}^{(1)}}{2}\left(1 + i(1 + i\kappa)v_4\right) b_2 H \left(1 + \sqrt{1 - \frac{4R_5}{R_6}}\right),$$

$$T_6 = \left(\bar{L}_1 + i(1 + i\kappa)\bar{L}_2\right)\mu_T^{(1)} \frac{\omega H}{\beta_1},$$

$$R_1 = 1 + \left(\frac{\mu_L^{(1)}}{\mu_T^{(1)}} - 1\right) a_z^2 + i(1 + i\kappa)\left(v_2 + \left(v_1 \frac{\mu_L^{(1)}}{\mu_T^{(1)}} - v_2\right) a_z^2\right),$$

$$R_2 = \left(b_1 R_1 - 2i(kH) a_x a_z \left(\frac{\mu_L^{(1)}}{\mu_T^{(1)}} - 1\right) + 2kHa_x a_z (1 + i\kappa)\left(v_1 \frac{\mu_T^{(1)}}{\mu_T^{(1)}} - v_2\right)\right)^2,$$

$$R_3 = R_1\left(-(kH)^2\left(1 + i(1 + i\kappa)v_2 - \xi_1\right) + \frac{\omega^2 H^2}{\beta_1^2}(1 + i\kappa)^2 \right.$$
$$\left. -(kH) a_x \left(\frac{\mu_L^{(1)}}{\mu_T^{(1)}} - 1\right)\left\{a_x (kH) + i a_z b_1 H\right\} + R_4\right),$$

$$R_4 = a_x kH(1 + i\kappa)\left(v_1 \frac{\mu_L^{(1)}}{\mu_T^{(1)}} - v_2\right)\left\{(b_1 H) a_z - i a_x (kH)\right\},$$

$$R_5 = -(kH)^2\left(1 + i(1 + i\kappa)v_3\right) + (kH)^2 \xi_2 (1 + i\kappa)^2 \frac{\omega^2}{\beta_2^2} H^2,$$

$$R_6 = b_2^2 H^2\left(\left(1 + i(1 + i\kappa)v_4\right)\right)\frac{C_{44}^{(1)}}{C_{66}^{(1)}}.$$

REFERENCES

1. Peltier W. The impulse response of a Maxwell Earth. Reviews of Geophysics. 1974; 12(4):649–669.
2. Osinski Z. Damping of vibrations. CRC Press; 2018.
3. Carcione J. Wave propagation in anisotropic linear viscoelastic media: theory and simulated wavefields. Geophysical Journal International. 1990; 101(3):739–750.
4. Mazzotti M, Marzani A, Bartoli I, et al. Guided waves dispersion analysis for prestressed viscoelastic waveguides by means of the safe method. International Journal of Solids and Structures. 2012; 49(18):2359–2372.
5. Sumbatyan M, Zampoli V, Vaccaro M. Baseline from a wave of a viscoelastic layer. Bulletin of Samara State University. 2007; 4:372–386.

6. Singh MK, Alam P. Surface wave analysis in orthotropic composite structure with irregular interfaces. International Journal of Applied and Computational Mathematics. 2020 Feb; 6:1–21.
7. Alam P, Singh MK. Study of Love waves in a clamped viscoelastic medium with irregular boundaries. Journal of Solid Mechanics. 2019 Jul 1; 11(3).
8. Sahu SA, Singh MK, Pankaj KK. Analysis of torsional waves in a prestressed composite structure with loosely bonded and corrugated boundaries. Mechanics of Composite Materials. 2018 Jul; 54:321–332.
9. Majhi S, Pal P, Kumar S. Propagation of SH waves in a visco-elastic layer overlying an inhomogeneous isotropic half-space. Ain Shams Engineering Journal. 2016. https://doi.org/10.1016/j.asej.2016.03.011.
10. Ren D, Shen X, Li C, et al. The fractional Kelvin-Voigt model for Rayleigh surface waves in viscoelastic FGM infinite half space. Mechanics Research Communications. 2018; 87:53–58.
11. Sato H, Fehler MC, Maeda T. Heterogeneity in the lithosphere. In: Seismic wave propagation and scattering in the heterogeneous Earth: Second edition. Springer; 2012. pp. 13–61.
12. Meissner R. The little book of planet Earth. Springer Science & Business Media; 2002.
13. Biot MA. The influence of initial stress on elastic waves. Journal of Applied Physics. 1940; 11(8):522–530.
14. Abd-Alla A, Mahmoud S, Abo-Dahab S, et al. Propagation of S-wave in a non-homogeneous anisotropic incompressible and initially stressed medium under influence of gravity field. Applied Mathematics and Computation. 2011; 217(9):4321–4332.
15. Garg N. Effect of initial stress on harmonic plane homogeneous waves in viscoelastic anisotropic media. Journal of Sound and Vibration. 2007; 303(3–5):515–525.
16. Shams M. Effect of initial stress on Love wave propagation at the boundary between a layer and a half-space. Wave Motion. 2016; 65:92–104.
17. Singh A, Parween Z, Chatterjee M, et al. Love-type wave propagation in a pre-stressed viscoelastic medium influenced by smooth moving punch. Waves in Random and Complex Media. 2015; 25(2):268–285.
18. Kumari P, Kumar Sharma V, Modi C. Modeling of magnetoelastic shear waves due to point source in a viscoelastic crustal layer over an inhomogeneous viscoelastic half space. Waves in Random and Complex Media. 2016; 26(2):101–120.
19. Kundu S, Pandit DK, Gupta S, et al. Love wave propagation in a fiber-reinforced medium sandwiched between an isotropic layer and gravitating half-space. Journal of Engineering Mathematics. 2016; 100(1):109–119.
20. Singh MK, Rahul AK, Saha S, Paul S, Tiwari R, Nandi S. On generalized Rayleigh waves in a pre-stressed piezoelectric medium. International Journal of Modern Physics C. 2024; 35(2). https://doi.org/10.1142/S0129183124400060.
21. Chattopadhyay A, Gupta S, Sahu S, et al. Torsional surface waves in a self-reinforced medium over a heterogeneous half space. International Journal of Geomechanics. 2011; 12(2):193–197.
22. Acharya D, Roy I. Magnetoelastic surface waves in electrically conducting fibre-reinforced media. Bulletin of the Institute of Mathematics Academia Sinica (New Series). 2009; 4(3):333–352.
23. Belfield A, Rogers T, Spencer A. Stress in elastic plates reinforced by fibres lying in concentric circles. Journal of the Mechanics and Physics of Solids. 1983; 31(1):25–54.
24. Hashin Z, Rosen BW. The elastic moduli of fiber-reinforced materials. Journal of Applied Mechanics. 1964; 31(2):223–232.

25. Singh B. Wave propagation in thermally conducting linear fibre-reinforced composite materials. Archive of Applied Mechanics. 2006; 75(8–9):513–520.
26. Ebrahimian H, Kohler M, Massari A, et al. Parametric estimation of dispersive viscoelastic layered media with application to structural health monitoring. Soil Dynamics and Earthquake Engineering. 2018; 105:204–223.
27. Srivastava A, Chattopadhyay A, Singh AK. Impact of inhomogeneous fiber-reinforced layer with frictional interface on Rayleigh-type wave propagation. Journal of Engineering Mathematics. 2019; 114(1):159–176.
28. Singh B, Sindhu R, Singh J. Rayleigh waves in a rotating orthotropic micropolar elastic solid half-space. International Journal of Geophysics. 2013; 1.
29. Vinh PC, Anh VTN, Linh NTK. Exact secular equations of Rayleigh waves in an orthotropic elastic half-space overlaid by an orthotropic elastic layer. International Journal of Solids and Structures. 2016; 83:65–72.
30. Yu J, Zhang C. Effects of initial stress on guided waves in orthotropic functionally graded plates. Applied Mathematical Modelling. 2014; 38(2):464–478.
31. Miller R. An approximate method of analysis of the transmission of elastic waves through a frictional boundary. Journal of Applied Mechanics. 1977; 44(4):652–656.
32. Miller RK. An estimate of the properties of Love-type surface waves in a frictionally bonded layer. Bulletin of the Seismological Society of America. 1979; 69(2):305–317.
33. Li J, Li H, Zhao J. An improved equivalent viscoelastic medium method for wave propagation across layered rock masses. International Journal of Rock Mechanics and Mining Sciences. 2015; 73:62–69.
34. Di Bartolomeo M, Massi F, Baillet L, et al. Wave and rupture propagation at frictional bimaterial sliding interfaces: From local to global dynamics, from stick-slip to continuous sliding. Tribology International. 2012; 52:117–131.
35. Singh MK, Sahu SA, Singhal A, Chaudhary S. Approximation of surface wave velocity in smart composite structure using Wentzel–Kramers–Brillouin method. Journal of Intelligent Material Systems and Structures. 2018 Nov; 29(18):3582–3597.
36. Singh MK, Rahul AK, Tiwari R, Paul S, Saha S. Propagation of surface wave in a composite material. Mechanics of Solids. 2023 Dec; 58(9):3226–3243.
37. Kaur T, Sharma SK, Singh AK. Influence of imperfectly bonded micropolar elastic half-space with non-homogeneous viscoelastic layer on propagation behavior of shear wave. Waves in Random and Complex Media. 2016; 26(4):650–670.
38. Biot MA. Mechanics of incremental deformations. John Wiley & Sons; 1965.
39. Spencer AJM. Deformations of fibre-reinforced materials. Clarendon Press; 1972.
40. Yu J, Ratolojanahary F, Lefebvre J. Guided waves in functionally graded viscoelastic plates. Composite Structures. 2011; 93(11):2671–2677.
41. Iwan WD. A generalization of the concept of equivalent linearization. International Journal of Non-Linear Mechanics. 1973; 8(3):279–287.
42. Kanai K. A new problem concerning surface waves. Earthquake Research Institute, the University of Tokyo. 1961; 39:359–366.
43. Love AEH. Some problems of geodynamics. Cambridge University Press; 1911.
44. Chattopadhyay A, Choudhury S. Propagation, reflection and transmission of magnetoelastic shear waves in a self-reinforced medium. International Journal of Engineering Science. 1990; 28(6):485–495.

Chapter 2

Understanding Rayleigh-type wave dynamics in complex medium

Pulkit Kumar and Moumita Mahanty

2.1 INTRODUCTION

Rayleigh wave is a kind of seismic surface wave causing the ground to shake in an elliptical motion, manifesting as undulating ground motion reminiscent of waves traversing across a lake or an ocean, i.e. the ground moves up and down and side-to-side in the same direction as Rayleigh wave propagation. These waves propagate along the Earth's surface with a characteristic elliptical motion, where particles move both vertically and horizontally in the direction of wave propagation. These surface waves, first theorized by Lord Rayleigh in 1885, exhibit both longitudinal and transverse characteristics, with their amplitude diminishing exponentially with depth which makes them distinct from other seismic wave types. Despite their surface-bound nature, Rayleigh waves exert significant influence during seismic events, often contributing substantially to the ground shaking experienced. The retrograde motion of Rayleigh waves refers to the elliptical trajectory followed by particles as the wave passes through the medium. Unlike the simple back-and-forth motion observed in some seismic waves, Rayleigh waves induce complex particle movements that result in both vertical and horizontal displacement. This retrograde motion imparts a characteristic rolling sensation to the ground. While the amplitude of Rayleigh waves decreases exponentially with depth, their influence on surface structures and geological formations can be profound, making them a subject of extensive study across various scientific disciplines. The study of Rayleigh wave holds significance across diverse fields, including earthquake engineering, seismology, geophysics, acoustics, telecommunications, and materials science. Understanding Rayleigh wave behaviour is particularly vital in seismology, aiding in the assessment of structural damage and earthquake impact estimation for construction purposes. Notable researchers such as Buchwald [1], Dutta [2], and Chattopadhyay et al. [3] have extensively explored Rayleigh wave propagation across various mediums, enriching our understanding of seismic phenomena.

DOI: 10.1201/9781003470885-2

Our planet, comprising diverse materials, exhibits inherent anisotropic properties, resulting from variations in elastic characteristics and the presence of initial stress. Understanding the propagation of seismic waves in such anisotropic mediums is crucial for comprehending the complexities of Earth's dynamics. Among the various types of anisotropic media studied extensively by researchers, transversely isotropic mediums stand out prominently. Transversely isotropic materials exhibit the properties that are symmetric with respect to an axis normal to a plane of isotropy. Within this transverse plane, material properties remain consistent in all directions, giving rise to infinite planes of symmetry. These materials, known as polar anisotropic materials, exhibit hexagonal symmetry. Examples include hexagonal crystals such as beryl, zinc, cadmium, and cobalt, as well as numerous fibre-reinforced composites. Extensive research by researchers such as Acharya et al. [4], Chattopadhyay [5], and Singh et al. [6] has provided valuable insights into the behaviour of transversely isotropic materials. Their contributions have enriched our understanding of the intricate dynamics governing wave propagation in such mediums, shedding light on fundamental principles underlying seismic phenomena.

In addition to anisotropy, the Earth is subject to significant initial stresses, which are inherent inelastic forces present within the Earth's structure even in the absence of external forces. These initial stresses stem from a multitude of factors, including temperature differentials, atmospheric pressure, overburden layers, gravitational forces, manufacturing processes, slow creep, hydrostatic tension or compression, and various geological processes such as quenching and cold working. The accumulation of these diverse influences results in a complex network of initial stresses distributed throughout the Earth's interior. The presence of initial stresses profoundly influences the response of solid materials to external loads, often leading to behaviours distinct from those observed in the absence of initial stresses. Furthermore, the dispersion of waves can be significantly affected by the initial friction of dry sandy layers within the Earth's structure. It is essential to consider the Earth's soil layers as predominantly sandy rather than purely elastic. A dry sandy layer is primarily composed of sand particles, with the sandiness parameter, denoted by η, by playing a crucial role in characterizing the properties of these layers. Weiskopf [7] explored the mechanics of dry sandy soils and proposed that the relationship $E/\mu = 2\eta(1+\sigma)$ could be applicable to such materials, where $\eta > 1$, E is the Young's modulus of elasticity, μ is the rigidity, and σ is the Poisson's ratio. Biot [8] explored how acoustic wave propagation behaves differently when there is initial stress, compared to the scenario without stress. A comprehensive analysis of this perspective can be found in Biot's [9], Guz's [10], and Chatterjee et al.'s [11] works. The Earth's crust, particularly its uppermost layers, exhibits a greater degree of heterogeneity compared to the deeper mantle, significantly influencing the propagation of surface waves. Bullen [12] demonstrated that density changes at discrete rates across Earth's layers, particularly in the heterogeneous crust. Further insights into the effects of heterogeneity on seismic wave propagation have been provided by researchers such as Sato [13], Dutta [14], and Bhattacharya [15] to name a few.

This chapter focuses on the propagation of Rayleigh-type waves in a semi-infinite medium characterized by pre-existing (initial) stress, heterogeneity, and transverse isotropy, with the additional consideration of dry sandy parameters. Employing analytical methods, a wave velocity relation is established specific to Rayleigh-type wave propagation in such complex environments. In the absence of anisotropy, heterogeneity, initial stress, and sandiness, the derived wave velocity relation closely aligns with established models of Rayleigh wave propagation. Through numerical computations and graphical representations, the influences of various parameters, including initial stress, heterogeneity, anisotropy, and sandiness, are explored extensively on the dimensionless phase velocity of Rayleigh-type wave propagation. Notably, the study addresses a gap in existing literature by examining the propagation of Rayleigh-type waves in a transversely isotropic, semi-infinite medium subjected to the combined effects of initial stress, material heterogeneity, and dry sandy parameters, thereby contributing novel insights to the field.

2.2 FORMULATION AND SOLUTION OF THE PROBLEM

In this study, the propagation of Rayleigh-type waves within a structure comprising heterogeneous, transversely isotropic, initially stressed sandy semi-infinite medium bounded by an x_1x_2-plane, where x_1-axis is horizontal and the x_2-axis is oriented downwards vertically is investigated. The medium experiences initial stress S_{11} along the x_1-axis (illustrated in Figure 2.1), with the elastic coefficients, density, and initial stress varying vertically with depth.

The displacement vector $\vec{u}$ at any time t is given as follows:

$$\vec{u} = u_i\left(x_1, x_2, x_3, t\right), \ i = 1,2,3 \tag{2.1}$$

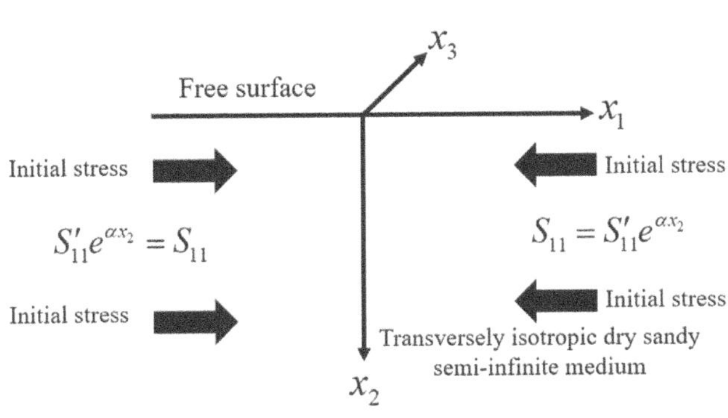

Figure 2.1 Schematic diagram of the problem.

where u_1, u_2, u_3 are the displacement components along the direction of x_1, x_2, x_3, respectively.

The stress–strain relationship for the transversely isotropic [16] sandy medium is defined as follows:

$$\begin{bmatrix} \tau_{11} \\ \tau_{22} \\ \tau_{33} \\ \tau_{12} \\ \tau_{13} \\ \tau_{23} \end{bmatrix} = \begin{bmatrix} c_{11} & c_{12} & c_{13} & 0 & 0 & 0 \\ c_{12} & c_{11} & c_{13} & 0 & 0 & 0 \\ c_{13} & c_{13} & c_{33} & 0 & 0 & 0 \\ 0 & 0 & 0 & \frac{c_{11}-c_{12}}{2} & 0 & 0 \\ 0 & 0 & 0 & 0 & c_{44} & 0 \\ 0 & 0 & 0 & 0 & 0 & c_{44} \end{bmatrix} \begin{bmatrix} e_{11} \\ e_{22} \\ e_{33} \\ e_{12} \\ e_{13} \\ e_{23} \end{bmatrix} \tag{2.2}$$

where c_{ij} are the elastic stiffness constants, τ_{ij} are the stress-tensor components, $e_{ij} = \frac{1}{2}\left(u_{i,j} + u_{j,i}\right)$ are strain-tensor components, and if $i \neq j, e_{ii} = u_{i,i}$, and η is the sandy parameter.

According to Biot [8], in the absence of external forces, the equilibrium equations for an unbounded medium with initial stresses S_{ij} in the Cartesian coordinate system (x_1, x_2, x_3) are given by

$$\frac{\partial \tau_{ij}}{\partial x_j} + \frac{\partial}{\partial x_j}\left[S_{kj}\omega_{ik} + S_{ij}e - S_{ik}e_{kj}\right] = \rho \frac{\partial^2 u_i}{\partial t^2}, \qquad i, j, k = 1,2,3 \tag{2.3}$$

where $\omega_{ik} = \left(u_{i,k} - u_{k,i}\right)/2$ is the rotational transform, $u_{i,j} = \partial u_i / \partial x_j$, $i, j, k = 1,2,3, e = \sum_{i=1}^{3} \frac{\partial u_i}{\partial x_i}$ is the cubical dilatation, S_{ij} is the initial stress in the medium, ρ is the density of the medium, and t is time.

Due to the presence of exponential heterogeneity in initial stress, density, and elastic constants, these parameters vary with depth and are considered as follows:

$$\left.\begin{aligned} & c_{11} = c'_{11}e^{\alpha x_2}, c_{12} = c'_{12}e^{\alpha x_2}, c_{13} = c'_{13}e^{\alpha x_2}, \\ & c_{33} = c'_{33}e^{\alpha x_2}, c_{44} = c'_{44}e^{\alpha x_2}, \rho = \rho_1 e^{\alpha x_2}, S_{11} = S'_{11}e^{\alpha x_2} \end{aligned}\right\} \tag{2.4}$$

where $c'_{11}, c'_{12}, c'_{13}, c'_{33}, c'_{44}$ are the elastic constants at $x_2 = 0, \alpha$ represents the material heterogeneity parameter in the medium with the dimension as inverse of length, ρ and S_{11} are the density and initial stress, respectively, whereas ρ_1, S'_{11} are constant for the anisotropic medium.

Using Eqs. (2.1), (2.2), and (2.3) in Eq. (2.4), the equations of motion are obtained in terms of displacement components and may be written as follows:

$$
\begin{aligned}
&c'_{11}\frac{\partial^2 u_1}{\partial x_1^2}+c'_{12}\frac{\partial^2 u_2}{\partial x_1\,\partial x_2}+c'_{13}\frac{\partial^2 u_3}{\partial x_1\,\partial x_3}+\alpha\frac{c'_{11}-c'_{12}}{2}\left(\frac{\partial u_1}{\partial x_2}+\frac{\partial u_2}{\partial x_2}\right)\\
&\quad+\frac{c'_{11}-c'_{12}}{2}\left(\frac{\partial^2 u_1}{\partial x_2^2}+\frac{\partial^2 u_2}{\partial x_1\,\partial x_2}\right)+c'_{44}\left(\frac{\partial^2 u_1}{\partial x_3^2}+\frac{\partial^2 u_3}{\partial x_1\,\partial x_3}\right)\\
&\quad+\frac{S'_{11}}{\eta}\left(\frac{\partial^2 u_2}{\partial x_1\,\partial x_2}+\frac{\partial^2 u_3}{\partial x_1\,\partial x_3}\right)-\alpha\frac{S'_{11}}{2\eta}\left(\frac{\partial u_1}{\partial x_2}+\frac{\partial u_2}{\partial x_1}\right)\\
&\quad-\frac{S'_{11}}{2\eta}\left(\frac{\partial^2 u_1}{\partial x_2^2}+\frac{\partial^2 u_2}{\partial x_1\,\partial x_2}\right)-\frac{S'_{11}}{2\eta}\left(\frac{\partial^2 u_1}{\partial x_3^2}+\frac{\partial^2 u_3}{\partial x_1\,\partial x_3}\right)=\frac{\rho_1}{\eta}\frac{\partial^2 u_1}{\partial t^2},
\end{aligned}
\tag{2.5}
$$

$$
\begin{aligned}
&\frac{c'_{11}-c'_{12}}{2}\left(\frac{\partial^2 u_1}{\partial x_1\,\partial x_2}+\frac{\partial^2 u_2}{\partial x_1^2}\right)+\alpha\left(c'_{12}\frac{\partial u_1}{\partial x_1}+c'_{11}\frac{\partial u_2}{\partial x_2}+c'_{13}\frac{\partial u_3}{\partial x_3}\right)\\
&\quad+\left(c'_{12}\frac{\partial^2 u_1}{\partial x_1\,\partial x_2}+c'_{11}\frac{\partial^2 u_2}{\partial x_2^2}+c'_{13}\frac{\partial^2 u_3}{\partial x_3\,\partial x_2}\right)\\
&\quad+c'_{44}\left(\frac{\partial^2 u_2}{\partial x_3^2}+\frac{\partial^2 u_3}{\partial x_2\,\partial x_3}\right)+\frac{S'_{11}}{2\eta}\left(\frac{\partial^2 u_1}{\partial x_1\,\partial x_2}-\frac{\partial^2 u_2}{\partial x_1^2}\right)=\frac{\rho_1}{\eta}\frac{\partial^2 u_2}{\partial t^2},
\end{aligned}
\tag{2.6}
$$

$$
\begin{aligned}
&c'_{44}\left(\frac{\partial^2 u_1}{\partial x_1\,\partial x_3}+\frac{\partial^2 u_3}{\partial x_1^2}\right)+\alpha c'_{44}\left(\frac{\partial u_2}{\partial x_3}+\frac{\partial u_3}{\partial x_2}\right)+c'_{44}\left(\frac{\partial^2 u_2}{\partial x_2\,\partial x_3}+\frac{\partial^2 u_3}{\partial x_2^2}\right)\\
&\quad+c'_{13}\left(\frac{\partial^2 u_1}{\partial x_1\,\partial x_3}+\frac{\partial^2 u_2}{\partial x_2\,\partial x_3}\right)+c'_{33}\frac{\partial^2 u_3}{\partial x_3^2}+\frac{S'_{11}}{2\eta}\left(\frac{\partial^2 u_1}{\partial x_1\,\partial x_3}-\frac{\partial^2 u_3}{\partial x_1^2}\right)=\frac{\rho_1}{\eta}\frac{\partial^2 u_3}{\partial t^2}.
\end{aligned}
\tag{2.7}
$$

The displacement components are characterized by the nature of the propagation of Rayleigh-type wave and are described as follows:

$$
u_1=u_1(x_1,x_2,t),u_2=u_2(x_1,x_2,t),u_3=0\text{ and }\frac{\partial(\bullet)}{\partial x_3}\equiv 0.
\tag{2.8}
$$

In view of Eq. (2.8), Eq. (2.7) vanishes identically, and nonvanishing coupled equations of motions for the propagation of Rayleigh-type wave Eqs. (2.5) and (2.6) result in the following:

$$
\begin{aligned}
&c'_{11}\frac{\partial^2 u_1}{\partial x_1^1}+c'_{12}\frac{\partial^2 u_2}{\partial x_1\,\partial x_2}+\alpha\frac{c'_{11}-c'_{12}}{2}\left(\frac{\partial u_1}{\partial x_2}+\frac{\partial u_2}{\partial x_1}\right)+\frac{c'_{11}-c'_{12}}{2}\left(\frac{\partial^2 u_1}{\partial x_2^2}+\frac{\partial^2 u_2}{\partial x_1\,\partial x_2}\right)\\
&\quad-\frac{\alpha}{2}\frac{S'_{11}}{\eta}\left(\frac{\partial u_1}{\partial x_2}+\frac{\partial u_2}{\partial x_1}\right)-\frac{S'_{11}}{2\eta}\left(\frac{\partial^2 u_1}{\partial x_2^2}-\frac{\partial^2 u_2}{\partial x_1\,\partial x_2}\right)=\frac{\rho_1}{\eta}\frac{\partial^2 u_1}{\partial t^2}
\end{aligned}
\tag{2.9}
$$

and

$$\frac{c'_{11}-c'_{12}}{2}\left(\frac{\partial^2 u_1}{\partial x_1\,\partial x_2}+\frac{\partial^2 u_2}{\partial x_2^2}\right)+\alpha\left(c'_{12}\frac{\partial u_1}{\partial x_1}+c'_{11}\frac{\partial u_2}{\partial x_2}\right)+\left(c'_{12}\frac{\partial^2 u_1}{\partial x_1\,\partial x_2}+c'_{11}\frac{\partial^2 u_2}{\partial x_2^2}\right)$$
$$-\frac{S'_{11}}{2\eta}\left(\frac{\partial^2 u_1}{\partial x_1\,\partial x_2}-\frac{\partial^2 u_2}{\partial x_1^2}\right)=\frac{\rho_1}{\eta}\frac{\partial^2 u_2}{\partial t^2}. \tag{2.10}$$

For the solution of Eqs. (2.9) and (2.10), consider the following substitution:

$$\left.\begin{aligned} u_1(x_1,x_2,t)&=P(x_2)e^{ik(x_1-ct)},\\ u_2(x_1,x_2,t)&=Q(x_2)e^{ik(x_1-ct)}\end{aligned}\right\} \tag{2.11}$$

where P and Q are the undetermined functions of x_2, c is the common phase velocity, k is the wave number, and $i=\sqrt{-1}$.

With the aid of Eq. (2.11), Eqs. (2.9) and (2.10) yield the following:

$$\left[\left(\frac{c'_{11}-c'_{12}}{2}-\frac{S'_{11}}{2\eta}\right)D^2+\alpha\left(\frac{c'_{11}-c'_{12}}{2}-\frac{S'_{11}}{2\eta}\right)D+k^2\left(\frac{\rho}{\eta}c^2-c'_{11}\right)\right]P$$
$$+ik\left[\left(\frac{c'_{11}+c'_{12}}{2}+\frac{S'_{11}}{2\eta}\right)D+\alpha\left(\frac{c'_{11}-c'_{12}}{2}-\frac{S'_{11}}{2\eta}\right)\right]Q=0 \tag{2.12}$$

and

$$ik\left[\left(\frac{c'_{11}+c'_{12}}{2}-\frac{S'_{11}}{2\eta}\right)D+\alpha c'_{12}\right]P+\left[c'_{11}D^2+\alpha c'_{11}D+k^2\left(\frac{\rho}{\eta}c^2+\frac{S'_{11}}{2\eta}-\frac{c'_{11}-c'_{12}}{2}\right)\right]Q=0 \tag{2.13}$$

where $D^2\equiv\frac{d^2}{dx_2^2}$, $D\equiv\frac{d}{dx_2}$.

Eqs. (2.12) and (2.13) require exponential solution for the surface waves, where u_1 and u_2 approach to zero as x_2 increases. Thus, horizontal and vertical displacements in the medium, due to the Rayleigh-type waves' propagation, can be expressed as follows:

$$\left.\begin{aligned} u_1(x_1,x_2,t)&=\left\{K_1e^{-im_1\eta x_2}+K_2e^{-im_2\eta x_2}\right\}e^{ik(x_1-ct)},\\ u_2(x_1,x_2,t)&=\left\{K_3e^{-im_1\eta x_2}+K_4e^{-im_2\eta x_2}\right\}e^{ik(x_1-ct)}\end{aligned}\right\} \tag{2.14}$$

where $m_j\,(j=1,2)$ are the roots of

$$m^4+\frac{2i\alpha}{k}m^3-\left[\left(\frac{c^2}{\eta\alpha_1^2}+\varepsilon\frac{c'_{11}-c'_{12}}{c'_{11}}-\frac{c'_{11}-c'_{12}}{2c'_{11}}\right)+\frac{\alpha^2}{k^2}+\left(\frac{c^2}{\eta\beta_1^2}-\frac{2c'_{11}}{c'_{11}-c'_{12}}\right)\frac{1}{1-\varepsilon}\right.$$
$$\left.+\left(\frac{c'_{11}+c'_{12}}{2c'_{11}}+\varepsilon\frac{c'_{11}-c'_{12}}{2c'_{11}}\right)\left(\frac{c'_{11}+c'_{12}}{c'_{11}-c'_{12}}+\varepsilon\right)\frac{1}{1-\varepsilon}\right]m^2$$
$$-\frac{i\alpha}{k}\left[\left(\frac{c^2}{\eta\alpha_1^2}+2\varepsilon\frac{c'_{11}-c'_{12}}{2c'_{11}}-\frac{c'_{11}-c'_{12}}{2c'_{11}}\right)+\left(\frac{c^2}{\eta\beta_1^2}-\frac{2c'_{11}}{c'_{11}-c'_{12}}\right)\frac{1}{1-\varepsilon}\right. \tag{2.15}$$
$$\left.+\left(\frac{c'_{11}+c'_{12}}{c'_{11}-c'_{12}}+\varepsilon\right)\frac{c'_{12}}{c'_{11}(1-\varepsilon)}+\left(\frac{c'_{11}+c'_{12}}{2c'_{11}}-\varepsilon\frac{c'_{11}-c'_{12}}{2c'_{11}}\right)\right]m$$
$$+\left[\left(\frac{c^2}{\eta\alpha_1^2}-1\right)\left(\frac{c^2}{\eta\beta_1^2(1-\varepsilon)}-1+2\varepsilon\right)+\frac{\alpha^2}{k^2}\frac{c'_{12}}{c'_{11}}\right]=0$$

with positive real part and $K_3=r_1K_1$, $K_4=r_2K_2$, $\alpha_1=\sqrt{\frac{c'_{11}}{\rho}}$ is the transverse wave velocity, $\beta_1=\sqrt{\frac{c'_{11}-c'_{12}}{2\rho}}$ is the longitudinal wave velocity, $\varepsilon=\frac{S'_{11}}{c'_{11}-c'_{12}}$ is the dimensionless initial stress, and where values of r_j are defined as follows:

$$r_j=\frac{(1-\varepsilon)m_j^2+\frac{i\alpha}{k}(1-\varepsilon)m_j-2\left(\frac{c^2}{\eta\beta_1^2}-\frac{2c'_{11}}{c'_{11}-c'_{12}}\right)}{\left(\frac{c'_{11}+c'_{12}}{c'_{11}-c'_{12}}+\varepsilon\right)m_j+\frac{i\alpha}{k}(1-\varepsilon)},\quad j=1,2. \tag{2.16}$$

The boundary conditions at the free surface $x_2=0$ of the semi-infinite medium require the traction-free condition which is defined mathematically as follows:

$$\left.\begin{aligned}&e^{\alpha x_2}\left[c'_{12}\frac{\partial u_1}{\partial x_1}+c'_{11}\frac{\partial u_2}{\partial x_2}\right]=0,\\&\frac{c'_{11}-c'_{12}}{2}\left[\frac{\partial u_1}{\partial x_2}+\frac{\partial u_2}{\partial x_1}\right]=0.\end{aligned}\right\} \tag{2.17}$$

By applying the boundary conditions mentioned above and removing K_1 and K_2, a closed-form relation is obtained as follows:

$$(m_1-r_1)(c'_{12}-r_2m_2c'_{11})=(m_2-r_2)(c'_{12}-r_1m_1c'_{11}). \tag{2.18}$$

This represents the wave velocity relationship for Rayleigh-type waves in a heterogeneous transversely isotropic medium under initial stress, with the values of m_1 and m_2 derived from Eq. (2.15).

2.3 PARTICULAR CASE

Case 1: When the semi-infinite medium is homogeneous, free from initial stress and sandy parameters, i.e. $\alpha = 0$, $\varepsilon = 0$, and $\eta = 1$, the wave velocity expression (18) for the propagation of Rayleigh-type wave in transversely isotropic semi-infinite medium simplifies to the following:

$$\left(m_1 - r_1\right)\left(c'_{12} - r_2 m_2 c'_{11}\right) = \left(m_2 - r_2\right)\left(c'_{12} - r_1 m_1 c'_{11}\right) \tag{2.19}$$

where m_j $(j = 1,2)$ are the roots of the equation

$$m^4 - \left[\left(\frac{c^2}{\alpha_1^2} - \frac{c'_{11} - c'_{12}}{2c'_{11}}\right) + \left(\frac{c^2}{\beta_1^2} - \frac{2c'_{11}}{c'_{11} - c'_{12}}\right) + \frac{\left(c'_{11} + c'_{12}\right)^2}{2c'_{11}\left(c'_{11} - c'_{12}\right)}\right] m^2 + \left[\left(\frac{c^2}{\alpha_1^2} - 1\right)\left(\frac{c^2}{\beta_1^2} - 1\right)\right] = 0 \tag{2.20}$$

with positive real part and values of $r_j = \dfrac{m_j^2 - 2\left(\frac{c^2}{\beta_1^2} - \frac{2c'_{11}}{c'_{11} - c'_{12}}\right)}{\frac{c'_{11} + c'_{12}}{c'_{11} - c'_{12}} m_j}$, $j = 1,2.$

Case 2: When the semi-infinite medium is homogeneous, isotropic and free from initial stress and sandy parameters, i.e. $\alpha = 0$, $\varepsilon = 0, c'_{11} = \lambda + 2\mu$, $c'_{12} = \lambda$ (λ and μ are Lame's constants), and $\eta = 1$, the wave velocity expression (18) for Rayleigh-type wave propagation in an isotropic medium transforms to the following:

$$\left(m_1 - r_1\right)\left(c'_{12} - r_2 m_2 c'_{11}\right) = \left(m_2 - r_2\right)\left(c'_{12} - r_1 m_1 c'_{11}\right) \tag{2.21}$$

where m_j $(j = 1,2)$ are the roots of the following:

$$m^4 - \left[\left(\frac{c^2}{\alpha_1^2} - \frac{\mu}{\lambda + 2\mu}\right) + \left(\frac{c^2}{\beta_1^2} - \frac{\lambda + 2\mu}{\mu}\right) + \frac{(\lambda + \mu)^2}{\mu(\lambda + 2\mu)}\right] m^2 + \left[\left(\frac{c^2}{\alpha_1^2} - 1\right)\left(\frac{c^2}{\beta_1^2} - 1\right)\right] = 0 \tag{2.22}$$

with positive real part and

$$k_3 = \alpha_2 k_1,\ k_4 = \alpha_3 k_2,\ \alpha_1^2 = \frac{\lambda + 2\mu}{\rho},\ \beta_1^2 = \frac{\mu}{\rho} \text{ and}$$

$$\alpha_j = \frac{m_j^2 - 2\left(\dfrac{c^2}{\alpha_4^2} - \dfrac{\lambda + 2\mu}{\mu}\right)}{\dfrac{\lambda + \mu}{\mu} m_j}, \quad j = 2,3.$$

The relation derived in Eq. (2.21) aligns closely with the findings of Chatterjee et al. [11] who investigated the removal of the heterogeneity parameter in their analysis of a semi-infinite medium.

Case 3: In the case of an isotropic semi-infinite medium with a Poisson's ratio of 0.25, which results in equal Lamé constants (i.e. $\lambda = \mu$), the phase velocity of the Rayleigh wave is found to be 0.9194 times the shear wave velocity of the isotropic elastic semi-infinite medium. This outcome is consistent with the results obtained by Knopoff [17].

2.4 NUMERICAL CALCULATION AND DISCUSSION

This section presents numerical computations and graphical representations concerning the derived wave velocity relation for the propagation of Rayleigh-type waves in a semi-infinite medium characterized by initial stress, transverse isotropy, heterogeneity, and sandy parameters. The impact of influential factors such as initial stress, heterogeneity, anisotropy, and sandy parameters on the phase velocity of Rayleigh-type wave propagation is investigated and illustrated graphically. To facilitate the numerical analysis, transversely isotropic (beryl) material data from Payton [16] and isotropic data from Gubbins [18] are utilized.

For beryl material

$$c_{11} = 28.2 \times 10^{10} N/m^2, \quad c_{12} = 9.94 \times 10^{10} N/m^2,$$
$$c_{13} = 6.95 \times 10^{10} N/m^2, \quad c_{33} = 24.8 \times 10^{10} N/m^2,$$
$$c_{44} = 6.86 \times 10^{10} N/m^2, \quad \rho = 2700\ kg/m^3.$$

For isotropic material

$$\mu = 7.38 \times 10^{10}\ N/m^2,\ \ \lambda = 5.4 \times 10^{10}\ N/m^2,\ \ \rho = 3323 kg/m^3.$$

Figures 2.2(a), 2.3(a), and 2.4(a) illustrate the propagation of Rayleigh-type waves in a complex structure comprising an initially stressed, transversely isotropic, heterogeneous, dry, sandy, semi-infinite medium, whereas

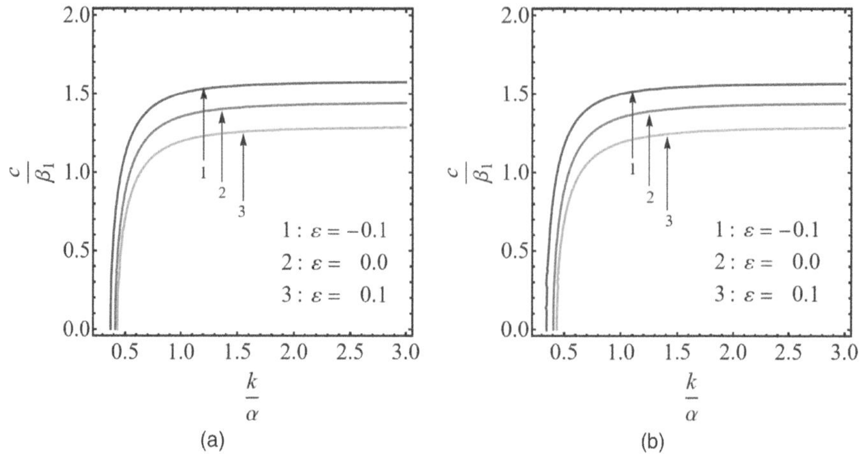

Figure 2.2 Variation of dimensionless phase velocity c/β_1 against dimensionless wave number k/α for different values of initial stress parameter $\varepsilon = S'_{11}/(c'_{11} - c'_{12})$ when medium is (a) transversely isotropic, (b) isotropic.

Figures 2.2(b), 2.3(b), and 2.4(b) depict the propagation of Rayleigh-type waves in a structure constituting an isotropic, heterogeneous, initially stressed, dry, sandy, semi-infinite medium. These figures demonstrate the relationship between dimensionless phase velocity (c/β_1) and wave number (k/α) for the propagation of Rayleigh-type wave, considering the influence of several parameters such as initial stress, sandiness, heterogeneity, and anisotropy. It is observed that the phase velocity of Rayleigh-type waves increases as the wave number increases.

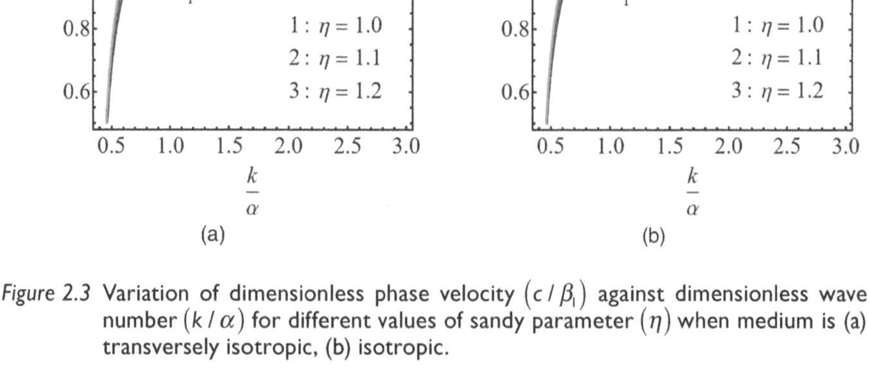

Figure 2.3 Variation of dimensionless phase velocity (c/β_1) against dimensionless wave number (k/α) for different values of sandy parameter (η) when medium is (a) transversely isotropic, (b) isotropic.

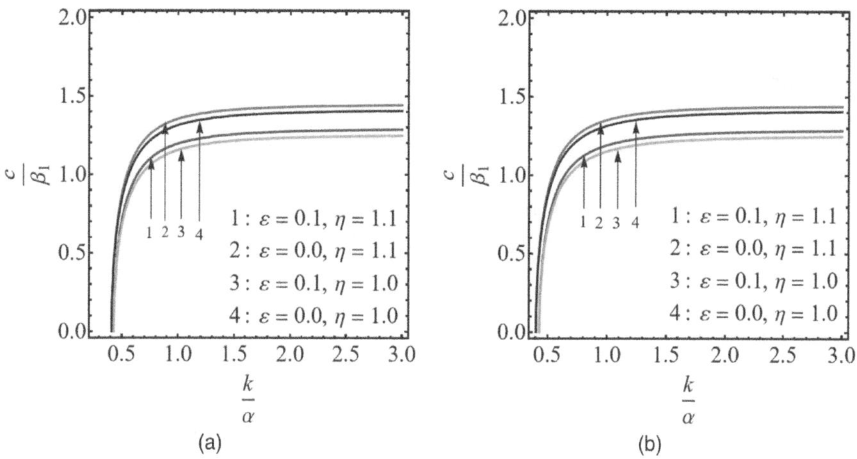

Figure 2.4 Variation of dimensionless phase velocity (c/β_1) against dimensionless wave number (k/α) for different cases of initial stress and sandiness, when medium is (a) transversely isotropic, (b) isotropic.

Figure 2.2 depicts the variation of the phase velocity of Rayleigh-type waves against the dimensionless wave number for two scenarios: a semi-infinite medium composed of beryl material (Figure 2.2(a)) and an isotropic material (Figure 2.2(b)). In both cases, curve 1 represents scenarios with tensile initial stress, while curve 3 represents scenarios with compressive initial stress applied to the heterogeneous, semi-infinite medium with dry sandiness. Additionally, curve 2 represents scenarios where the medium is dry sandy and heterogeneous but without the initial stress. These curves reveal that as the initial stress parameter increases, the phase velocity of Rayleigh-type waves decreases. The horizontal initial stress applied to the semi-infinite medium increases the vertical inter-particle distance, consequently enlarging the voids within the medium. This expansion discourages the distortional component of Rayleigh-type wave propagation, leading to a decrease in phase velocity. This phenomenon suggests that the initial stress parameter associated with the semi-infinite medium hinders the phase velocity of Rayleigh-type waves.

In Figure 2.3, the curves are plotted to demonstrate the effect of sandiness parameter on the phase velocity of Rayleigh-type wave in a transversely isotropic semi-infinite medium (Figure 2.3(a)) and in isotropic semi-infinite medium (Figure 2.3(b)) under the consideration of compressive initial stress. Curve 1 is sketched for the case when the semi-infinite medium is free from the sandiness, whereas curves 2 and 3 are with the sandiness in the semi-infinite medium. It has been observed that the variation of phase velocity increases with the increase of sandiness in the structure. The presence of sandiness may alter the energy dissipation mechanisms within the medium. Sands typically have higher damping characteristics compared to solid materials. As a result, the energy loss during wave propagation may be reduced in a sandy medium, allowing Rayleigh waves to travel faster.

Figure 2.4 presents a comparative analysis of the influence of compressive initial stress and sandiness on the phase velocity of Rayleigh-type waves for four distinct scenarios. Curve 1 illustrates the case where the structure undergoes compressive initial stress and sandiness, while curve 2 depicts the scenario without initial stress but with sandiness. Curve 3 corresponds to the condition with initial stress but without sandiness, and curve 4 represents the absence of both initial stress and sandiness. These analyses are conducted for both transversely isotropic (Figure 2.4(a)) and isotropic (Figure 2.4(b)) semi-infinite mediums. The findings indicate that when the semi-infinite medium is devoid of compressive initial stress but includes sandiness, the phase velocity of Rayleigh-type waves reaches its maximum. Conversely, the presence of initial stress without sandiness results in the minimum phase velocity of the propagating wave, regardless of whether the semi-infinite medium is transversely isotropic or isotropic. Notably, the presence of sandiness appears to enhance wave propagation, while the imposition of initial stress exerts a dampening effect, leading to reduced wave velocities.

Figure 2.5 illustrates the variation of dimensionless phase velocity against dimensionless wave number to examine the influence of anisotropy. The comparison is made across situation when the semi-infinite medium is with heterogeneity, initial stress, and sandiness, considering the presence (curve 1) or absence (curve 2) of anisotropy. It is clear from the figure that the anisotropy enhances the phase velocity of the propagating wave. It is because the wave may encounter regions of the medium where the elastic moduli are higher or where the material is stiffer, allowing it to propagate faster compared

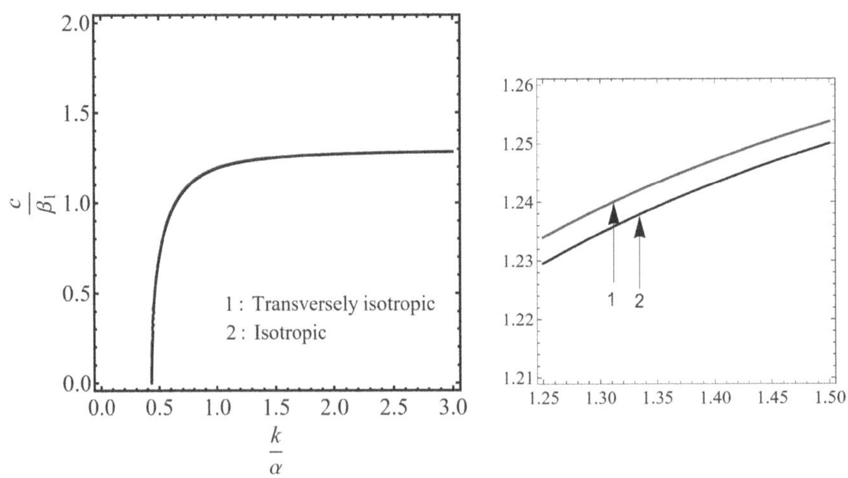

Figure 2.5 Variation of dimensionless phase velocity (c/β_1) against dimensionless wave number (k/α) for the comparison of anisotropy.

to isotropic medium where the elastic moduli are uniform in all directions. Moreover, it is noted from all the figures that heterogeneity also adversely affects the phase velocity of Rayleigh-type waves, regardless of whether the semi-infinite medium is composed of anisotropic or isotropic material. The presence of heterogeneity introduces spatial variations in material properties, leading to wavefront dispersion and scattering, which ultimately decrease the phase velocity of Rayleigh-type waves compared to homogeneous medium.

2.5 CONCLUSIONS

This chapter examined the propagation of Rayleigh-type waves in an initially stressed, heterogeneous, transversely isotropic, sandy, semi-infinite medium. A closed-form wave velocity relation was derived analytically for this structure, and special cases were discussed in the absence of heterogeneity, sandiness, anisotropy, and initial stress parameters. Numerical calculations were conducted to explore the influence of these parameters on the phase velocity of Rayleigh-type wave propagation, supported by graphical demonstrations.

The key findings of this study can be summarized as follows:

- Initial stress was found to adversely affect the phase velocity of Rayleigh-type wave propagation with respect to wave number, while sandy parameters exhibited a favourable effect on phase velocity.
- The impact of initial stress on phase velocity was observed to be more significant than that of sandy parameters.
- Phase velocity consistently increased with wave number across all scenarios.
- The obtained results were in good agreement with existing literature, validating the analysis.
- Comparative analysis revealed that the presence of sandiness without compressive initial stress resulted in maximum phase velocity, while the presence of compressive initial stress without sandiness led to minimum phase velocity, irrespective of the semi-infinite medium's isotropy or anisotropy.
- Anisotropy was found to enhance the phase velocity of the propagating wave, as indicated by the results.

In summary, this research offers valuable insights into the intricate relationships among various parameters affecting the phase velocity of Rayleigh-type waves, enhancing our understanding of wave behaviour in heterogeneous mediums. The investigations into initially stressed elastic mediums remain crucial and warrant ongoing attention, given their significance across multiple disciplines such as earthquake engineering, seismology, and geophysics. The exploration of surface wave in heterogeneous geological mediums holds considerable promise for geophysical prospecting and improving our comprehension of earthquake causality and damage estimation. The systematic examination of these surface waves not only promotes human safety but also satisfies our curiosity about the structure and evolution of the Earth. Furthermore, the use of artificially generated seismic waves provides valuable insights into rock layer configurations for oil exploration and offers crucial information about shallow layer rigidity for engineering applications.

ACKNOWLEDGEMENTS

Pulkit Kumar expresses gratitude to the CNDE at Iowa State University, Ames, Iowa, USA, and Moumita Mahanty extends thanks to the Centre for Data Science at the ITER, Siksha 'O' Anusandhan (Deemed to be University), Bhubaneswar, Odisha, India for providing research facilities.

REFERENCES

1. V. T. Buchwald, "Rayleigh waves in transversely isotropic media," *Q. J. Mech. Appl. Math.*, vol. 14, no. 3, pp. 293–318, 1961, doi: 10.1093/qjmam/14.3.293.
2. S. Dutta, "Rayleigh wave propagation in a two-layer anisotropic media," *Pure Appl. Geophys. PAGEOPH*, vol. 60, no. 1, pp. 51–60, Dec. 1965, doi: 10.1007/BF00874806.
3. A. Chattopadhyay, N. P. Mahata, and A. Keshri, "Rayleigh waves in a medium under initial stresses," *Acta Geophys. Pol.*, vol. 34, no. 1, pp. 57–62, 1986.
4. D. P. Acharya, I. Roy, and S. Sengupta, "Effect of magnetic field and initial stress on the propagation of interface waves in transversely isotropic perfectly conducting media," *Acta Mech.*, vol. 202, nos. 1–4, pp. 35–45, Jan. 2009, doi: 10.1007/s00707-008-0027-5.
5. A. Chattopadhyay, "On the strong SH motion in a transversely isotropic layer lying over an isotropic elastic material due to a momentary point source," *Ind. J. Pure Appl. Maths.*, Springer, vol. 9, pp. 886–892, 1978, Accessed: Dec. 20, 2017. [Online]. Available: http://www.insa.nic.in/writereaddata/UpLoadedFiles/IJPAM/20005a67_528.pdf
6. A. K. Singh, N. Kumari, A. Chattopadhyay, and S. A. Sahu, "Smooth moving punch in an initially stressed transversely isotropic magnetoelastic medium due to shear wave," *Mech. Adv. Mater. Struct.*, vol. 23, no. 7, pp. 774–783, Jul. 2016, doi: 10.1080/15376494.2015.1029161.
7. W. H. Weiskopf, "Stresses in soils under a foundation," *J. Franklin Inst.*, vol. 239, no. 6, pp. 445–465, Jun. 1945, doi: 10.1016/0016-0032(45)90189-X.
8. M. A. Biot, "The influence of initial stress on elastic waves," *J. Appl. Phys.*, vol. 11, no. 8, pp. 522–530, 1940, doi: 10.1063/1.1712807.
9. M. A. Biot, "Surface instability in finite anisotropic elasticity under initial stress," *Proc. R. Soc. London. Ser. A. Math. Phys. Sci.*, vol. 273, no. 1354, pp. 329–339, May 1963, doi: 10.1098/rspa.1963.0092.
10. A. N. Guz, "Elastic waves in bodies with initial (residual) stresses," *Int. Appl. Mech.*, vol. 38, no. 1, pp. 23–59, 2002, doi: 10.1023/A:1015379824503.
11. M. Chatterjee, S. Dhua, A. Chattopadhyay, and S. A. Sahu, "Seismic waves in heterogeneous crust-mantle layers under initial stresses," *J. Earthq. Eng.*, vol. 20, no. 1, pp. 39–61, 2016, doi: 10.1080/13632469.2015.1038371.

12. K. E. Bullen and B. A. Bolt, *An introduction to the theory of seismology*. Fourth Edition, Cambridge University Press, New York, 1985. doi: 10.1193/1.1585435
13. Y. Sato, "Love waves propagated upon heterogeneous medium," *Bull. Earthq. Res. Inst. Univ. Tokyo*, vol. 30, pp. 1–12, 1952.
14. S. Dutta, "Rayleigh waves in a two-layer heterogeneous medium," *Bull. Seismol. Soc. Am.*, vol. 53, no. 3, pp. 517–526, Apr. 1963. doi: 10.1785/bssa0530030517.
15. J. Bhattacharya, "The possibility of the propagation of Love type waves in an intermediate heterogeneous layer lying between two semi-infinite isotropic homogeneous elastic layers," *Pure Appl. Geophys. PAGEOPH*, vol. 72, no. 1, pp. 61–71, 1969, doi: 10.1007/BF00875693.
16. R. G. Payton, *Elastic wave propagation in transversely isotropic media*. First Edition, Kluwer Academic Publishers Group, Dordrecht, The Netherlands, 1983. DOI: 10.1007/978-94-009-6866-0.
17. L. Knopoff, "On Rayleigh wave velocities," *Bull. Seism. Soc. Am.*, vol. 42, no. 4, pp. 307–308, 1952.
18. D. Gubbins, *Seismology and plate tectonics*. Cambridge, Cambridge University Press, 1990. doi: 10.1111/j.1365-246X.1990.tb05689.x.

Chapter 3

Effects of attenuation coefficient and dissipation factors on the dispersion phenomenon of Rayleigh waves through the piezo-viscoelastic medium

Venkatesan P, Parvez Alam, Tabinda Nahid, and Sumit Kumar Vishwakarma

3.1 INTRODUCTION

Rayleigh wave propagation has been extensively studied due to its significance in non-destructive testing, material characterization, and seismology. Bluestein [1], Cheng [2], Fang [3], and Wang [4] have extensively studied Rayleigh wave movement in piezoelectric media. Piezo-viscoelastic media are distinguished from other materials under study by exclusive blending of viscoelastic and piezoelectric characteristics. The ability to convert mechanical energy into electrical energy and vice versa essentially characterizes these media. In piezo-viscoelastic media, the attenuation coefficient is one of the key factors affecting Rayleigh wave behaviour. Because viscoelastic materials have built-in energy dissipation mechanisms, this coefficient measures the rate at which the wave amplitude diminishes as it moves through the medium. In conjunction with the attenuation coefficient, dissipation factors play a pivotal role. These factors represent the extent to which the medium converts mechanical energy into heat, thereby affecting the wave's velocity and amplitude. Singh and Alam [5] have discussed Rayleigh wave propagation in the compressed viscoelastic layer.

The dispersion phenomenon, where the wave velocity depends on its frequency, is significantly impacted by these factors. Understanding how the attenuation coefficient and dissipation factors influence dispersion can provide deeper insights into the material properties and the wave dynamics within piezo-viscoelastic mediums. This knowledge is crucial for improving the design and performance of devices that rely on Rayleigh wave propagation, such as sensors and actuators in structures and advanced materials. It's widely acknowledged that numerous authors such as Saha et al. [6], Singh [7], Chai and Wu [8], Simionescu-Panait [9], Liu [10], and Son [11] have examined how Rayleigh waves affect piezoelectric structures. Yang [12] has studied several problems pertaining to the propagation of Rayleigh wave in piezoelectric materials. By examining several facets of this wave's interactions with the medium, his research offers important new perspectives on

DOI: 10.1201/9781003470885-3

how they behave. Rayleigh wave propagation through a piezoelectric strip upon a piezomagnetic semi-infinite semi-space has been studied by Pang et al [13]. The effect of rotation on Rayleigh waves was studied by Clarke and Burdness [14]. Cao et al. [15] examined the characteristic of Rayleigh wave propagation in piezoelectric semiconductor half-space. Li et al. [16] investigated the behaviour and propagation characteristics of Rayleigh wave, emphasizing their complex dynamics and fundamental ideas. Cao et al. [17] have studied the effects of Rayleigh wave in layered piezoelectric structures. The existence of Rayleigh waves on a half-space with a piezoelectric orthotropic substratum was discovered by Chaudhary et al. [18]. Singh and Sindhu [19] have discussed the propagation of Rayleigh wave in a micropolar piezoelectric medium. Paswan et al. [20] studied the influence of Rayleigh wave propagations within composite structures, analysing its impact under various conditions.

This study aims to investigate the dispersion behaviour of Rayleigh waves travelling along the free surface of a piezo-viscoelastic half-space medium, which is analysed in relation to the attenuation coefficient, dissipation factor, piezoelectric and dielectric constants. We comparatively investigated the electrically open and short conditions for this examination. The outcomes of this research will not only contribute to the theoretical framework but also have practical implications in the design and optimization of piezo-viscoelastic systems.

3.2 FORMULATION OF THE PROBLEM AND SOLUTIONS

We consider a Rayleigh wave propagation behaviour at the surface of a piezo-viscoelastic material (PVEM) half-space, as shown in Figure 3.1. The PVEM half-space is free of traction within the o-xyz coordinate system, where the x-axis follows the wave movement direction, and the z-axis extends downwards in a vertical orientation. Let us assume the displacement vector components can be described as follows:

$$u = u(x,z,t),\ v = 0,\ w = w(x,z,t),\ \varphi = \varphi(x,z,t) \text{ and } \frac{\partial(.)}{\partial y} = 0, \tag{3.1}$$

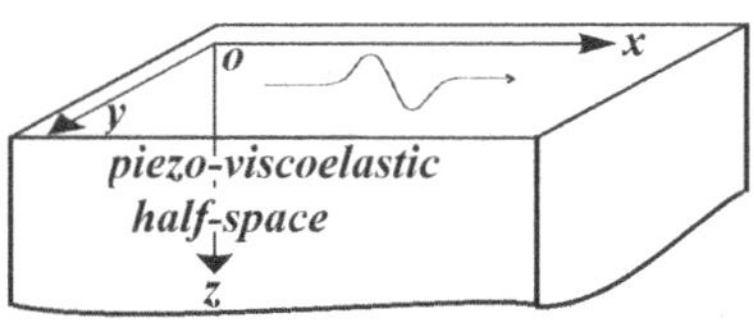

Figure 3.1 Geometry of the problem.

The equilibrium equations on electrical displacement and the equation of motion (Chaudhary et al. [18]) can be written as follows:

$$\tau_{ij,j} = \rho \ddot{u}_i, \tag{3.2}$$

$$D_{i,i} = 0,\ i, j = 1,2,3, \tag{3.3}$$

where ρ represents the mass density, u_i denotes the mechanical displacement components, while dot symbolizes differentiation in relation to time and the comma signifies differentiation in relation to spatial coordinates.

The fundamental governing relations for the PVEM medium are given below:

$$\tau_{xx} = C_{11}\frac{\partial u}{\partial x} + C_{12}\frac{\partial v}{\partial y} + C_{13}\frac{\partial w}{\partial z} + e_{31}\frac{\partial \varphi}{\partial z}, \tag{3.4}$$

$$\tau_{yy} = C_{12}\frac{\partial u}{\partial x} + C_{12}\frac{\partial v}{\partial y} + C_{13}\frac{\partial w}{\partial z} + e_{31}\frac{\partial \varphi}{\partial z}, \tag{3.5}$$

$$\tau_{zz} = C_{13}\frac{\partial u}{\partial x} + C_{13}\frac{\partial v}{\partial y} + C_{33}\frac{\partial w}{\partial z} + e_{33}\frac{\partial \varphi}{\partial z}, \tag{3.6}$$

$$\tau_{yz} = C_{44}\left(\frac{\partial v}{\partial z} + \frac{\partial w}{\partial y}\right) + e_{15}\frac{\partial \varphi}{\partial y}, \tag{3.7}$$

$$\tau_{xz} = C_{44}\left(\frac{\partial w}{\partial x} + \frac{\partial u}{\partial z}\right) + e_{15}\frac{\partial \varphi}{\partial x}, \tag{3.8}$$

$$\tau_{xy} = \left(\frac{C_{11} - C_{12}}{2}\right)\left(\frac{\partial v}{\partial x} + \frac{\partial u}{\partial y}\right), \tag{3.9}$$

$$D_x = e_{15}\left(\frac{\partial w}{\partial x} + \frac{\partial u}{\partial z}\right) - \epsilon_{11}\frac{\partial \varphi}{\partial x}, \tag{3.10}$$

$$D_y = e_{15}\left(\frac{\partial v}{\partial z} + \frac{\partial w}{\partial y}\right) - \epsilon_{11}\frac{\partial \varphi}{\partial y}, \tag{3.11}$$

$$D_z = e_{31}\frac{\partial u}{\partial x} + e_{31}\frac{\partial v}{\partial y} + e_{33}\frac{\partial w}{\partial z} - \epsilon_{11}\frac{\partial \varphi}{\partial z}. \tag{3.12}$$

where the elastic, piezoelectric, and dielectric constants of PVEM are C_{ij}, e_{ij}, and ϵ_{ij}. These constant expressions are as follows:

$$C_{ij} = C_{ij}^{*} + i\omega C_{ij}', \tag{3.13}$$

where i denotes complex number, ω is angular frequency, and C_{ij}^{*} and C_{ij}' represent the elasctic moduli and elastic loss moduli of the considered material.

The governing equation motions for piezo-viscoelastic medium are as follows:

$$C_{11}\frac{\partial^2 u}{\partial x^2} + (C_{13} + C_{44})\frac{\partial^2 w}{\partial x \partial z} + C_{44}\frac{\partial^2 u}{\partial z^2} + (e_{31} + e_{15})\frac{\partial^2 \varphi}{\partial x \partial z} = \rho\frac{\partial^2 u}{\partial t^2}, \tag{3.14}$$

$$C_{44}\frac{\partial^2 w}{\partial x^2} + (C_{13} + C_{44})\frac{\partial^2 u}{\partial x \partial z} + C_{33}\frac{\partial^2 w}{\partial z^2} + e_{15}\frac{\partial^2 \varphi}{\partial x^2} + e_{33}\frac{\partial^2 \varphi}{\partial z^2} = \rho\frac{\partial^2 w}{\partial t^2}, \tag{3.15}$$

$$e_{15}\frac{\partial^2 w}{\partial x^2} + (e_{15} + e_{31})\frac{\partial^2 u}{\partial x \partial z} + e_{33}\frac{\partial^2 w}{\partial z^2} - \epsilon_{11}\frac{\partial^2 \varphi}{\partial x^2} - \epsilon_{33}\frac{\partial^2 \varphi}{\partial z^2} = 0. \tag{3.16}$$

From Eqs. (3.14)–(3.16) can be written as follows:

$$u = Ue^{ik(x-ct)},\ w = We^{ik(x-ct)},\ \varphi = \Phi e^{ik(x-ct)}. \tag{3.17}$$

Substituting Eq. (3.17) into Eqs. (3.14)–(3.16), we get the following:

$$C_{44}U'' + (C_{13} + C_{44})ikW' + (e_{31} + e_{15})ik\Phi' + (\rho c^2 - C_{11})k^2 U = 0, \tag{3.18}$$

$$C_{33}W'' + (C_{13} + C_{44})ikU' + (\rho c^2 - C_{44})k^2 U + e_{33}\Phi'' - e_{15}k^2\Phi = 0, \tag{3.19}$$

$$(e_{15} + e_{31})ikU' + e_{33}W'' - k^2 e_{15}W + \epsilon_{11}k^2\Phi = 0. \tag{3.20}$$

Solving Eqs. (3.18)–(3.20), we obtain the following:

$$(D^6\alpha_1 + D^4\alpha_2 + D^2\alpha_3 + \alpha_4)(U, W, \Phi) = 0, \tag{3.21}$$

where α_1, α_2, α_3, α_4, v_1, and v_2 are given in Appendix.

Let us consider

$$U = A_1 e^{-\alpha z},\ W = A_3 e^{-\alpha z},\ \Phi = A_\Phi e^{-\alpha z}, \tag{3.22}$$

where A_1, A_3, and A_Φ represent coefficients, while α characterizes the decay rate originating from the free surface at $z = 0$.

Substituting Eq. (3.22) into Eqs. (3.18)–(3.20), we get the following:

$$\left(C_{44}\alpha^2+\left(\rho c^2-C_{11}\right)k^2\right)A_1-ik\alpha\left(C_{13}+C_{44}\right)A_3-ik\alpha\left(e_{31}+e_{15}\right)A_\Phi=0, \quad (3.23)$$

$$-ik\alpha\left(C_{13}+C_{44}\right)A_1+\left(C_{44}\alpha^2+\left(\rho c^2-C_{44}\right)k^2\right)A_3+\left(e_{33}\alpha^2-e_{15}k^2\right)A_\Phi=0, \quad (3.24)$$

$$-ik\alpha\left(e_{31}+e_{15}\right)A_1+\left(e_{33}\alpha^2-e_{15}k^2\right)A_3+\left(\epsilon_{11}k^2-\epsilon_{33}\alpha^2\right)A_\Phi=0, \quad (3.25)$$

There is a non-trivial solution to the homogeneous linear equations in Eqs. (3.23)–(3.25), given to the determinant of the coefficient matrix is equal to zero.

$$\begin{vmatrix} C_{44}\alpha^2+\left(\rho c^2-C_{11}\right)k^2 & -ik\alpha\left(C_{13}+C_{44}\right) & -ik\alpha\left(e_{31}+e_{15}\right) \\ -ik\alpha\left(C_{13}+C_{44}\right) & \left(C_{44}\alpha^2+\left(\rho c^2-C_{44}\right)k^2\right) & \left(e_{33}\alpha^2-e_{15}k^2\right) \\ -ik\alpha\left(e_{31}+e_{15}\right) & \left(e_{33}\alpha^2-e_{15}k^2\right) & \left(\epsilon_{11}k^2-\epsilon_{33}\alpha^2\right) \end{vmatrix}=0. \quad (3.26)$$

Eq. (3.26) is a sixth-order equation concerning. We should get three origins with a negative real component if we remove the origin with a positive real component, which can be expressed as follows:

$$\begin{cases} U(z)=\sum_{j=1}^{3}A_je^{-\alpha_jz} \\ W(z)=\sum_{j=1}^{3}m_jA_je^{-\alpha_jz} \\ \Phi(z)=\sum_{j=1}^{3}n_jA_je^{-\alpha_jz} \end{cases} \quad (3.27)$$

Solutions of Eq. (3.21) with the help of Eq. (3.27) can be obtained as follows:

$$u=\left(A_1e^{-\alpha_1z}+A_2e^{-\alpha_2z}+A_3e^{-\alpha_3z}\right)e^{ik(x-\omega t)}, \quad (3.28)$$

$$w=\left(m_1A_1e^{-\alpha_1z}+m_2A_2e^{-\alpha_2z}+m_3A_3e^{-\alpha_3z}\right)e^{ik(x-\omega t)}, \quad (3.29)$$

$$\varphi=\left(n_1A_1e^{-\alpha_1z}+n_2A_2e^{-\alpha_2z}+n_3A_3e^{-\alpha_3z}\right)e^{ik(x-\omega t)}. \quad (3.30)$$

3.3 BOUNDARY CONDITION

The following is a mathematical formulation of the boundary conditions at $z = 0$ for the current problem:

For open case

$$(a)\tau_{xz} = 0, \tag{3.31}$$

$$(b)\tau_{zz} = 0, \tag{3.32}$$

$$(c)D_z = 0, \tag{3.33}$$

For short case

$$(c)\tau_{xz} = 0, \tag{3.34}$$

$$(d)\tau_{zz} = 0, \tag{3.35}$$

$$(e)\varphi = 0, \tag{3.36}$$

3.4 DISPERSION RELATION

Using Eqs. (3.28)–(3.30), we derive a system of homogeneous algebraic equations characterized by indeterminate variables. These variables are then integrated into the boundary conditions through the corresponding components of electrical displacement.

3.4.1 Dispersion relation for open circuit case

Eqs. (3.31)–(3.33), which describe electrically open boundary conditions, can be written as follows:

$$\left(\begin{array}{ccc|c} X_{11} & X_{12} & X_{13} & A_1 \\ X_{21} & X_{22} & X_{23} & A_2 \\ X_{31} & X_{32} & X_{33} & A_3 \end{array}\right) = 0, \tag{3.37}$$

where $X_{11} = ikC_{44}m_1 - C_{44}\alpha_1 - n_1\alpha_1 e_{15}$, $X_{12} = ikC_{44}m_2 - C_{44}\alpha_2 - n_2\alpha_2 e_{15}$, $X_{13} = ikC_{44}m_3 - C_{44}\alpha_3 - n_3\alpha_3 e_{15}$, $X_{21} = ikC_{13} - \alpha_1 m_1 - n_1\alpha_1 e_{33}$, $X_{22} = ikC_{13} - \alpha_2 m_2 - n_2\alpha_2 e_{33}$, $X_{23} = ikC_{13} - \alpha_3 m_3 - n_3\alpha_3 e_{33}$, $X_{31} = ike_{31} - e_{33}\alpha_1 - n_1\alpha_1\epsilon_{11}$, $X_{32} = ike_{31} - e_{33}\alpha_2 - n_2\alpha_2\epsilon_{11}$, $X_{33} = ike_{31} - e_{33}\alpha_3 - n_3\alpha_3\epsilon_{11}$, and the values of m_1, m_2, m_3, n_1, n_2, and n_3 are given in Appendix.

The dispersion equation of Rayleigh waves is obtained from Eq. (3.37) for the presence of a significant result, which requires that the factor associated with the coefficient array must also be zero and its real part of the matrix be zero.

3.4.2 Dispersion relation for short circuit case

With the help of boundary conditions described in Eqs. (3.34)–(3.36), they can be expressed as follows:

$$\left(\begin{array}{ccc|c} Y_{11} & Y_{12} & Y_{13} & A_1 \\ Y_{21} & Y_{22} & Y_{23} & A_2 \\ Y_{31} & Y_{32} & Y_{33} & A_3 \end{array}\right) = 0, \tag{3.38}$$

where $Y_{11} = ikC_{44}m_1 - C_{44}\alpha_1 - n_1\alpha_1e_{15}$, $Y_{12} = ikC_{44}m_2 - C_{44}\alpha_2 - n_2\alpha_2e_{15}$, $Y_{13} = ikC_{44}m_3 - C_{44}\alpha_3 - n_3\alpha_3e_{15}$, $Y_{21} = ikC_{13} - \alpha_1m_1 - n_1\alpha_1e_{33}$, $Y_{22} = ikC_{13} - \alpha_2m_2 - n_2\alpha_2e_{33}$, $Y_{23} = ikC_{13} - \alpha_3m_3 - n_3\alpha_3e_{33}$, $Y_{31} = n_1$, $Y_{32} = n_2$, $Y_{33} = n_3$.

The necessary and sufficient condition for the dispersion equation of Rayleigh waves is obtained from Eq. (3.38) for the existence of a significant solution, which requires that the determinant associated with the coefficient matrix must also be zero and its real part of the matrix be zero.

Considering the complex wave number $k = k_1 + ik_2$, we obtained the following:

$$\begin{cases} k = k_1(1+i\delta),\ C_{11} = C_{11}^*(1+iQ_1),\ C_{13} = C_{13}^*(1+iQ_2),\ C_{33} = C_{33}^*(1+iQ_3), \\ C_{44} = C_{44}^*(1+iQ_4), \end{cases} \tag{3.39}$$

where k_1, k_2 are real, and the dimensionless attenuation coefficient is $\delta = \frac{k_1}{k_2}$; $Q_1 = \frac{C'_{11}}{C^*_{11}}$, $Q_2 = \frac{C'_{13}}{C^*_{13}}$ $Q_3 = \frac{C'_{33}}{C^*_{33}}$, and $Q_4 = \frac{C'_{44}}{C^*_{44}}$, are the piezo-viscoelastic constant parameters.

3.5 NUMERICAL COMPUTATIONS AND DISCUSSION

In order to demonstrate the behaviour of the Rayleigh wave in the considered geometry. We have taken the following dimension parameters (Table 3.1): ZnO are chosen to be (Cao et al. [15]) to show the effects of all the material parameters used in the analysis, which are outlined below:

$C_{11} = 21\times10^{10}\ Nm^{-2}$, $C_{13} = 10.5\times10^{10}\ Nm^{-2}$, $C_{13} = 21.1\times10^{10}\ Nm^{-2}$,

Table 3.1 Fixed values of parameters

Parameters	$\delta = (k_2/k_1)$	$Q_1 = (C'_{11}/C^*_{11})$	$Q_2 = (C'_{13}/C^*_{13})$	$Q_3 = (C'_{33}/C^*_{33})$	$Q_4 = (C'_{44}/C^*_{44})$	$e_{15}(C/m^2)$	$\epsilon_{11}(F/m)$
Figure 3.2	–	1500	1000	1300	1500	–0.48	6.71×10^{-11}
Figure 3.3	1000	–	1000	1300	1500	–0.48	6.71×10^{-11}
Figure 3.4	1000	1500	–	1300	1500	–0.48	6.71×10^{-11}
Figure 3.5	1000	1500	1000	–	1500	–0.48	6.71×10^{-11}
Figure 3.6	1000	1500	1000	1300	–	–0.48	6.71×10^{-11}
Figure 3.7	1000	1500	1000	1300	1500	–	6.71×10^{-11}
Figure 3.8	1000	1500	1000	1300	1500	–0.48	–

"–" denotes the parameter variation in the associated figure.

$C_{44} = 4.3\times10^{10}\ Nm^{-2},\ e_{31} = -0.57\ Cm^{-2},,\ e_{33} = 1.32\ Cm^{-2}\ ,\ e_{15} = -0.48\ Cm^{-2},$

$\epsilon_{11} = 7.61\times10^{-11}Fm^{-1},\ \epsilon_{33} = 8.85\times10^{-11}Fm^{-1},\ \rho = 5700\ kgm^{-3}.$

Figure 3.2 reveals that the phase velocity of Rayleigh waves is affected by the attenuation coefficient (δ), which results from the complex spatial frequency. Figure 3.2(a) and (b) shows that phase velocity decreases and attenuation coefficient's magnitude enhances.

Figure 3.3 shows the phase velocity contrary to spatial frequency for distinct magnitude of dissipation factor (Q_1). It exhibits that when wave number (k_1) rises, so does phase velocity. Figure 3.3(a) and (b) shows that changes in

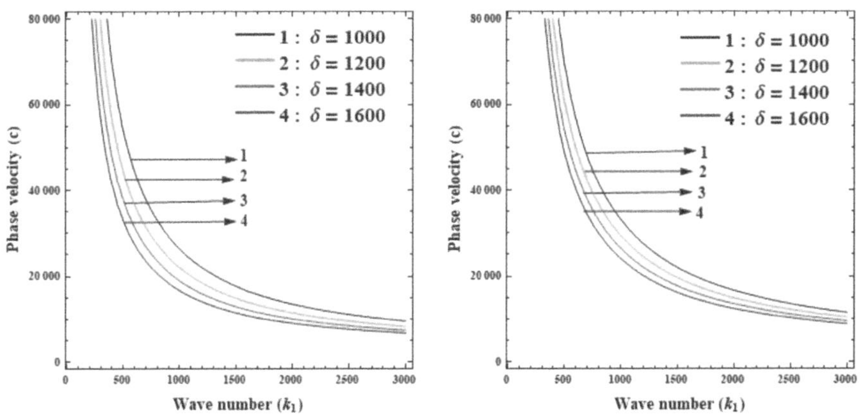

(a) (b)

Figure 3.2 The change of phase velocity (c) as a function of wave number (k_1) for distinct magnitudes of attenuation coefficient (δ): (a) represents electrically open; (b) represents electrically short.

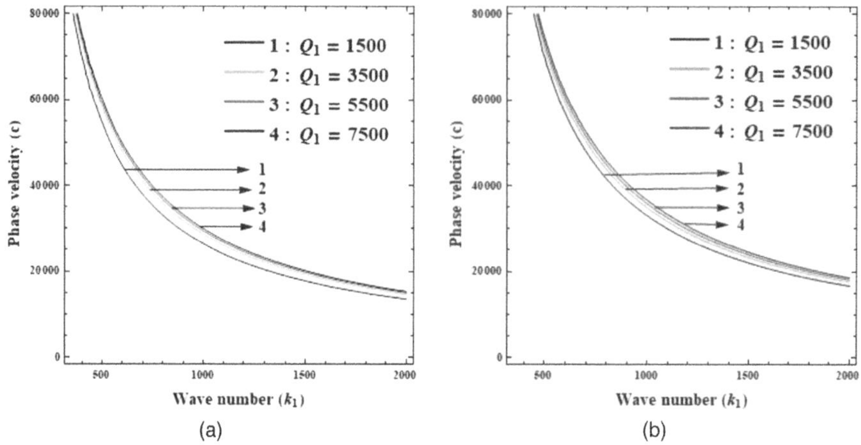

Figure 3.3 The change of phase velocity (c) as a function of real wave number (k_1) for distinct magnitudes of dissipation factor (Q_1): (a) represents electrically open; (b) represents electrically short.

the elastic constant have less of an effect in situations with electrically open and short circumstances.

The impact of dissipation factor (Q_2) on the Rayleigh wave phase velocity is shown in Figure 3.4. A thorough analysis shows that when the wave is electrically open and short, its phase velocity increases as the elastic constant increases.

Figure 3.5 illustrates the significance of piezo-viscoelastic medium's dissipation factor parameter (Q_3) on the graph. Figure 3.5(a) and (b) shows that the Rayleigh wave phase velocity increases as the elastic constant increase.

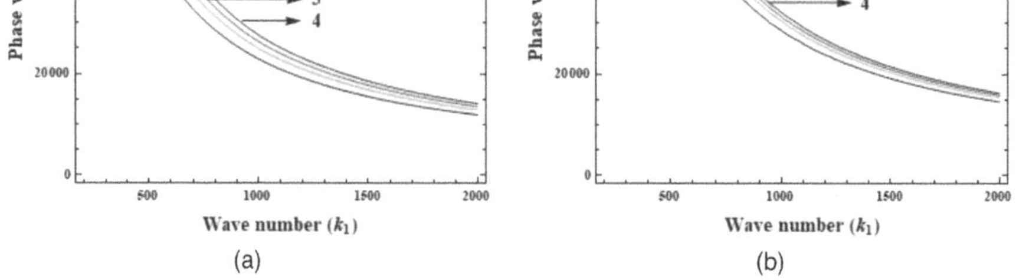

Figure 3.4 The change of phase velocity (c) as a function of real wave number (k_1) for distinct magnitudes of dissipation factor (Q_2): (a) represents electrically open; (b) represents electrically short.

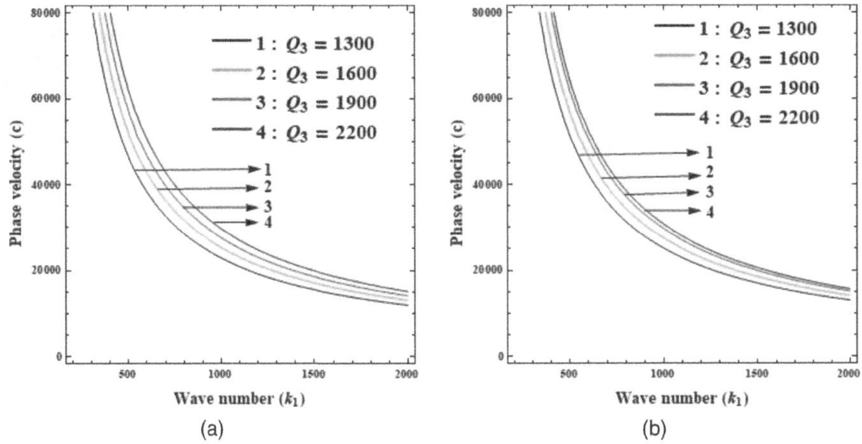

Figure 3.5 The change of phase velocity (c) against real wave number (k_l) for different magnitudes of dissipation factor $\left(Q_3\right)$: (a) represents electrically open case; (b) represents electrically short case.

Figure 3.6 shows the impacts of the dissipation factor (Q_4) on the Rayleigh wave's phase velocity. Figure 3.6(a) and (b) makes it evident that phase velocity diminishes as the elastic constant's magnitude increases.

The impact of the piezoelectric coefficient (e_{15}) on the phase velocity of the Rayleigh waves is considered. The dispersion curves of Rayleigh waves for various piezoelectric coefficient values for electrically open cases are shown in

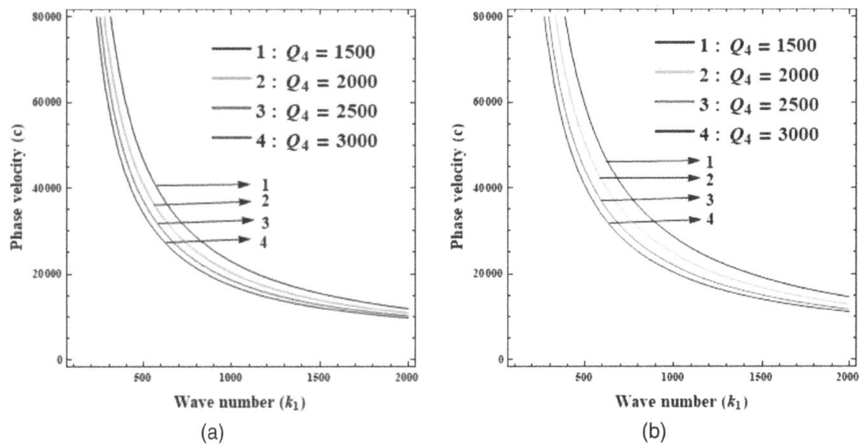

Figure 3.6 The change of phase velocity (c) as a function of real wave number (k_l) for distinct magnitudes of dissipation factor $\left(Q_4\right)$: (a) represents electrically open; (b) represents electrically short.

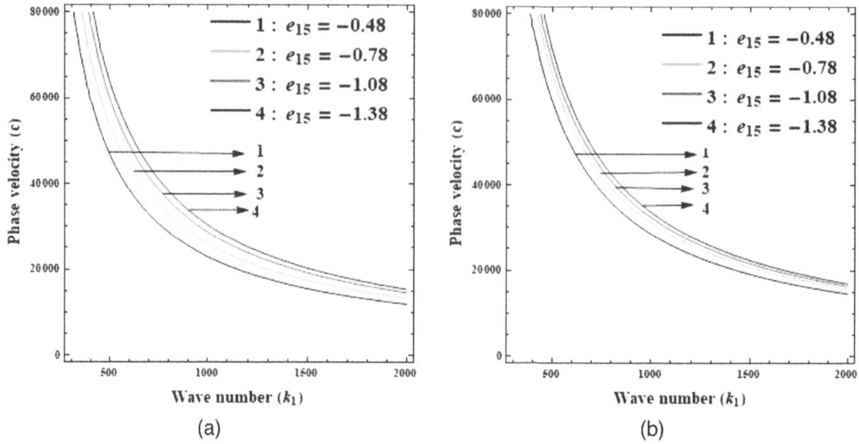

Figure 3.7 The change of phase velocity (c) as a function of real wave number (k_l) for distinct magnitudes of piezoelectric constant $\left(e_{15}\right)$: (a) represents electrically open; (b) represents electrically short.

Figure 3.7. According to the figure, in electrically open and short cases, phase velocity rises as piezoelectric coefficient values rise.

Figure 3.8 reveals the effect of dielectric coefficient $\left(\epsilon_{11}\right)$ on the wave's phase velocity. Curves on the graph indicate that phase velocity decreases as we increase the values of dielectric coefficient in both cases.

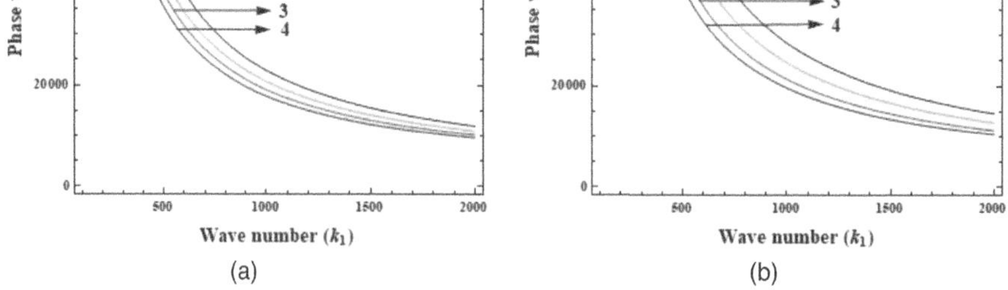

Figure 3.8 The change of phase velocity (c) as a function of real wave number (k_l) for distinct magnitudes of dielectric constant $\left(\epsilon_{11}\right)$: (a) represents electrically open; (b) represents electrically short.

3.6 CONCLUSION

In a piezo-viscoelastic half-space, the propagation of coupled Rayleigh surface waves is examined. An electrically open and short case of the dispersion relations is formed in the matrix determinant form, and the material is handled as exponential in the medium. Numerical calculations and graphical representations have been used to show the parametric influences on the wave's velocity profile. From the discussion and numerical facts presented above, the following conclusions can be made:

- The attenuation coefficients, elastic constants, piezoelectric constant, and dielectric coefficient all have an impact on the wave's phase velocity in both electrical cases.
- In both electrically open and short cases, it is found that phase velocity rises as the values of dissipation factors Q_1, Q_2, Q_3, and piezoelectric constant increase.
- The substance of the piezo-viscoelastic medium is found to possess a considerable impact on the wave's phase velocity. Figure (a) and (b) shows that as the attenuation coefficient, dissipation factor (Q_4), and dielectric constant increase, the phase velocity decreases.

3.7 APPENDIX

$$\alpha_1 = \frac{-\left(\epsilon_{33}\left(C_{33}+e_{33}^2\right)\right)}{e_{33}\left(C_{33}\right)},\ v_1 = \frac{C_{11}}{C_{44}},\ v_2 = \frac{C_{44}}{C_{33}},$$

$$\alpha_2 = \frac{2e_{15}}{C_{33}} + \frac{e_{33}}{C_{33}} + \frac{\epsilon_{33}}{e_{33}}\left(v_1+v_2\right) + \frac{\epsilon_{11}}{e_{33}} - \frac{\epsilon_{33}\left(C_{13}+C_{44}\right)^2}{e_{33}C_{33}C_{44}}$$
$$-\frac{\left(e_{15}+e_{31}\right)\left(C_{13}+C_{44}\right)}{e_{33}C_{44}} + \frac{\left(e_{15}+e_{31}\right)^2}{e_{33}C_{44}},$$

$$\alpha_3 = \frac{\epsilon_{11}}{e_{33}}\left(v_1+v_2\right) - \frac{e_{15}^2}{e_{33}\left(C_{13}+C_{44}\right)} - \frac{\epsilon_{11}\left(C_{13}+C_{44}\right)^2}{e_{33}C_{33}C_{44}}$$
$$+\frac{2\left(e_{15}+e_{31}\right)\left(C_{13}+C_{44}\right)e_{15}}{e_{33}C_{44}C_{33}} - \frac{\left(e_{15}+e_{31}\right)^2 v_2}{e_{33}C_{44}},$$

$$\alpha_4 = \frac{\epsilon_{11}v_1v_2}{e_{33}} + \frac{e_{15}^2 v_1}{e_{33}C_{33}},$$

$$m_1 = \frac{-\left(e_{31}+e_{15}\right)\left(C_{13}+C_{44}\right)k^2\alpha_1^2 - \left(C_{44}\alpha_1^2+\left(\rho c^2-C_{11}\right)k^2\right)\left(e_{33}\alpha_1^2-e_{15}k^2\right)}{\left(-i\left(C_{13}+C_{44}\right)k\alpha_1\right)\left(e_{33}\alpha_1^2-e_{15}k^2\right)+\left(i\left(e_{31}+e_{15}\right)k\alpha_1\right)\left(C_{33}\alpha_1^2+\left(\rho c^2-C_{44}\right)k^2\right)},$$

$$m_2 = \frac{-(e_{31}+e_{15})(C_{13}+C_{44})k^2\alpha_2^2 - \left(C_{44}\alpha_2^2 + (\rho c^2 - C_{11})k^2\right)(e_{33}\alpha_2^2 - e_{15}k^2)}{\left(-i(C_{13}+C_{44})k\alpha_2\right)(e_{33}\alpha_2^2 - e_{15}k^2) + \left(i(e_{31}+e_{15})k\alpha_2\right)\left(C_{33}\alpha_2^2 + (\rho c^2 - C_{44})k^2\right)},$$

$$m_3 = \frac{-(e_{31}+e_{15})(C_{13}+C_{44})k^2\alpha_3^2 - \left(C_{44}\alpha_3^2 + (\rho c^2 - C_{11})k^2\right)(e_{33}\alpha_3^2 - e_{15}k^2)}{\left(-i(C_{13}+C_{44})k\alpha_3\right)(e_{33}\alpha_3^2 - e_{15}k^2) + \left(i(e_{31}+e_{15})k\alpha_3\right)\left(C_{33}\alpha_3^2 + (\rho c^2 - C_{44})k^2\right)},$$

$$n_1 = \frac{\left(C_{44}\alpha_1^2 + (\rho c^2 - C_{11})k^2\right)\left(C_{33}\alpha_1^2 + (\rho c^2 - C_{44})k^2\right) - \left(i(C_{13}+C_{44})k\alpha_1\right)^2}{-\left(i(C_{13}+C_{44})k\alpha_1\right)(e_{33}\alpha_1^2 - e_{15}k^2) + \left(i(e_{31}+e_{15})k\alpha_1\right)\left(C_{33}\alpha_1^2 + (\rho c^2 - C_{44})k^2\right)},$$

$$n_2 = \frac{\left(C_{44}\alpha_2^2 + (\rho c^2 - C_{11})k^2\right)\left(C_{33}\alpha_2^2 + (\rho c^2 - C_{44})k^2\right) - \left(i(C_{13}+C_{44})k\alpha_2\right)^2}{-\left(i(C_{13}+C_{44})k\alpha_2\right)(e_{33}\alpha_2^2 - e_{15}k^2) + \left(i(e_{31}+e_{15})k\alpha_2\right)\left(C_{33}\alpha_2^2 + (\rho c^2 - C_{44})k^2\right)},$$

$$n_3 = \frac{\left(C_{44}\alpha_3^2 + (\rho c^2 - C_{11})k^2\right)\left(C_{33}\alpha_3^2 + (\rho c^2 - C_{44})k^2\right) - \left(i(C_{13}+C_{44})k\alpha_3\right)^2}{-\left(i(C_{13}+C_{44})k\alpha_3\right)(e_{33}\alpha_3^2 - e_{15}k^2) + \left(i(e_{31}+e_{15})k\alpha_3\right)\left(C_{33}\alpha_3^2 + (\rho c^2 - C_{44})k^2\right)}.$$

ACKNOWLEDGEMENT

The author Venkatesan P (Ph.D. Reg. No. 22PHD0124) and Dr. Parvez Alam (corresponding author) thank Vellore Institute of Technology, Vellore for providing "VIT SEED Grant – RGEMS Fund (Sanction order No.: SG20220003)" for carrying out this research work.

REFERENCES

1. Bluestein, J.L.: A new surface wave in piezoelectric medium. Appl. Phys. Lett. 13, 412–413 (1968).
2. Cheng, N.C., Sun, C.T.: Wave propagation in two-layered piezoelectric plates. J. Acoust. Soc. Am. 57, 632–638 (1974).
3. Fang, H., Yang, J., Jiang, Q.: Surface acoustic waves propagating over a rotating piezoelectric half-space. IEEE Trans. Ultrason. Ferroelectr. Freq. Control. 48, 998–1004 (2001).
4. Wang, Q.: Wave propagation in a piezoelectric coupled solid medium. J. Appl. Mech. 69, 819–824 (2002).
5. Singh, M.K., Alam, P. Attenuation and dispersion characteristic of Rayleigh waves in a compressed viscoelastic strip: A comparative study. Bol. Soc. Mat. Mex. 26, 1333–1340 (2020).
6. Saha, S., Chattopadhyay, A., Singh, A.K.: Propagation of Rayleigh type wave in an initially stressed Voigt type viscoelastic layer. Procedia Eng. 173, 1162–1168 (2017).
7. Singh, B.: Wave propagation in a pre-stressed piezoelectric half-space. Acta Mech. 211, 337–344 (2010).

8. Chai, J.F., Wu, T.T.: Propagation of surface waves in a pre-stressed piezoelectric material. J. Acoust. Soc. Am. 100, 2112–2122 (1996).
9. Simionescu-Panait, O.: The influence of initial fields on wave propagation in piezoelectric crystals. Int. J. Appl. Electromag. Mech. 12, 241–251 (2000).
10. Liu, H., Kuang, Z.B., Cai, Z.M.: Propagation of surface acoustic waves in prestressed anisotropic layered piezoelectric structures. Acta Mech. Solida Sin. 16, 16–23 (2003).
11. Son, M.S., Kang, Y.J.: The effect of initial stress on the propagation behavior of SH waves in piezoelectric coupled plates. Ultrasonics 51, 489–495 (2011).
12. Yang, J.S.: A review of a few topics in piezoelectricity. Appl. Mech. Rev. 59, 335–345 (2006).
13. Pang, Y., Liu, J.X., Wang, Y.S., Zhao, X.F.: Propagation of Rayleigh-type surface waves in a transversely isotropic piezoelectric layer over a piezomagnetic half-space. J. Appl. Phys. 103, 074901 (2008).
14. Clarke, N.S., Burdness, J.S.: Rayleigh waves on a rotating surface. ASME J. Appl. Mech. 61, 724–726 (1994).
15. Cao, X., Hu, S., Liu, J., Shi, J.: Generalized Rayleigh surface waves in a piezoelectric semiconductor half space. Meccanica 54, 271–281 (2019).
16. Li, K., Jing, S., Yu, J., Zhang, X., Zhang, B.: The complex Rayleigh waves in a functionally graded piezoelectric half-space: An improvement of the Laguerre polynomial approach. Materials 13(10), 2320 (2020).
17. Cao, X., Jin, F., Wang, Z.: On dispersion relations of Rayleigh waves in a functionally graded piezoelectric material (FGPM) half-space. Acta Mech. 200, 247–261 (2008).
18. Chaudhary, S., Sahu, S.A., Singhal, A.: Analytic model for Rayleigh wave propagation in piezoelectric layer overlaid orthotropic substratum. Acta Mech. 228, 495–529 (2017).
19. Singh, B., Sindhu, R.: On propagation of Rayleigh type surface wave in a micropolar piezoelectric medium. Open J. Acoust. 6(4), 35–44 (2016).
20. Paswan, B., Singh, P., Sahu, S.A.: Mathematical study for the Rayleigh wave propagation in a composite structure with piezoelectric material. J. Solid Mech. 15(2), 144–159 (2023).

Chapter 4

Investigation on Love waves scattering in an orthotropic composite structure with surface irregularity

Siddhartha Koley, Anil Negi, and Abhishek Kumar Singh

4.1 INTRODUCTION

The phenomenon of Love waves propagation in layered structures has consistently attracted the widespread interest of various researchers across the globe. Wang and Yang [1] used the analytical procedure to examine the dispersion characteristics corresponding to the Love waves propagation in a layered fluid-saturated transversely isotropic poroelastic structure. Li et al. [2] analyzed the characteristics of Love waves propagation in a functionally graded-layered piezoelectric structure.

The propagation of Love waves in special class of anisotropic structures exhibit certain complicated properties, which are not analyzed much. The investigation on the Love waves propagation in these types of anisotropic structures can provide crucial information that can be beneficial in real-world scenarios. In yesteryears, the anisotropic materials were widely used in many engineering applications. Orthotropic materials are certain types of anisotropic structures that exhibit diverse material characteristics in two mutually perpendicular directions. The engineering structures having orthotropic symmetry have substantial applications in geotechnical and civil engineering. The different kinds of engineering materials viz. fibrous composites, laminated composites, and rolled sheets, can be considered as orthotropic materials. These types of materials are widely used in the design and development of many engineering structures such as bridge decks, highway bridges, and railway bridges because of their high specific strength, thickness, and resistance to cracking [3–5].

The phenomenon of surface acoustic wave propagation in orthotropic composites can be useful in the structural health monitoring and non-destructive testing of engineering structures. Therefore, it provides substantial applications for the development of engineering structures. Due to this, the investigation of surface waves propagation in orthotropic structures has attracted many researchers. Destrade [6] investigated the propagation characteristics of surface acoustic waves in an orthotropic incompressible half-space. Destrade [7] also developed a mathematical model for the seismic surface waves

DOI: 10.1201/9781003470885-4

propagation in an orthotropic elastic half-space having heterogeneity in the form of an exponentially graded function. The constitutive relation to model the shock waves in orthotropic material structures was proposed by Vignjevic et al. [8]. Biswas [9] examined the Rayleigh surface waves in a thermoelastic orthotropic structure. Abd-Alla and Ahmed [10] investigated the Love waves propagation in a pre-stressed non-homogeneous orthotropic elastic layer lying over orthotropic elastic space. Wang et al. [11] employed the exponential and generalized power law model to examine the dispersion characteristics associated with the Love waves propagation in an inhomogeneous orthotropic structure. Yu and Zhang [12] analyzed the impact of pre-stress on the elastic waves propagation in functionally graded material (FGM) plates having orthotropic symmetry.

The orthotropic composites may contain some structural defects due to sudden high impact loads, structural design, manufacturing process, and erosion [13]. The surface irregularity generated from the structural defects can significantly influence the dynamic performance of the orthotropic structures, which has drawn the attention of some researchers. Singh et al. [14] observed the efficacy of surface irregularity and initial stress on the dynamic characteristics of orthotropic material structure under the moving line load. Singh et al. [13] accomplished the investigation on the induced incremental stress caused by a line load moving on the different orthotropic half-spaces.

The propagation of Love-type waves over an irregular surface can lead to the scattering of these waves, which may further result in significant strength loss of the composite structure. Due to this, it is essential to deal with the scattering behavior of guided waves in a layered composite structure. Owing to this, several researchers made significant efforts to analyze the scattering phenomena of elastic waves in different types of layered structures. Slavin and Wolf [15] discussed the Love waves scattering due to the interaction of Love waves with an irregularity in the elastic layer overlying a rigid elastic half-space. Wolf [16] examined the Love waves scattering due to the irregular upper surface of a layered isotropic elastic structure. Chen et al. [17] obtained the scattering relations for elastic waves propagation in functionally graded isotropic elastic plate. Anagnostopoulos [18] performed analytical investigation on the two-dimensional direct scattering problem associated with the propagation of time-harmonic elastic waves in an orthotropic elastic structure. Chattopadhyay et al. [19] analyzed the propagation of horizontally polarized shear waves (SH-waves) encountering the irregular upper portion of isotropic viscoelastic structure. The scattering behavior of Love-type wave due to the presence of irregularity in the upper surface layer in a transversely isotropic piezoelectric composite structure has been studied by Negi and Singh [20].

However, no effort has been accomplished until date to examine the Love waves scattering in a layered orthotropic composite elastic structure. Based on the above-mentioned reasons, the authors have made an attempt to deal with the scattering characteristics of Love-type wave (one type of guided waves) in a layered orthotropic composite structure containing irregularity in the upper surface.

The present analysis conferred the Love waves propagation and hence its scattering caused by the presence of irregularity in the upper surface of a layered orthotropic composite elastic structure. The solution methodology employed in this investigation contains the usage of double-sided Fourier transforms and contour integral method. Surface irregularities of various shapes, such as rectangular, parabolic, and triangular notches, have been considered in the upper layer of the structure. The expressions for the dispersion relation and group velocity have been derived analytically. The effect of anisotropic parameter on the group and phase velocities has also been examined. The deduction of reflected displacement has also been done for all the considered cases of surface irregularity. The influences of the vertical irregularity and anisotropic parameters for the upper layer on the reflected displacement have also been discussed for all shapes of irregularity in the upper surface.

4.2 PROBLEM FORMULATION

Let us consider the Love waves propagation in the orthotropic elastic layer having finite thickness H nested on the orthotropic elastic half-space. In the Cartesian coordinate system for problem's geometry (Figure 4.1), it has been considered that the y-axis denotes the wave propagation direction while x-axis is pointing in the vertical downward direction. An irregularity of thickness b

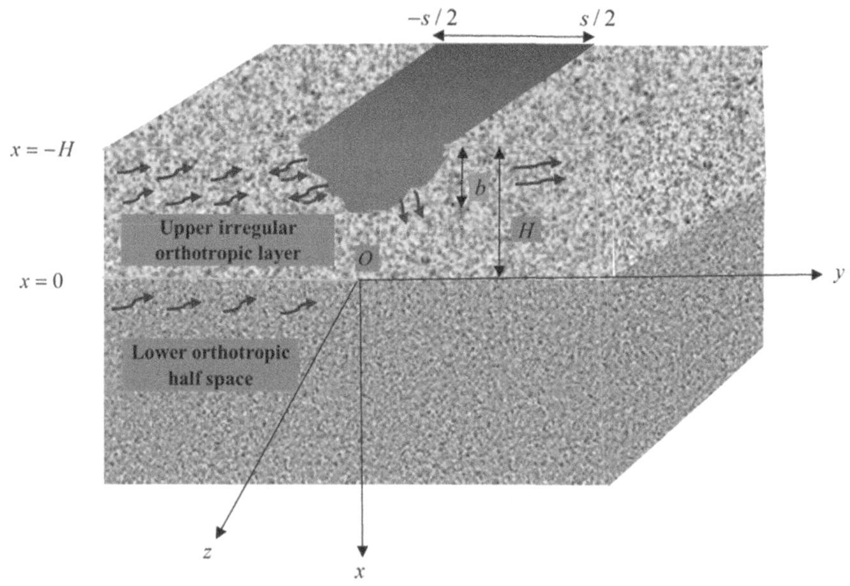

Figure 4.1 Geometry of the problem.

and span S has been assumed in the upper surface layer of the aforementioned elastic structure. At the interface $(x = 0)$ of the said orthotropic structure, the origin is considered, which is depicted in Figure 4.1.

The equation of the upper irregular surface of the orthotropic layer can be written as follows:

$$x = x_B = -H + bh(y), \tag{4.1}$$

where

$$h(y) = \begin{cases} 0, & \text{for } y \le -\frac{s}{2} \text{ and } y \ge \frac{s}{2} \\ f(y), & \text{for } -\frac{s}{2} \le y \le \frac{s}{2} \end{cases} \tag{4.2}$$

in which b ($\ll 1$) signifies the vertical irregularity parameter of the surface irregularity and the function $f(y)$ corresponds to the shape of the irregularity.

Let us consider (u_1, v_1, w_1) and (u_2, v_2, w_2) denote the displacement components associated with the upper layer and lower half-space in the direction of x, y, and z-axis, respectively. In view of the propagation properties of the Love waves, the displacement components for the upper layer and lower half-space can be expressed as follows:

$$u_1 = v_1 = 0,\ w_1 = w_1(x, y, t), \text{ and } u_2 = v_2 = 0,\ w_2 = w_2(x, y, t). \tag{4.3}$$

The governing equations for the propagation of Love waves in the orthotropic medium can be written as follows:

$$\tau_{ij,j} = \rho \ddot{u}_i \left(i, j = 1, 2, 3\right), \tag{4.4}$$

where ρ is the mass density of the orthotropic material, u_i $(i = 1, 2, 3)$ signifies the displacement components along the coordinate axes, and τ_{ij} $(i, j = 1, 2, 3)$ denote the components corresponding to the stress tensor.

The constitutive equation associated with the orthotropic elastic medium can be expressed as follows:

$$\left.\begin{aligned} \tau_{11} &= c_{11}e_{11} + c_{12}e_{22} + c_{13}e_{33}, \\ \tau_{22} &= c_{12}e_{11} + c_{12}e_{22} + c_{33}e_{33}, \\ \tau_{33} &= c_{13}e_{11} + c_{23}e_{22} + c_{33}e_{33}, \\ \tau_{23} &= 2c_{44}e_{23}, \\ \tau_{13} &= 2c_{55}\varepsilon_{13}, \\ \tau_{12} &= 2c_{66}e_{12}, \end{aligned}\right\} \tag{4.5}$$

where $c_{ij}\,(i,j=1,2,...6)$ represent the elastic constants and $e_{ij}\,(i,j=1,2,3)$ refer to the components of strain.

The displacement and strain components can be related as follows:

$$e_{11}=\frac{\partial u}{\partial x},e_{22}=\frac{\partial v}{\partial y},e_{33}=\frac{\partial w}{\partial z},e_{23}=\frac{1}{2}\left(\frac{\partial w}{\partial y}+\frac{\partial v}{\partial z}\right),e_{31}=\frac{1}{2}\left(\frac{\partial w}{\partial x}+\frac{\partial u}{\partial z}\right),$$
$$e_{12}=\frac{1}{2}\left(\frac{\partial u}{\partial y}+\frac{\partial v}{\partial x}\right). \tag{4.6}$$

In view of Eqs. (4.3) and (4.6), the constitutive Eq. (4.5) provides

$$\tau_{11}=0,\tau_{22}=0,\tau_{33}=0,\tau_{12}=0,\tau_{23}=c_{44}\frac{\partial w}{\partial y},\tau_{13}=c_{55}\frac{\partial w}{\partial x}. \tag{4.7}$$

Using Eq. (4.7) in Eq. (4.4), we get

$$\frac{\partial^2 w}{\partial x^2}+\frac{1}{\gamma^2}\frac{\partial^2 w}{\partial y^2}=\frac{1}{\beta^2}\frac{\partial^2 w}{\partial t^2}, \tag{4.8}$$

where $\gamma=\sqrt{\frac{c_{44}}{c_{55}}}$ represents the anisotropy parameter, and $\beta=\sqrt{\frac{c_{44}}{\rho}}$ denotes the velocity of the shear wave.

In view of Eq. (4.8), the non-vanishing equation of motion for Love waves propagation in the upper orthotropic layer can be expressed as the following:

$$\frac{\partial^2 w_1}{\partial x^2}+\frac{1}{\gamma_1^{\,2}}\frac{\partial^2 w_1}{\partial y^2}=\frac{1}{\beta_1^{\,2}}\frac{\partial^2 w_1}{\partial t^2}, \tag{4.9}$$

where $\gamma_1=\sqrt{\frac{c_{44}^1}{c_{55}^1}}$; and $\beta_1=\sqrt{\frac{c_{44}^1}{\rho_1}}$.

Similarly, non-zero governing equation for Love waves propagation in the lower orthotropic half-space can be written as the following:

$$\frac{\partial^2 w_2}{\partial x^2}+\frac{1}{\gamma_2^{\,2}}\frac{\partial^2 w_2}{\partial y^2}=\frac{1}{\beta_2^{\,2}}\frac{\partial^2 w_2}{\partial t^2}, \tag{4.10}$$

where $\gamma_2=\sqrt{\frac{c_{44}^2}{c_{55}^2}}$ and $\beta_2=\sqrt{\frac{c_{44}^2}{\rho_2}}$.

Let us assume the solution of Eqs. (4.9) and (4.10) as the following:

$$w_p=W_p(x,y)e^{i\omega t}, \tag{4.11}$$

where ω represents the angular frequency, $p=1,2$ corresponds to the upper orthotropic layer and lower orthotropic half-space, respectively.

Using Eq. (4.11) in Eqs. (4.9) and (4.10), we have

$$\frac{\partial^2 W_1}{\partial y^2} + \frac{1}{\gamma_1^2}\frac{\partial^2 W_1}{\partial x^2} + L_1^2 W_1 = 0, \tag{4.12}$$

$$\frac{\partial^2 W_2}{\partial y^2} + \frac{1}{\gamma_2^2}\frac{\partial^2 W_2}{\partial x^2} + L_2^2 W_2 = 0, \tag{4.13}$$

where $L_1^2 = \frac{\omega^2}{\beta_1^2}$ and $L_2^2 = \frac{\omega^2}{\beta_2^2}$.

4.3 CONDITIONS AT THE BOUNDARIES

i. The traction free boundary condition for the upper irregular surface i.e. at $x = -H + bh(y)$ can be written as the following:

$$\tau_{32}^1 - \gamma_1^2 bh' \tau_{31}^1 = 0. \tag{4.14}$$

ii. The continuity conditions at the interface between the layer and half-space i.e. at $x = 0$ are
   a. $w_1 = w_2$ (4.15)
   b. $\tau_{13}^1 = \tau_{31}^2$. (4.16)

4.4 SOLUTION TO THE PROBLEM

In view of Eqs. (4.12) and (4.13), the solution for the incident wave field can be written as the following [16, 19]:

$$w_{1,in} = A\cos\gamma_1 S_1 (x + H) e^{-i\eta y}, \tag{4.17}$$

$$w_{2,in} = B e^{-\gamma_2 S_2 y} e^{-i\eta y}, \tag{4.18}$$

where $S_1^2 = L_1^2 - \eta^2, S_2^2 = \eta^2 - L_2^2$; $\eta = \frac{\omega}{c_{ph}}$ signifies the angular wave number; c_{ph} denotes the phase velocity for Love waves; K_1 and K_2 are wave numbers; A and B signify the arbitrary coefficients associated with the incident wave field.

The solution of Love waves corresponding to the scattered wave field in $\upsilon-$plane has been considered as follows [16, 19]:

$$w_{1,scatt} = \int_c \left\{ C(\upsilon) e^{-i\gamma_1 \xi_1 x} + D(\upsilon) e^{i\gamma_1 \xi_1 x} \right\} e^{-i\upsilon y} d\upsilon, \tag{4.19}$$

$$w_{2,scatt} = \int_c E(\upsilon) e^{-\gamma_2 \xi_2 x} e^{-i\eta y} d\upsilon, \tag{4.20}$$

where $\xi_1^2 = L_1^2 - \upsilon^2$, $\xi_2^2 = L_2^2 - \upsilon^2$, $\upsilon = \frac{\omega}{c_{ph}}$ signifies the angular wave number; c_{ph} signifies the phase velocity; K_1 and K_2 represent the wave numbers corresponding to scattered wave field; $C(\upsilon), D(\upsilon),$ *and* $E(\upsilon)$ represent the arbitrary constants corresponding to the scattered wave field. In Eqs. (4.17)–(4.20), $w_{p,\, in}$ and $w_{p,\, scatt}$ are the displacement functions for incident and scattered wave fields related to the upper layer $(p = 1)$ and to the lower half-space $(p = 2)$ for an orthotropic medium.

Therefore, the total mechanical displacement function associated with the Love waves propagation in the studied medium can be represented as the following:

$$w_1 = w_{1,in} + w_{1,scatt} = A\cos\gamma_1 S_1(x+H)e^{-i\eta y} + \int_c \left\{C(\upsilon)e^{-i\gamma_1\xi_1 x} + D(\upsilon)e^{i\gamma_1\xi_1 x}\right\}, \quad (4.21)$$

$$w_2 = w_{2,in} + w_{2,scatt} = Be^{-\gamma_2 S_2 y}e^{-i\eta y} + \int_c E(\upsilon)e^{-\gamma_2\xi_2 x}e^{-i\eta y}d\upsilon. \quad (4.22)$$

Using Eqs. (4.21) and (4.22) into boundary condition Eq. (4.15), we get

$$A\cos\gamma_1 S_1 H = B, \quad (4.23)$$

$$C(\upsilon) + D(\upsilon) = E(\upsilon). \quad (4.24)$$

Putting the value of Eq. (4.23) in Eq. (4.22), we obtain

$$w_2 = A\cos\gamma_1 S_1 He^{-\gamma_2 S_2 y}e^{-i\eta y} + \int_c E(\upsilon)e^{-\gamma_2\xi_2 x}e^{-i\eta y}d\upsilon. \quad (4.25)$$

Again applying Eq. (4.21) and Eq. (4.25) in the boundary condition Eq. (4.16), we have

$$C(\upsilon) - D(\upsilon) = \frac{E(\upsilon)}{\lambda}, \quad (4.26)$$

where $\lambda = \frac{i\xi_1\gamma_1 c_{55}^1}{\xi_2\gamma_2 c_{55}^2}$.

Solving Eq. (4.24) and Eq. (4.26), we get

$$\frac{2\lambda}{\lambda+1}C(\upsilon) = E(\upsilon), \quad \frac{C(\upsilon)(\lambda-1)}{(\lambda+1)} = D(\upsilon). \quad (4.27)$$

Using Eq. (4.27) in Eq. (4.21) and Eq. (4.25), we get the following expressions:

$$w_1 = A\cos\gamma_1 S_1(x+H)e^{-i\eta y} + \int_c \frac{2C}{(\lambda+1)}\left\{\lambda\cos\gamma_1\xi_1 x - i\sin\gamma_1\xi_1 x\right\}e^{-i\upsilon y}d\upsilon, \quad (4.28)$$

$$w_2 = A\cos\gamma_1 S_1 H e^{-\gamma_2 S_2 x} e^{-i\eta y} + \int_c C\frac{2\lambda}{(\lambda+1)} e^{-\gamma_2\xi_2 x} e^{-i\upsilon y} d\upsilon. \tag{4.29}$$

Since the irregularity is very small as $(b \ll 1)$, the arbitrary coefficients, cosine and sine functions associated with the displacement functions can be expressed as the following:

$$C(\upsilon) = bC_1(\upsilon),\ \sin S_1 bh \cong S_1 bh,\ \cos S_1 bh \cong 1. \tag{4.30}$$

On substituting the expressions of w_1 from Eq. (4.28) and the concept from Eq. (4.30) in the boundary condition Eq. (4.14), we get

$$A\left[i\eta h' - S_1^2 h\right]\gamma_1 e^{-i\eta y} + \int_c \frac{2C_1(\upsilon)\xi_1}{(\lambda+1)}(-i\cos\xi_1\gamma_1 H + \lambda\sin\xi_1\gamma_1 H)e^{-i\upsilon y} d\upsilon = 0. \tag{4.31}$$

On inverting Eq. (4.31), we have

$$C_1(\upsilon) = \frac{A(\lambda+1)}{4\pi(-i\cos\xi_1\gamma_1 H + \lambda\sin\xi_1\gamma_1 H)\xi_1}\int_{-\infty}^{\infty}\left[i\eta h' - S_1^2 h\right]\gamma_1 e^{ix(\upsilon-\eta)} dx. \tag{4.32}$$

Using Eq. (4.32) in Eqs. (4.28) and (4.29), we obtain

$$w_1 = A\cos\gamma_1 S_1(x+H)e^{-i\eta y} + \int_{-\infty}^{\infty}\frac{Ab}{2\pi}\left[i\eta h' - S_1^2 h\right]\gamma_1 e^{-i\eta x} dx \times \int_c T_0(\upsilon) d\upsilon, \tag{4.33}$$

$$w_2 = A\cos\gamma_1 S_1 H e^{-\gamma_2 S_2 x} e^{-i\eta y} + \frac{Ab}{2\pi}\int_{-\infty}^{\infty}(i\eta h' - S_1^2 h)\gamma_1 e^{-i\eta x} dx \times \int_c T_{00}(\upsilon) d\upsilon. \tag{4.34}$$

where

$$T_0(\upsilon) = \frac{\{\lambda\cos\gamma_1\xi_1 x - i\sin\gamma_1\xi_1 x\}}{(i\cos\xi_1\gamma_1 H - \lambda\sin\xi_1\gamma_1 H)\xi_1} e^{i\upsilon x - i\upsilon y}, \tag{4.35}$$

$$T_{00}(\upsilon) = \frac{\lambda e^{-\gamma_2\xi_2 x} e^{-i\upsilon y + i\upsilon x}}{(i\cos\xi_1\gamma_1 H - \lambda\sin\xi_1\gamma_1 H)\xi_1}. \tag{4.36}$$

The poles corresponding to contour integrals in Eqs. (4.33) and (4.34) can be determined as follows:

$$(i\cos\xi_1\gamma_1 H - \lambda\sin\xi_1\gamma_1 H)\xi_1 = 0. \tag{4.37}$$

Putting the values of ξ_1 and γ_1 into Eq. (4.37), we have

$$\tan\left(\sqrt{\frac{c_{44}^1}{c_{55}^1}}\sqrt{(L_1^2-\upsilon^2)}\right)H=\frac{c_{55}^2\sqrt{\frac{c_{44}^2}{c_{55}^2}}\sqrt{(\upsilon^2-L_2^2)}}{c_{55}^1\sqrt{\frac{c_{44}^1}{c_{55}^1}}\sqrt{(L_1^2-\upsilon^2)}}. \tag{4.38}$$

Eq. (4.38) describes the generalized periodic equations for the Love waves that are for both the scattered and incident wave fields in the considered orthotropic composite medium. In Eq. (4.38), υ_m associated with real $\upsilon-$axis indicates the equation's roots. The graph corresponding to the function (Figure 4.2) indicates regarding N_0 roots exist for Eq. (4.38), where N_0 represents the integral part of the number $\left[\left\{(L_1^2-L_2^2)\frac{H}{\pi}\right\}+1\right]$ [16].

On taking the expressions of L_1^2, L_2^2 and ω from Eqs. (4.13–4.18), Eq. (4.38) provides

$$\tan\left(\sqrt{\frac{c_{44}^1}{c_{55}^1}}\sqrt{\left(\frac{c_{ph}^2}{\beta_1^2}-1\right)}\right)\upsilon H=\frac{c_{44}^2\sqrt{\frac{c_{44}^2}{c_{55}^2}}\sqrt{\left(1-\frac{c_{ph}^2}{\beta_2^2}\right)}}{c_{44}^1\sqrt{\frac{c_{44}^1}{c_{55}^1}}\sqrt{\left(\frac{c_{ph}^2}{\beta_1^2}-1\right)}}. \tag{4.39}$$

Eq. (4.39) denotes the dispersion relation associated with the scattered field of the Love waves propagation in layered orthotropic composite structure.

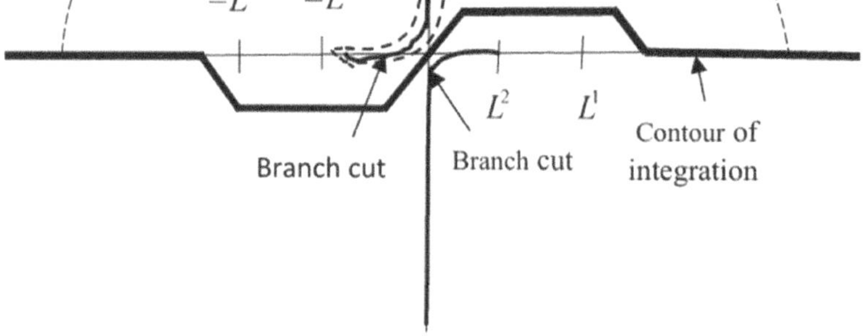

Figure 4.2 Contour of integration in υ plane.

The relation between the group velocity and phase velocity in the aforesaid medium [2, 21–23] can be expressed as the following:

$$c_g = c_{ph} + \upsilon \frac{dc_{ph}}{d\upsilon}, \tag{4.40}$$

where c_g is the group velocity of Love waves.

With the help of Eqs. (4.39) and (4.40), we get

$$\frac{c_g}{\beta_1} = \frac{K_2\left[\left(\frac{c_{ph}^2}{\beta_1^2}-1\right)\frac{c_{ph}^2}{\beta_2^2\upsilon}+\left(1-\frac{c_{ph}^2}{\beta_2^2}\right)\frac{c_{ph}^2}{\beta_1^2\upsilon}\right]+K_1\left[-\delta\left\{\gamma_1 H\left(\frac{c_{ph}^2}{\beta_1^2}-1\right)-\frac{c_{ph}^2 H}{\beta_1^2}\right\}\right]}{K_2\left[\left(\frac{c_{ph}^2}{\beta_1^2}-1\right)\frac{c_{ph}\beta_1^2}{\beta_1\beta_2^2\upsilon}+\left(1-\frac{c_{ph}^2}{\beta_2^2}\right)\frac{c_{ph}}{\beta_1\upsilon}\right]+K_1\left[\frac{\delta\gamma_1\upsilon H c_{ph}}{\beta_1\upsilon}\right]}, \tag{4.41}$$

where $K_2 = c_{55}^2\gamma_2, K_1 = c_{55}^1\gamma_1;\ \delta = \left(\frac{c_{ph}^2}{\beta_1^2}-1\right)\sqrt{1-\frac{c_{ph}^2}{\beta_2^2}}\sec^2\left(\gamma_1\sqrt{\left(\frac{c_{ph}^2}{\beta_1^2}-1\right)}\upsilon H\right)$.

Eq. (4.41) represents the expression of group velocity related to the scattered field of Love waves propagation in the said orthotropic elastic structure.

On applying the residue theorem, the expressions in integral form of $T_0(\upsilon)$ and $T_{00}(\upsilon)$ containing in Eqs. (4.33) and (4.34) and using the concept defined in Figure 4.2, we have

$$\int_c T_0(\upsilon)d\upsilon = 2\pi i\sum \mathrm{Res}\,T_0 - \int_{\substack{Branch\\ line}} T_0(\upsilon)d\upsilon - \int_{c_\infty} T_0(\upsilon)d\upsilon, \tag{4.42}$$

$$\int_c T_{00}(\upsilon)d\upsilon = 2\pi i\sum \mathrm{Res}\,T_{00} - \int_{\substack{Branch\\ line}} T_{00}(\upsilon)d\upsilon - \int_{c_\infty} T_{00}(\upsilon)d\upsilon. \tag{4.43}$$

The residues of the integral T_0 and T_{00} at the poles υ_m are obtained as

$$\mathrm{Res}\,T_0 = \frac{\cos\gamma_1\xi_{1m}(x+H)}{\gamma_1(\upsilon_m H)}e^{i\upsilon_m(x-y)}, \tag{4.44}$$

$$\mathrm{Res}\,T_{00} = \frac{\cos\gamma_1\xi_{1m}He^{-\gamma_2\xi_{2m}x}}{\gamma_1(\upsilon_m H)}e^{i\upsilon_m(x-y)}. \tag{4.45}$$

The integrals present in Eqs. (4.42) and (4.43) can be solved using the methodology given by Sezawa [24]. In view of this, in the upper half-plane, the contour consisting of a real axis along with a semi-circle having the large radius has been assumed. On owing to the characteristics of Love-type wave propagation, we consider K_1, K_2 in the form of branch points intersect the real axis along integral's contour. Because of having big values of y and x,

Eqs. (4.42) and (4.43) having the term $\frac{1}{x^{3/2}}$ can be neglected in the field of branch line integral. Therefore, the integrals contained in Eqs. (4.42) and (4.43) in the arcs of a semicircle at infinity also vanish.

Therefore, with the help of the expressions mentioned in Eqs. (4.44) and (4.45) and also the condition $z > x$ and $\upsilon_m < 0$, Eqs. (4.42) and (4.43) can be written as follows:

$$\int_c T_0(\upsilon)d\upsilon = 2\pi i \sum_{m=1}^{N^o} \frac{\cos\gamma_1\xi_{1m}(x+H)}{\gamma_1(\upsilon_m H)} e^{i\upsilon_m(x-y)}, \tag{4.46}$$

$$\int_c T_{00}(\upsilon)d\upsilon = 2\pi i \sum_{m=1}^{N^0} \frac{\cos\gamma_1\xi_{1m}He^{-\gamma_2\xi_{2m}x}}{\gamma_1(\upsilon_m H)} e^{i\upsilon_m(x-y)}. \tag{4.47}$$

Similarly, for $y < x$ and $\upsilon_m > 0$, we obtain

$$\int_c T_0(\upsilon)d\upsilon = -2\pi i \sum_{m=1}^{N^o} \frac{\cos\gamma_1\xi_{1m}(x+H)}{\gamma_1(\upsilon_m H)} e^{i\upsilon_m(x-y)}, \tag{4.48}$$

$$\int_c T_{00}(\upsilon)d\upsilon = -2\pi i \sum_{m=1}^{N^0} \frac{\cos\gamma_1\xi_{1m}He^{-\gamma_2\xi_{2m}x}}{\gamma_1(\upsilon_m H)} e^{i\upsilon_m(x-y)}, \tag{4.49}$$

where N^0 indicates the total no. of roots.

Using Eqs. (4.46)–(4.49) and $\upsilon_m > 0$, Eqs. (4.33) and (4.34) are simplified to

$$\begin{aligned} w_1 &= A\cos\gamma_1 S_1(x+H)e^{-i\eta y} - ibA\sum_{m=1}^{N^0} \frac{\cos\gamma_1\xi_{1m}(x+H)}{(\upsilon_m H)} \\ &\times \left\{ e^{-i\upsilon_m y}\int_{-\infty}^{x}(i\eta h' - S_1^2 h)e^{i(\upsilon_m-\eta)x}dx + e^{i\upsilon_m y}\int_{x}^{\infty}(i\eta h' - S_1^2 h)e^{-i(\upsilon_m+\eta)x}dx \right\}, \end{aligned} \tag{4.50}$$

$$\begin{aligned} w_2 &= A\cos\gamma_1 S_1 He^{-\gamma_2 S_2 x}e^{-i\eta y} - ibA\sum_{m=1}^{N^0} \frac{\cos\gamma_1\xi_{1m}He^{-\gamma_2\xi_{2m}x}}{(\upsilon_m H)} \\ &\times \left\{ e^{-i\upsilon_m y}\int_{-\infty}^{x}(i\eta h' - S_1^2 h)e^{i(\upsilon_m-\eta)x}dx + e^{i\upsilon_m y}\int_{x}^{\infty}(i\eta h' - S_1^2 h)e^{-i(\upsilon_m+\eta)x}dx \right\} \end{aligned} \tag{4.51}$$

In view of Eq. (4.2), Eqs. (4.50) and (4.51) provide

$$\begin{aligned} w_1 &= A\cos\gamma_1 S_1(x+H)e^{-i\eta y} + ibA\sum_{m=1}^{N^0} \frac{\cos\gamma_1\xi_{1m}(x+H)}{(\upsilon_m H)} \\ &\times \left\{ e^{i\upsilon_m y}(\eta\upsilon_m + \eta^2 + S_1^2)\int_{-\frac{s}{2}}^{\frac{s}{2}} he^{-i(\upsilon_m+\eta)x}dx \right\}, \end{aligned} \tag{4.52}$$

$$w_2 = A\cos\gamma_1 S_1 He^{-\gamma_2 S_2 x}e^{-i\eta y} + ibA\sum_{m=1}^{N^0}\frac{\cos\gamma_1\xi_{1m}He^{-\gamma_2\xi_{2m}x}}{(\upsilon_m H)}$$
$$\times\left\{e^{i\upsilon_m y}(\eta\upsilon_m + \eta^2 + S_1^2)\int_{-\frac{s}{2}}^{\frac{s}{2}} he^{-i(\upsilon_m+\eta)x}dx\right\}. \tag{4.53}$$

where Eqs. (4.52) and (4.53) provide the total displacement components of Love waves within the layer and the half-space of the said orthotropic structure, respectively.

4.5 SPECIAL CASES

Let us consider the three practical scenarios viz. (I) parabolic-shaped irregularity, (II) triangular notch-shaped irregularity, and (III) rectangular-shaped irregularity in the upper surface of studied orthotropic structure.

4.5.1 Case I: Parabolic-shaped irregularity present in the upper surface layer

The equation for the parabolic-shaped irregularity present in upper layer is defined as follows (as displayed by Figure 4.3):

$$h(y) = \begin{cases} 0, & y > \frac{s}{2}, \text{ and } y < \frac{-s}{2} \\ \left(1 - \frac{4y^2}{s^2}\right), & \frac{-s}{2} \le y \le \frac{s}{2}. \end{cases} \tag{4.54}$$

The upper surface in the orthotropic material is considered as it allows the propagation of only primary waves. This wave satisfies the dispersion Eq. (4.39).

Using Eq. (4.54) and concept $\eta = \upsilon_1$, $S_1 = \xi_1$; $m = 1$ (for first mode) in Eq. (4.52), we get

$$\frac{w_{1,ref}^{P}}{A} = ib\frac{\cos\gamma_1\xi_{11}(x+H)}{(\upsilon_1 H)}e^{i\upsilon_1 y}(2\upsilon_1^{\,2} + \xi_{11}^2)\left(\frac{2\sin\upsilon_1 s}{s^2\upsilon_1^{\,3}} - \frac{2\cos\upsilon_1 s}{s\upsilon_1^{\,2}}\right). \tag{4.55}$$

Eq. (4.55) represents the component of reflected displacement induced in the upper layer consisting of a parabolic-shaped surface irregularity present in the studied orthotropic composite structure influenced by Love waves.

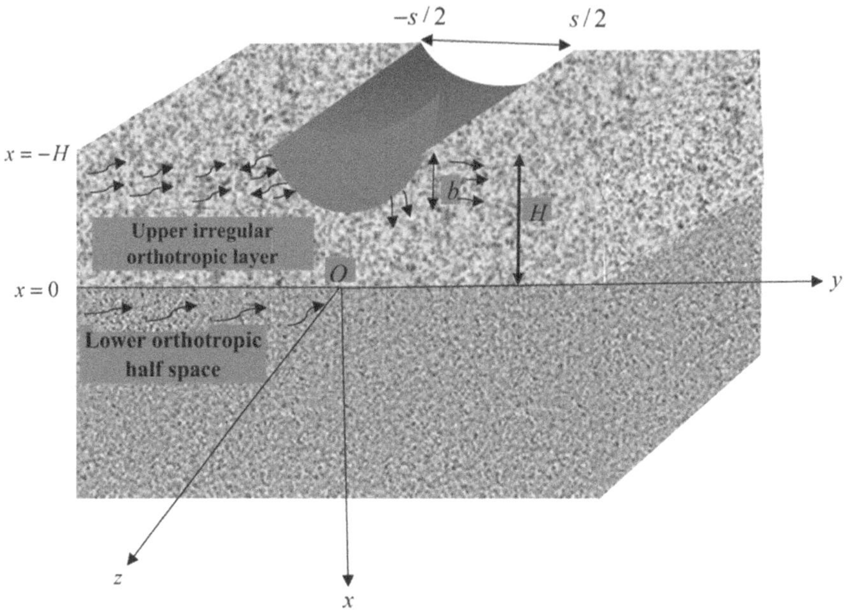

Figure 4.3 Geometry of parabolic-shaped surface irregularity.

4.5.2 Case II. Triangular notch-shaped irregularity present in the upper surface layer

The equation expressing the surface irregularity of the type triangular notch in upper layer is defined as follows (as shown in Figure 4.4):

$$h(y) = \begin{cases} 0, & y > \dfrac{s}{2}, y < \dfrac{-s}{2} \\ 1 + \dfrac{2y}{s}, & \dfrac{-s}{2} \leq y \leq 0 \\ 1 - \dfrac{2y}{s}, & 0 \leq y \leq \dfrac{s}{2}. \end{cases} \tag{4.56}$$

Applying the same procedure as mentioned for the physical scenario of irregularity of parabolic shape present in upper surface layer and using Eq. (4.56) in Eq. (4.52), we obtain the non-vanishing component of reflected displacement for the physical scenario when irregularity of triangular notch shape present in the upper surface layer as the following:

$$\frac{w^T_{1,ref}}{A} = ib\frac{\cos\gamma_1\xi_{11}(x+H)}{(\upsilon_1 H)} \times e^{i\upsilon_1 y}(2\upsilon_1^{\,2} + \xi_{11}^2)\left(\frac{2\sin^2\frac{\upsilon_1 s}{2}}{s\upsilon_1^{\,2}}\right). \tag{4.57}$$

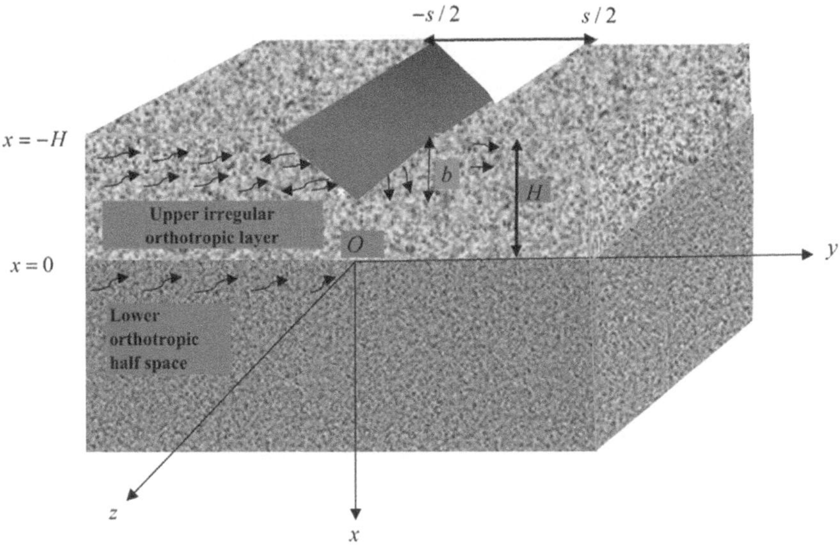

Figure 4.4 Geometry of triangular notch-shaped surface irregularity.

Eq. (4.57) represents the component of reflected displacement induced in the upper layer for the physical scenario of triangular notch-shaped irregularity present in upper surface of the studied orthotropic composite structure influenced by Love waves.

4.5.3 Case III. Rectangular-shaped irregularity present in the upper surface layer

The equation representing the surface irregularity of rectangular shape present on upper surface can be expressed as follows (as shown in Figure 4.5):

$$h(y) = \begin{cases} 0, & y > \dfrac{s}{2} \text{ and } y < \dfrac{-s}{2}, \\ 1, & \dfrac{-s}{2} \le y \le \dfrac{s}{2}. \end{cases} \tag{4.58}$$

On employing a similar approach as mentioned for the physical scenario of parabolic shape irregularity present in the upper surface and using Eq. (4.58) in Eq. (4.52), we obtain

$$\frac{w_{1,ref}^{R}}{A} = ib\frac{\cos\gamma_1\xi_{11}(x+H)}{(\upsilon_1 H)} \times e^{i\upsilon_1 y}(2\upsilon_1{}^2 + \xi_{11}^2)\left(\frac{2\sin\upsilon_1 s}{\upsilon_1}\right). \tag{4.59}$$

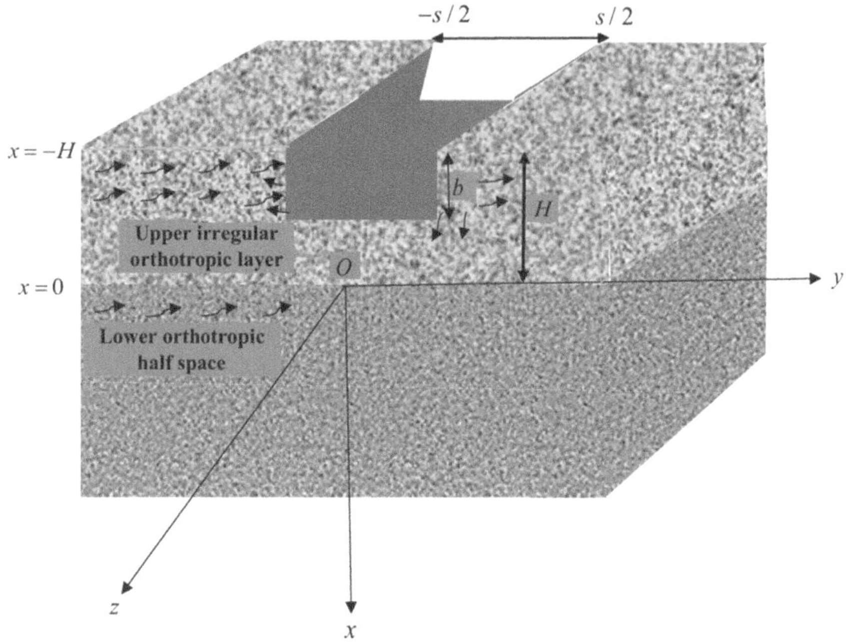

Figure 4.5 Geometry of rectangular-shaped surface irregularity.

Eq. (4.59) adduces the non-vanishing reflected displacement component induced in the upper layer having the rectangular-shaped irregularity present in the upper surface of the considered orthotropic elastic structure.

4.6 VALIDATION

4.6.1 Dispersion relation

If $c_{44}^1 = \mu^1, c_{55}^2 = \mu^2$ and $\gamma_1 = \sqrt{\frac{c_{44}^1}{c_{55}^1}} = 1,\ \gamma_2 = \sqrt{\frac{c_{44}^2}{c_{55}^2}} = 1$ in the dispersion relation Eq. (4.39), we have

$$\tan\sqrt{\left(\frac{c_{ph}^2}{\beta_1^2} - 1\right)}\upsilon H = \frac{\mu^2\sqrt{\left(1 - \frac{c_{ph}^2}{\beta_2^2}\right)}}{\mu^1\sqrt{\left(\frac{c_{ph}^2}{\beta_1^2} - 1\right)}}, \tag{4.60}$$

where μ^1 and μ^2 denote the rigidities related to the isotropic elastic layer and isotropic elastic half-space, respectively. Eq. (4.60) represents the dispersion relation related to the scattered Love waves in the layered isotropic elastic medium having irregular surface. The obtained expression is found to match the classical equation given by Love [25].

4.6.2 Reflected displacement component when parabolic-shaped irregularity present in upper surface layer

For the isotropic elastic structure, i.e. $\gamma_1 = \sqrt{\frac{c_{44}^1}{c_{55}^1}} = 1$, the expression in Eq. (4.55) reduces to

$$\frac{w_{1,ref}^{P}}{A} = ib\frac{\cos\xi_{11}(x+H)}{(\upsilon_1 H)}e^{i\upsilon_1 y}(2\upsilon_1^{\,2}+\xi_{11}^2)\left(\frac{2\sin\upsilon_1 s}{s^2\upsilon_1^{\,3}}-\frac{2\cos\upsilon_1 s}{s\upsilon_1^{\,2}}\right). \tag{4.61}$$

Eq. (4.61) gives the expression for the reflected displacement associated with Love wave propagation in the upper surface layer of a layered isotropic elastic medium. This result is consistent with those reported by Wolf [16] and Chattopadhyay et al. [19] for the case of an isotropic elastic structure.

4.6.3 Reflected displacement component when rectangular-shaped irregularity present in upper surface layer

Similarly, if $\gamma_1 = \sqrt{\frac{c_{44}^1}{c_{55}^1}} = 1$, the expression in Eq. (4.57) provides

$$\frac{w_{1,ref}^{T}}{A} = ib\frac{\cos\xi_{11}(x+H)}{(\upsilon_1 H)}\times e^{i\upsilon_1 y}(2\upsilon_1^{\,2}+\xi_{11}^2)\left(\frac{2\sin^2\frac{\upsilon_1 s}{2}}{s\upsilon_1^{\,2}}\right). \tag{4.62}$$

Eq. (4.62) presents the expression for the reflected displacement during Love wave propagation in the upper surface of the studied material structure consisting of an irregularity of triangular notch shape. This result exactly matches those determined by Wolf [16] and Chattopadhyay et al. [19] for the case of isotropic elastic structure.

4.7 NUMERICAL SIMULATIONS AND RESULTS DISCUSSION

In view of numerical simulations and graphical manifestation, the dimensionless phase velocity, group velocity, and the non-vanishing component of reflected displacement corresponding to the scattered Love waves in the irregular-layered orthotropic composite medium; the following physio-mechanical properties have been taken into account (given in Table 4.1).

Table 4.1 Materials characteristic of orthotropic material [17]

Properties	c_{11}	c_{12}	c_{13}	c_{22}	c_{23}	c_{33}	c_{44}	c_{55}	c_{66}	ρ
Si_3N_4 (Upper layer)	433	195	127	433	127	574	108	108	108	3.2
Zinc (Lower substrate)	62.7	36.2	36.2	62.7	50.8	162.8	38.5	38.5	38.5	7.14

Unit: $c_{ij}\left(\times 10^9 N/m^2\right)$, $\rho\left(\times 10^3 kg/m^3\right)$.
Unless otherwise stated as follows:

$$(y+H)/H = 10, \quad x/H = 10, \quad s/H = 0.01, \quad b/H = 0.001.$$

The graphical demonstrations depicting the efficacy of the anisotropy parameter associated with the upper surface layer of the considered orthotropic structure on non-dimensional phase and group velocities for the scattered Love wave are accomplished through Figure 4.6(a) and (b), respectively. Figures 4.7(a) to 4.9(b) show the behavior of induced non-vanishing component of reflected displacement in orthotropic featuring surface irregularity in the form of parabolic, triangular notch, and rectangular shapes. Figure 4.10 illustrates the comparative analysis of reflected displacement of

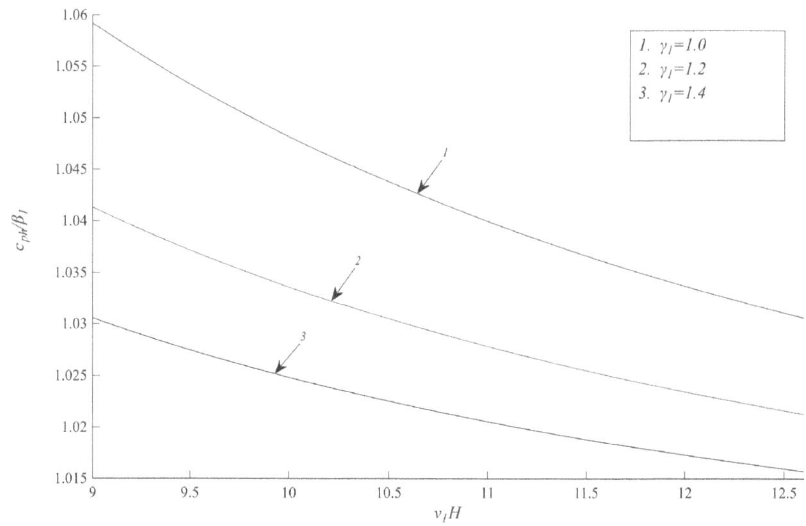

(a)

Figure 4.6 Non-dimensional (a) phase velocity $\left(\frac{c_{ph}}{\beta_1}\right)$ and (b) group velocity $\left(\frac{c_g}{\beta_1}\right)$ contrasted with non-dimensional wave number $\left(\upsilon_1 H\right)$ for diverse anisotropy parameter (γ_1) values of the upper layer in the orthotropic composite material structure. *(Continued)*

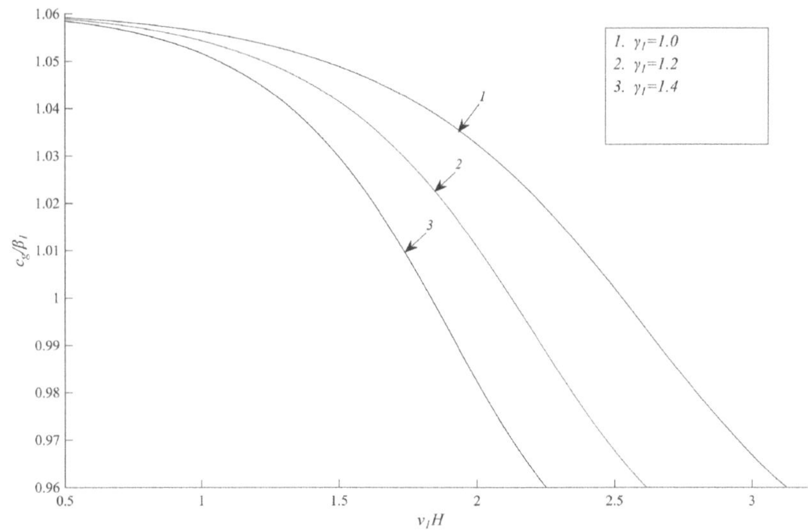

(b)

Figure 4.6 (Continued)

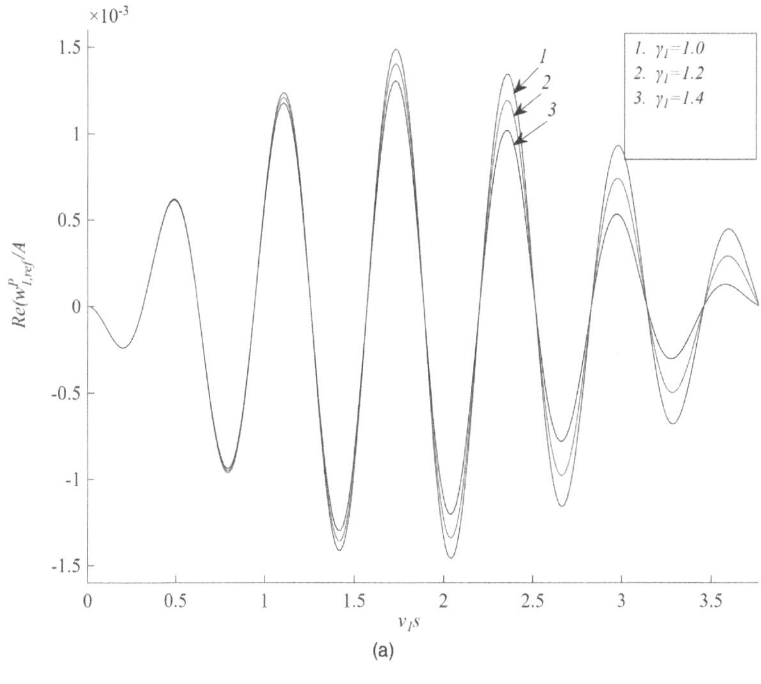

(a)

Figure 4.7 Non-dimensional reflected displacement $\text{Re}\left(w_1^P, ref / A\right)$ contrasted with non-dimensional width $\left(\upsilon_1 s\right)$ for diverse (a) anisotropy parameter (γ_1) values and (b) vertical irregularity parameter $\left(b / H\right)$ values associated with the irregularity of the parabolic shape present in the upper orthotropic layer of the studied orthotropic composite structure. (*Continued*)

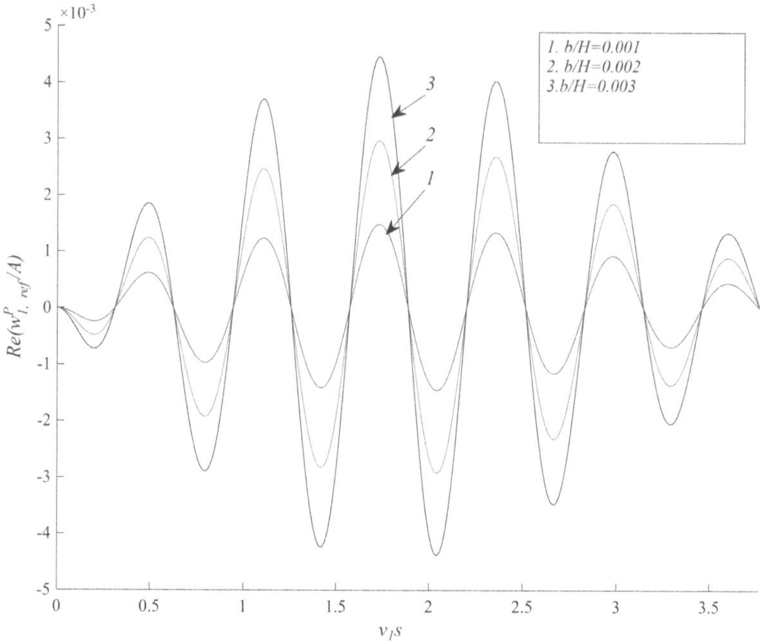

(b)

Figure 4.7 (Continued)

different shapes of irregularity viz. parabolic, triangular notch, and rectangular, in the upper surface layer in the studied orthotropic structure.

Figure 4.6(a) and (b) depict the phase and group velocities, respectively, with the influence of anisotropy parameter for shear rigidity associated with the upper orthotropic surface layer. The said anisotropy parameter is the ratio of two-shear rigidities in two perpendicular directions. The anisotropic parameter $\gamma_1 = 1$ implies that the case of an isotropic structure. It is reported from this figure that the phase and group velocities become minimum corresponding to the isotropic case. Moreover, these velocities decay as the anisotropy prevails in the orthotropic structure.

Figures 4.7(a), 4.8(a), and 4.9(a) illustrate the effect of vertical irregularity parameter on the reflected displacement in the studied orthotropic medium in which the upper surface layer contains distinct shapes of irregularity viz. parabolic, triangular notch, and rectangular, respectively. It is reported by the aforesaid figures that the enhancement of the anisotropy parameter in the considered orthotropic composite structure causes the decrease in the non-vanishing components of reflected displacement for all discussed physical scenarios of irregularity in the upper surface. The physical observation for the obtained pattern may lie with the fact that wave energy decreases as the anisotropy parameter increases (as discussed for Figure 4.6(a) and (b)).

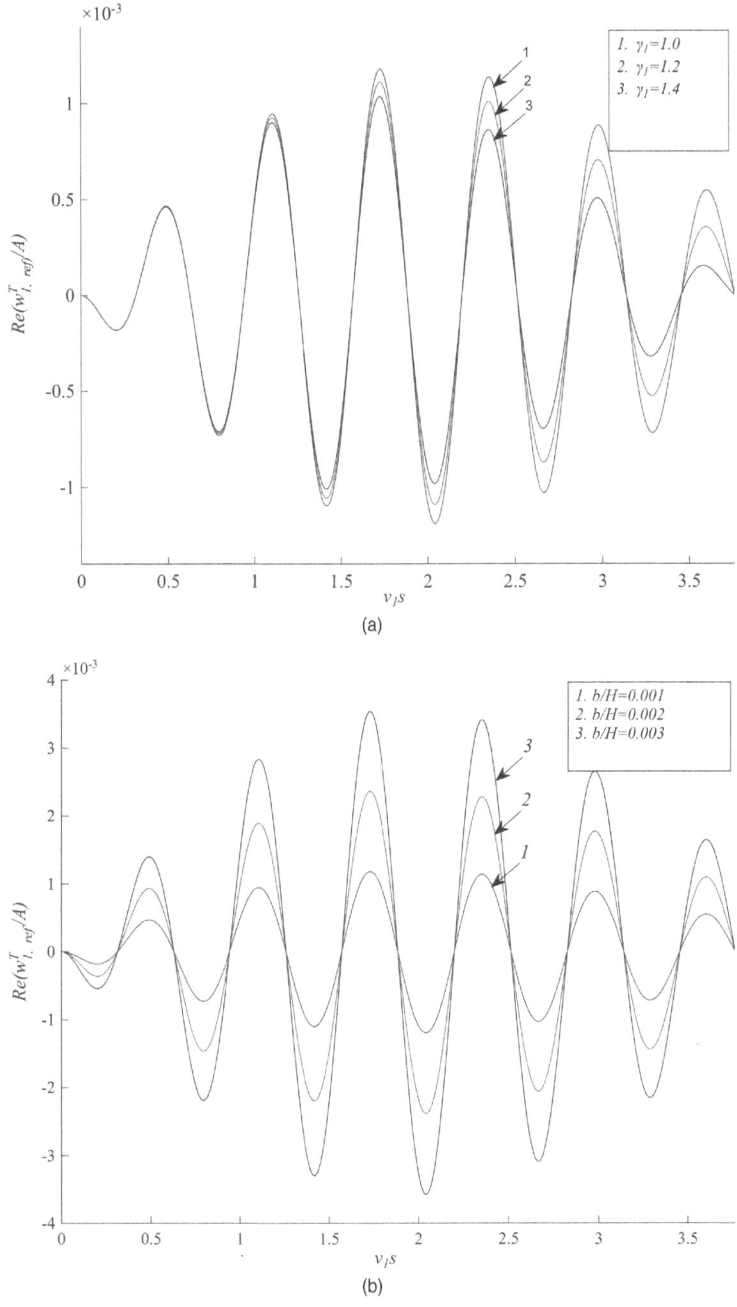

Figure 4.8 Non-dimensional component of reflected displacement $\mathrm{Re}\left(w_l^T, ref / A\right)$ contrasted with non-dimensional width $\left(\upsilon_l s\right)$ for diverse (a) anisotropy parameter $\left(\gamma_l\right)$ values and (b) vertical irregularity parameter $\left(b / H\right)$ values related to the irregularity with the shape of triangular notch present in upper surface layer in the studied orthotropic composite structure.

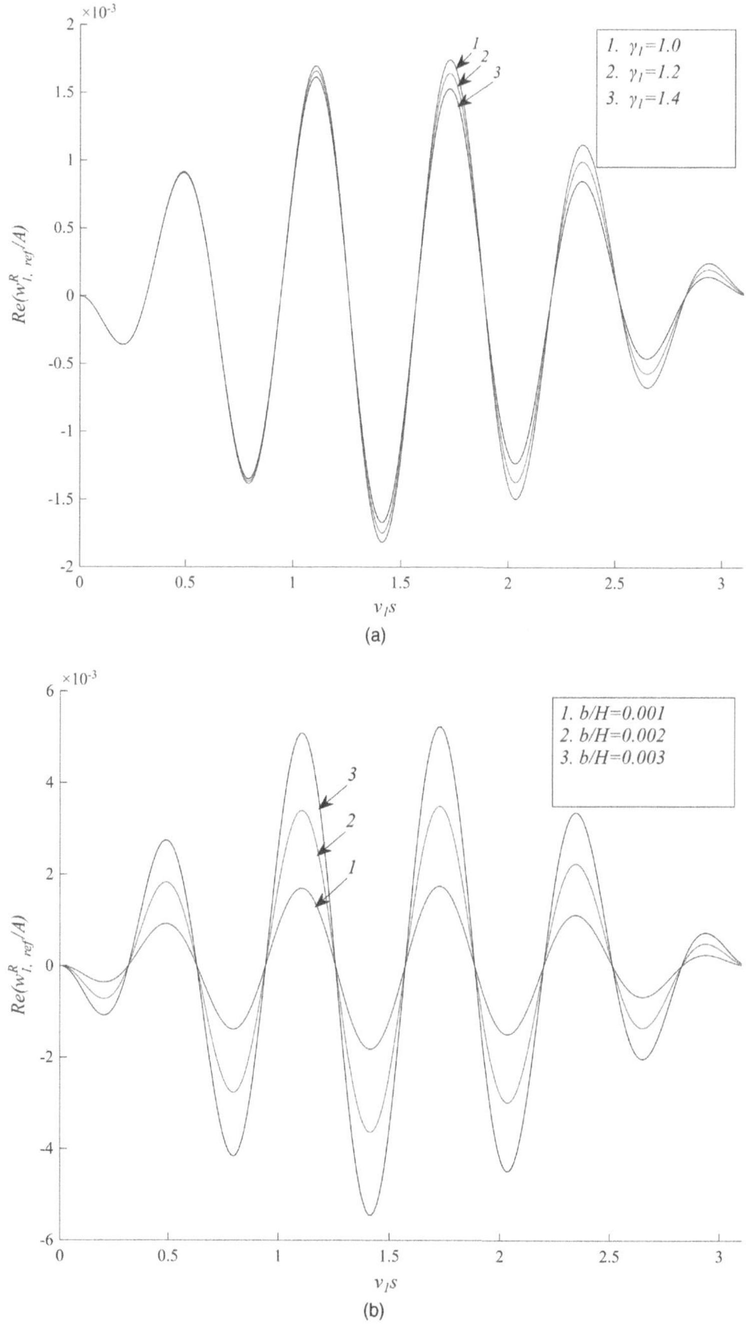

Figure 4.9 Non-dimensional component of reflected displacement $\mathrm{Re}\left(w_l^R, ref / A\right)$ contrasted with non-dimensional width $\left(\upsilon_l s\right)$ for diverse (a) anisotropy parameter (γ_l) values and (b) vertical irregularity parameter (b / H) values for the irregularity of rectangular shape present in the studied orthotropic composite structure.

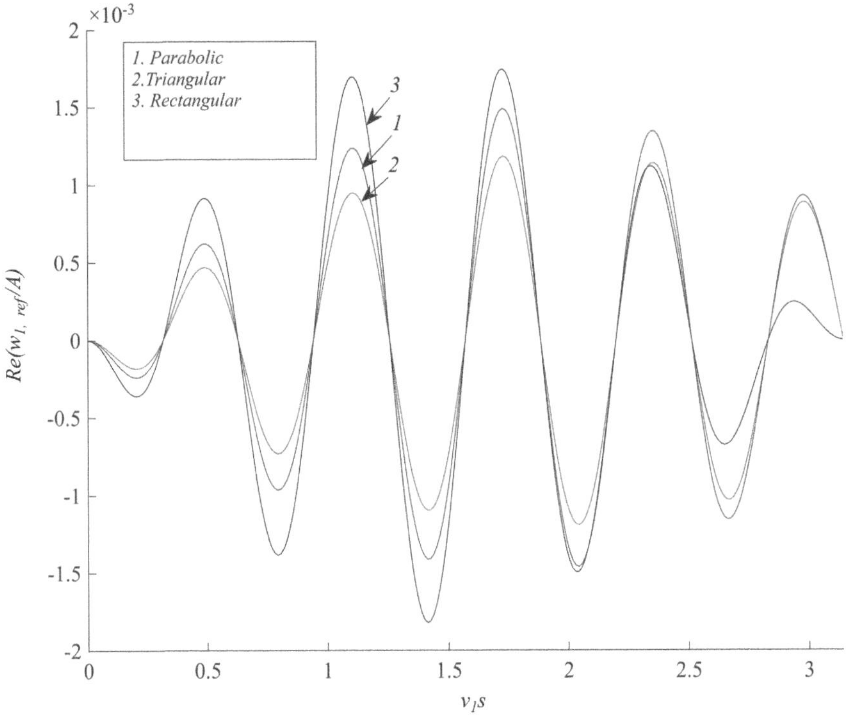

Figure 4.10 Non-dimensional component of reflected displacement $\mathrm{Re}\left(w_1, ref / A\right)$ contrasted with non-dimensional width $\left(v_1 s\right)$ for all three analyzed cases (viz. parabolic, triangular notch, rectangular) of irregularity present in the upper orthotropic surface of the studied orthotropic composite structure.

Due to this, the reflected mechanical displacement is reduced by the elevation of the anisotropy parameter.

Figures 4.7(b), 4.8(b), and 4.9(b) exhibit the effect of vertical irregularity parameter on the reflected displacement for the irregularity of three distinct shapes viz. parabolic, triangular notch, and rectangular in the upper surface. The above-mentioned figures provide that as the value of vertical irregularity parameter increases, the reflected displacement decreases. This may be attributed to the fact that the increment of vertical irregularity parameter indicates a greater depth of irregularity present in the upper surface. Further, the greater depth of surface irregularity may cause the greater number of wave components to encounter through the area of irregularity in the upper surface. It further leads to the rise in the studied reflected displacement.

Figure 4.10 imparts the comparison of induced non-vanishing component of reflected displacement among the irregularities of diverse shapes viz. parabolic (curve 1), triangular notch (curve 2), and rectangular (curve 3) in the upper surface. The observation of this figure provides that initially (for lower values of dimensional width), the reflected displacement corresponding to the scenario of surface irregularity having rectangular shape is maximum whereas the same is minimum for the scenario when irregularity of triangular notch shape present in the upper layer. Moreover, the reflected displacement when the surface irregularity is of parabolic shape falls between those of the rectangular and triangular notch cases. After a certain critical point, as the dimensional width attains higher value, the parabolic shape surface irregularity attains maximum value whereas rectangular shape irregularity attains minimum value.

4.8 CONCLUSIONS

The present investigation encapsulates the Love waves scattering due to presence of irregularity in the upper surface in a layered orthotropic composite structure. Surface irregularity with distinct shapes viz. parabolic, triangular notch, and rectangular have been considered to be present in upper surface of considered orthotropic composite structure which has been mathematically addressed using the perturbation technique. The expressions for dispersion relation and group velocity corresponding to Love waves for scattered field have been determined analytically. The efficacy of anisotropy parameter on scattered Love waves has also been analyzed. The relation associated with the induced non-vanishing component of reflected displacement has also been obtained with the help of analytical treatment. The influences of the vertical irregularity parameter and the anisotropy parameter on the reflected displacement in the upper layer of the specified orthotropic composite structure have been investigated. Furthermore, the following key points can be regarded as the primary outcomes of the conducted study:

1. The phase and group velocities of the scattered Love waves decrease as the anisotropy parameter of the upper layer in the studied orthotropic composite structure increases.
2. The reflected displacement induced by Love waves propagation has a favorable impact on the vertical irregularity parameter for all considered surface irregularity types viz. parabolic shape, triangular notch, and rectangular in the orthotropic composite structure.
3. The reflected displacement induced by the propagation of Love waves decreases as the anisotropy parameter corresponding to orthotropic layer increases, for all studied physical shapes of surface irregularity such as parabolic, triangular notch, and rectangular.
4. Among the considered surface irregularities in the layered orthotropic composite structure, the rectangular-shaped irregularity yields the highest reflected displacement, while the triangular notch-shaped

irregularity results in the lowest. The reflected displacement when the irregularity having parabolic shape present falls between these two extremes.

4.9 APPLIED ENGINEERING INSIGHTS

The current study may find its potential uses in the design and development of bridge decks, railway bridge, highway bridges, etc. The orthotropic materials such as fibrous composite, laminated composites are generally used in the design of such structures. As the phenomenon of an earthquake occurs, it causes the propagation of various seismic waves. Love waves are a type of seismic wave known for their destructive nature. Love waves passing through the irregularity present in upper surface may undergo scattering within the structure. An increase in the velocity of scattered waves may exacerbate their impact on structural integrity, making buildings more susceptible to damage. Scattering phenomena also give rise to a reflected wave field, which, if intensified, can potentially cause internal damage to the structure. Hence, it is essential to investigate scattering effects caused by surface irregularities. From the current investigation, it is reported that the upper layer of higher anisotropy can reduce the velocity (energy) of scattered wave and also can result in a lesser magnitude of the reflected displacement.

The present analysis also suggests that the reflected wave field reaches its maximum when the irregularity is of rectangular shape and its minimum when the irregularity is in the form of triangular notch in the upper surface. Owing to all the above facts, it is evident that the knowledge obtained from this analysis can be applicable to geotechnical and civil engineering.

ACKNOWLEDGMENT

The authors express their heartfelt gratitude to IIT(ISM), Dhanbad, and UPES, Dehradun for providing the necessary research facilities, which played a crucial role in the successful completion of this work.

DECLARATION OF INTEREST STATEMENT

On behalf of all the authors, the corresponding author affirms that there are no conflicts of interest.

DATA AVAILABILITY

All data used in the present study are included and referenced within the manuscript.

REFERENCES

1. Wang, Y. S., & Zhang, Z. M. (1998). Propagation of Love waves in a transversely isotropic fluid-saturated porous layered half-space. *The Journal of the Acoustical Society of America*, *103*(2), 695–701.
2. Li, X. Y., Wang, Z. K., & Huang, S. H. (2004). Love waves in functionally graded piezoelectric materials. *International Journal of Solids and Structures*, *41*(26), 7309–7328.
3. Zhu, X. Q., & Law, S. S. (2003). Dynamic behavior of orthotropic rectangular plates under moving loads. *Journal of Engineering Mechanics*, *129*(1), 79–87.
4. Maljaars, J., van Dooren, F., & Kolstein, H. (2012). Fatigue assessment for deck plates in orthotropic bridge decks. *Steel Construction*, *5*(2), 93–100.
5. Saha, A., Kundu, S., Gupta, S., & Kumar Vaishnav, P. (2017). SH wave propagation in a finite thicker layer of the void pore sandwiched by heterogeneous orthotropic media. *International Journal of Geomechanics*, *17*(5), 06016033.
6. Destrade, M. (2001). Surface waves in orthotropic incompressible materials. *The Journal of the Acoustical Society of America*, *110*(2), 837–840.
7. Destrade, M. (2007). Seismic Rayleigh waves on an exponentially graded, orthotropic half-space. *Proceedings of the Royal Society A: Mathematical, Physical and Engineering Sciences*, *463*(2078), 495–502.
8. Vignjevic, R., Campbell, J. C., Bourne, N. K., & Djordjevic, N. (2008). Modeling shock waves in orthotropic elastic materials. *Journal of Applied Physics*, *104*(4), 044904.
9. Biswas, S. (2020). Surface waves in porous nonlocal thermoelastic orthotropic medium. *Acta Mechanica*, *231*, 2741–2760.
10. Abd-Alla, A. M., & Ahmed, S. M. (1999). Propagation of Love waves in a non-homogeneous orthotropic elastic layer under initial stress overlying semi-infinite medium. *Applied Mathematics and Computation*, *106*(2–3), 265–275.
11. Wang, C. D., Chou, H. T., & Peng, D. H. (2017). Love-wave propagation in an inhomogeneous orthotropic medium obeying the exponential and generalized power law models. *International Journal of Geomechanics*, *17*(7), 04017003.
12. Yu, J., & Zhang, C. (2014). Effects of initial stress on guided waves in orthotropic functionally graded plates. *Applied Mathematical Modelling*, *38*(2), 464–478.
13. Singh, A. K., Pal, M. K., Negi, A., & Mistri, K. C. (2019). Analytical study on dynamic response due to a moving load on distinctly characterized orthotropic half-spaces under different physical conditions with comparative approach. *Arabian Journal for Science and Engineering*, *44*(5), 4863–4883.
14. Singh, A. K., Lakshman, A., & Chattopadhyay, A. (2016). Effect of irregularity and anisotropy on the dynamic response due to a shear load moving on an irregular orthotropic half-space under influence of gravity. *Multidiscipline Modeling in Materials and Structures*, *12*(1), 194–214.
15. Slavin, L. M., & Wolf, B. (1970). Scattering of Love waves in a surface layer with an irregular boundary for the case of a rigid underlying half-space. *Bulletin of the Seismological Society of America*, *60*(3), 859–877.
16. Wolf, B. (1970). Propagation of Love waves in layers with irregular boundaries. *Pure and Applied Geophysics*, *78*(1), 48–57.
17. Chen, W. Q., Wang, H. M., & Bao, R. H. (2007). On calculating dispersion curves of waves in a functionally graded elastic plate. *Composite Structures*, *81*(2), 233–242.
18. Anagnostopoulos, K. A., Charalambopoulos, A., & Massalas, C. V. (2005). On the investigation of elasticity equation for orthotropic materials and the solution of the associated scattering problem. *International Journal of Solids and Structures*, *42*(24–25), 6376–6408.

19. Chattopadhyay, A., Gupta, S., Sharma, V. K., & Kumari, P. (2010). Propagation of shear waves in viscoelastic medium at irregular boundaries. *Acta Geophysica*, *58*(2), 195–214.
20. Negi, A., & Kumar Singh, A. (2019). Analysis on scattering characteristics of Love-type wave due to surface irregularity in a piezoelectric structure. *The Journal of the Acoustical Society of America*, *145*(6), 3756–3783.
21. Graff, K. F. (2012). *Wave motion in elastic solids*. Courier Corporation.
22. Miklowitz, J. (2012). *The theory of elastic waves and waveguides*. Elsevier.
23 Novotny, O. (1999). Seismic surface waves. *Bahia, Salvador: Instituto de Geociencias*, *61*.
24. Sezawa, K. (1935). Love-waves generated from a source of a certain depth. *Bulletin of the Earthquake Research Institute, Tokyo Imperial University*, *13*(1), 1–17.
25 Love, A. E. H. (1944). *A treatise on the mathematical theory of elasticity*. Dover Publications.

Chapter 5

Love-type surface wave propagation in a layered anisotropic medium with sliding interface

Pranav P, Amit Kumar Rahul, and Manoj Kumar Singh

5.1 INTRODUCTION

Over the past few decades, research has been done on the seismic wave propagation in multilayer structures. This research strives to increase the knowledge of the Earth's internal mechanics. Natural occurrences such as earthquakes generate two categories of seismic waves: body waves, which traverse the Earth's interior, and surface waves, which move along its surface. Love-type waves, which are distinguished by their confinement close to the surface and horizontal shear motion, are an important family of surface waves. Love [1] was the first to anticipate the presence of Love waves in 1911, and they usually show attenuation as depth increases. Foundational concepts and prior studies related to seismic wave propagation, anisotropy, elasticity, and viscoelastic theory are extensively documented in the works of Achenbach, Ewing et al., Biot, and Gubbins [2–5].

An anisotropic material with mechanical properties that vary along three mutually perpendicular directions—typically aligned with material symmetry axes—is called an orthotropic medium. Orthotropic materials exhibit direction-dependent reactions because of their internal structure, layering, or preferred grain alignment, in contrast to isotropic media, which have attributes like stiffness or wave speed that are constant in all directions. With a wide range of applications, an orthotropic medium's wave propagation characteristics are highlighted in a number of publications. Early studies, such as Abd-Alla and Ahmed, examined Love waves under initial stress in a non-homogeneous orthotropic layer, establishing foundational dispersion relations [6]. Kundu et al. explored fibre-reinforced orthotropic media, demonstrating how material reinforcement enhances wave velocity [7]. Further advancements by Panja and Mandal, and Pandit et al. incorporated viscoelastic damping and porosity effects, revealing the critical role of energy dissipation in wave propagation [8, 9]. Recent studies have pushed the boundaries by investigating non-local elasticity and fractured porous media, highlighting the influence of microstructural defects and scale-dependent material behaviour [10, 11]. Additionally, some researchers have addressed irregular

 DOI: 10.1201/9781003470885-5

interfaces and multilayered composites, respectively, emphasizing the impact of geometric and stress-induced anisotropies [12, 13].

A unique class of anisotropic material that behaves isotropically in a plane perpendicular to a single axis of symmetry is known as transversely isotropic (TI) media. In these media, mechanical properties vary along the axis perpendicular to the symmetry plane but stay the same in all other directions within the symmetry plane. As a result, the TI model is simpler than fully anisotropic formulations but more general than isotropic. Over the years, researchers have explored various physical effects influencing wave propagation in TI media. Ahmad and Khan, and Singh investigated rotational and thermoelastic effects, respectively, in TI media [14, 15], while Singh and Acharya et al. studied the behaviour of wave in magnetoelastic TI medium [16, 17]. These studies demonstrate how coupling additional physical fields with anisotropy can significantly alter wave characteristics. More directly relevant to the present context, Alam and Kundu analysed the effects of initial stress and heterogeneity on Love-type wave propagation in a TI medium over an inhomogeneous substrate, revealing how material gradation and stress states influence dispersion [18]. Bayones et al. extended this investigation by incorporating rotational effects into a heterogeneous, initially stressed TI layer, further illustrating the interplay between anisotropy, rotation, and stress in seismic response [19]. In composite systems, some researchers examined wave propagation through layered media combining TI, orthotropic, isotropic, and porous components, emphasizing the impact of material mismatches and pre-stressing on wave behaviour [20–22]. Zhu et al. added the perspective of functional grading, which introduces spatial variation in elastic properties [23].

The interface condition between adjacent media is a critical factor in determining wave behaviour in the research of seismic and elastic wave propagation. The majority of analytical models have historically assumed a welded (completely bonded) contact, which implies that stresses and displacements are continuous throughout the boundary. However, interfaces may allow relative motion, resulting in what is known as a sliding contact, in many realistic geological and material systems, such as fault zones, fractured layers, or manufactured composites. Some of the authors have done research with sliding interface [24, 25].

In this study, we focus on the theoretical modelling of seismic wave transmission between two contrasting geological media, each possessing distinct mechanical characteristics. The present Earth model (see Figure 5.1) considers the Love-type wave propagation in a composite structure consisting of a viscoelastic orthotropic medium overlying a heterogeneous TI elastic half-space which are initially stressed. Unlike most of the existing literature that assumes welded contact between layers, this study incorporates a sliding interface to more accurately represent imperfect bonding. This unified approach provides a more realistic understanding of seismic surface wave behaviour in anisotropic, stressed, and imperfectly bonded geological settings.

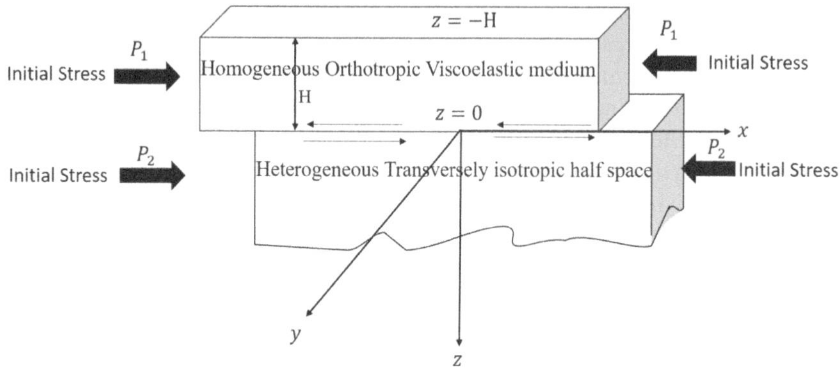

Figure 5.1 Geometrical model.

Nomenclature

Heterogeneous Transversely Isotropic Medium

A, C, N, G, S	Material constants	$\tilde{v}_2$	Displacement
P_2	Initial stress	k	Angular wave number
$\tilde{\sigma}^*_{ij}$	Stress component	ω	Angular frequency
ρ_2	Density	β	Heterogeneity parameter

Homogeneous Orthotropic Viscoelastic Medium

$\tilde{\sigma}_{ij}$	Stress component	e_{ij}	Strain component
G^*_{ij}	Elastic constants	$\tilde{v}_1$	Displacement
P_1	Initial stress	ρ_1	Density
$\bar{G}_{44}, \bar{G}_{66}$	Internal frictions		

5.2 PROBLEM FORMULATION

Consider the configuration illustrated in Figure 5.1. The Cartesian coordinate system's origin is marked as 0. The orthotropic viscoelastic medium occupying the domain $\{-H \leq z \leq 0\}$ is embedded over heterogeneous TI half-space with domain $\{0 \leq z < \infty\}$. Both media are subjected to initial stresses, labelled P_1 and P_2, respectively. The interface of the two media is a sliding interface. The Kelvin–Voigt viscoelasticity model is taken into account. Love-type waves are considered to propagate in the x-direction and the particle motion polarized in the y-direction. A quadratic variation of heterogeneity with depth is considered in the lower half-space, aligned with the z-direction.

The following presumptions are made regarding the propagation of Love-type waves:

$$\tilde{u}_i = 0, \;\; \tilde{v}_i = \tilde{v}_i(x, z, t), \tilde{w}_i = 0, \;\; \frac{\partial}{\partial y} \equiv 0, \tag{5.1}$$

where the quantities $\tilde{u}_i, \tilde{v}_i, \tilde{w}_i$ correspond to displacement along the x, y, and z directions, respectively.

5.2.1 For orthotropic homogeneous viscoelastic layer

The upper layer is subjected to a pre-applied initial stress P_1 oriented along the x-axis. Then, Biot [4] provides the motion equations in this upper medium without body forces and under initial stress.

$$\begin{cases} \dfrac{\partial \tilde{\sigma}_{xx}}{\partial x} + \dfrac{\partial \tilde{\sigma}_{xy}}{\partial y} + \dfrac{\partial \tilde{\sigma}_{xz}}{\partial z} - P_1\left(\dfrac{\partial w_z}{\partial y} - \dfrac{\partial w_y}{\partial z}\right) = \rho_1 \dfrac{\partial^2 \tilde{u}_1}{\partial t^2} \\ \dfrac{\partial \tilde{\sigma}_{yx}}{\partial x} + \dfrac{\partial \tilde{\sigma}_{yy}}{\partial y} + \dfrac{\partial \tilde{\sigma}_{yz}}{\partial z} - P_1 \dfrac{\partial w_z}{\partial x} = \rho_1 \dfrac{\partial^2 \tilde{v}_1}{\partial t^2} \\ \dfrac{\partial \tilde{\sigma}_{zx}}{\partial x} + \dfrac{\partial \tilde{\sigma}_{zy}}{\partial y} + \dfrac{\partial \tilde{\sigma}_{zz}}{\partial z} - P_1 \dfrac{\partial w_y}{\partial x} = \rho_1 \dfrac{\partial^2 \tilde{w}_1}{\partial t^2} \end{cases} \tag{5.2}$$

where $\tilde{\sigma}_{ij}$ are the incremental stress, $(\tilde{u}_1, \tilde{v}_1, \tilde{w}_1)$ are displacement components, ρ_1 is the density of the upper layer, and (w_x, w_y, w_z) denote the components of rotation along suffix x, y, z direction, respectively.

The sole non-vanishing equation, using Eq. (5.1) in Eq. (5.2), is

$$\frac{\partial \tilde{\sigma}_{xy}}{\partial x} + \frac{\partial \tilde{\sigma}_{yz}}{\partial z} - \frac{P_1}{2}\frac{\partial^2 \tilde{v}_1}{\partial x^2} = \rho_1 \frac{\partial^2 \tilde{v}_1}{\partial t^2}, \tag{5.3}$$

Within the layer, the stress components take the form:

$$\begin{cases} \tilde{\sigma}_{xx} = G^*_{11}e_{11} + G^*_{12}e_{22} + G^*_{13}e_{33} \\ \tilde{\sigma}_{yy} = G^*_{12}e_{11} + G^*_{22}e_{22} + G^*_{23}e_{33} \\ \tilde{\sigma}_{zz} = G^*_{13}e_{11} + G^*_{23}e_{22} + G^*_{33}e_{33} \\ \tilde{\sigma}_{yz} = 2G^*_{44}e_{23} \\ \tilde{\sigma}_{xz} = 2G^*_{55}e_{13} \\ \tilde{\sigma}_{xy} = 2G^*_{66}e_{12} \end{cases} \tag{5.4}$$

where $e_{ij} = \frac{1}{2}\left(\frac{\partial u_i}{\partial x_j} + \frac{\partial u_j}{\partial x_i}\right)$ is the strain components.

Using Eq. (5.1) and Eq. (5.4), stress–displacement relationships are given by

$$\tilde{\sigma}_{xy} = G^*_{66}\frac{\partial \tilde{v}_1}{\partial x}, \quad \tilde{\sigma}_{yz} = G^*_{44}\frac{\partial \tilde{v}_1}{\partial z}, \tag{5.5}$$

where $G_{66}^{*}=G_{66}+\bar{G}_{66}\frac{\partial}{\partial t}$, $G_{44}^{*}=G_{44}+\bar{G}_{44}\frac{\partial}{\partial t}$; G_{66},G_{44} are elastic coefficient, and $\bar{G}_{66},\bar{G}_{44}$ are internal frictions due to viscoelasticity of the medium.

The solution of Eq. (5.3) can be considered in the following form:

$$\tilde{v}_1=V_1(z)e^{i(kx-\omega t)}, \tag{5.6}$$

where V_1 is the unknown function.

Using Eqs. (5.5) and (5.6) in Eq. (5.3), the following differential equation is obtained:

$$\frac{d^2V_1}{dz^2}+\xi_1^2 V_1=0, \tag{5.7}$$

where $\xi_1=\sqrt{\frac{k^2G_{66}\left((-1+i\gamma_1)+\eta_1+\frac{c^2}{\alpha_1^2}\right)}{G_{44}(1-i\gamma_2)}}, \gamma_1=\omega\frac{\bar{G}_{66}}{G_{66}}, \gamma_2=\omega\frac{\bar{G}_{44}}{G_{44}}, \eta_1=\frac{P_1}{2G_{66}}, \alpha_1=\sqrt{\frac{G_{66}}{\rho_1}}$;

where γ_1,γ_2 are the viscoelastic parameters, and η_1 is the initial stress parameter.

Eq. (5.7) yields the expression below:

$$V_1(z)=A_1Cos(\xi_1 z)+A_2Sin(\xi_1 z), \tag{5.8}$$

From Eq. (5.6), the displacement component for the orthotropic is obtained as

$$\tilde{v}_1(z)=A_1Cos(\xi_1 z)+A_2Sin(\xi_1 z)e^{i(kx-\omega t)} \tag{5.9}$$

5.2.2 For the heterogeneous transversely isotropic (TI) half-space

Consider the component set of the medium's displacement vectors to be $(\tilde{u}_2,\tilde{v}_2,\tilde{w}_2)$. The dynamical equilibrium equations for media under initial stress [4] are as follows:

$$\begin{cases}\dfrac{\partial\tilde{\sigma}^{*}_{xx}}{\partial x}+\dfrac{\partial\tilde{\sigma}^{*}_{xy}}{\partial y}+\dfrac{\partial\tilde{\sigma}^{*}_{xz}}{\partial z}-P_2\left(\dfrac{\partial\phi_z}{\partial y}-\dfrac{\partial\phi_y}{\partial z}\right)=\rho_2\dfrac{\partial^2\tilde{u}_2}{\partial t^2}\\ \dfrac{\partial\tilde{\sigma}^{*}_{xy}}{\partial x}+\dfrac{\partial\tilde{\sigma}^{*}_{yy}}{\partial y}+\dfrac{\partial\tilde{\sigma}^{*}_{yz}}{\partial y}-P_2\left(\dfrac{\partial\phi_z}{\partial x}\right)=\rho_2\dfrac{\partial^2\tilde{v}_2}{\partial t^2}\\ \dfrac{\partial\tilde{\sigma}^{*}_{xz}}{\partial x}+\dfrac{\partial\tilde{\sigma}^{*}_{yz}}{\partial y}+\dfrac{\partial\tilde{\sigma}^{*}_{zz}}{\partial z}-P_2\left(\dfrac{\partial\phi_y}{\partial x}\right)=\rho_2\dfrac{\partial^2\tilde{w}_2}{\partial t^2}\end{cases} \tag{5.10}$$

The stress–displacement relationships are given by

$$\begin{cases} \tilde{\sigma}^*_{xx} = A\dfrac{\partial \tilde{u}_2}{\partial x} + (A-2N)\dfrac{\partial \tilde{v}_2}{\partial y} + S\dfrac{\partial \tilde{w}_2}{\partial z} \\ \tilde{\sigma}^*_{xy} = N\left(\dfrac{\partial \tilde{u}_2}{\partial y} + \dfrac{\partial \tilde{v}_2}{\partial x}\right) \\ \tilde{\sigma}^*_{yy} = (A-2N)\dfrac{\partial \tilde{u}_2}{\partial x} + A\dfrac{\partial \tilde{v}_2}{\partial y} + S\dfrac{\partial \tilde{w}_2}{\partial z} \\ \tilde{\sigma}^*_{yz} = G\left(\dfrac{\partial \tilde{w}_2}{\partial y} + \dfrac{\partial \tilde{v}_2}{\partial z}\right) \\ \tilde{\sigma}^*_{zz} = S\dfrac{\partial \tilde{u}_2}{\partial x} + S\dfrac{\partial \tilde{v}_2}{\partial y} + C\dfrac{\partial \tilde{w}_2}{\partial z} \\ \tilde{\sigma}^*_{xz} = G\left(\dfrac{\partial \tilde{u}_2}{\partial z} + \dfrac{\partial \tilde{w}_2}{\partial x}\right) \end{cases} \tag{5.11}$$

The components of rotation ϕ_x, ϕ_y, ϕ_z are represented by

$$\phi_x = \frac{1}{2}\left(\frac{\partial \tilde{w}_2}{\partial y} - \frac{\partial \tilde{v}_2}{\partial z}\right),\ \phi_y = \frac{1}{2}\left(\frac{\partial \tilde{u}_2}{\partial z} - \frac{\partial \tilde{w}_2}{\partial x}\right),\ \phi_z = \frac{1}{2}\left(\frac{\partial \tilde{v}_2}{\partial x} - \frac{\partial \tilde{u}_2}{\partial y}\right) \tag{5.12}$$

The material inhomogeneities within the half-space are defined by

$$\begin{cases} A = A^*(1+\beta z), S = S^*(1+\beta z), C = C^*(1+\beta z), N = N^*(1+\beta z) \\ P_2 = P_2^*(1+\beta z), \rho_2 = \rho_2^*(1+\beta z), G = G^*(1+\beta z) \end{cases} \tag{5.13}$$

Using Eq. (5.1) in Eq. (5.10), the only non-vanishing equation is

$$\frac{\partial \tilde{\sigma}^*_{xy}}{\partial x} + \frac{\partial \tilde{\sigma}^*_{yz}}{\partial z} - P_2\left(\frac{\partial \phi_z}{\partial x}\right) = \rho_2 \frac{\partial^2 \tilde{v}_2}{\partial t^2} \tag{5.14}$$

A wave-form solution corresponding to Eq. (5.14) is assumed as follows:

$$\tilde{v}_2(z) = V_2(z)e^{i(kx-\omega t)} \tag{5.15}$$

Inserting Eqs. (5.12), (5.13), and (5.15) into Eq. (5.14) yields

$$\frac{d^2V_2}{dz^2} + \frac{\beta}{(1+\beta z)}\frac{dV_2}{dz} + k^2\xi_2^2 V_2 = 0 \tag{5.16}$$

where $\xi_2^2 = \left(-\frac{N^*}{G^*} + \eta_2 + \frac{c^2}{\alpha_2^2}\right)$, $\eta_2 = \frac{P_2^*}{2G^*}$, $\alpha_2 = \sqrt{\frac{G^*}{\rho_2^*}}$.
where η_2 is the initial stress parameter of half-space.

Let's substitute $V_2(z) = \frac{M(z)}{\sqrt{1+\beta z}}$ in Eq. (5.16); we get

$$\frac{d^2M}{dz^2} + \chi^2 M = 0 \tag{5.17}$$

where $\chi = \sqrt{\frac{\beta^2}{4(1+\beta z)^2} + k^2\xi_2^2}$.

Assuming weak heterogeneity $(\beta z \ll 1)$, the depth-dependent parameter χ may be approximated as constant, which reduces the governing equation to a harmonic form.

Hence, Eq. (5.17) admits the following solution:

$$V_2(z) = \frac{A_3 e^{i\chi z} + A_4 e^{-i\chi z}}{\sqrt{1+\beta z}} \tag{5.18}$$

Then the displacement component we obtain is as follows:

$$\tilde{v}_2(z) = \frac{A_4 e^{-i\chi z}}{\sqrt{1+\beta z}} e^{i(kx-\omega t)} \tag{5.19}$$

5.3 BOUNDARY CONDITIONS

1. No external traction acts on the layer's upper surface

i.e., $\tilde{\sigma}_{yz} = 0$ at $z = -H$ (5.20)

2. The parameter δ governs the anticipated sliding of the interface. Consequently, the boundary limitations associated with displacement and traction at the interface are provided by
   - $\tilde{\sigma}^*_{yz} = (1-\delta)\tilde{\sigma}_{yz}$ $at\ z = 0$ (5.21)
   - $\delta\tilde{\sigma}_{yz} + (1-\delta)k\mathrm{F}\tilde{v}_1 = (1-\delta)k\mathrm{F}\tilde{v}_2$ $at\ z = 0$ (5.22)

where δ, the sliding parameter, which quantifies interfacial slip, is defined as follows: when $\delta = 0$, it indicates a welded contact interface; when $\delta = 1$, it indicates a smooth contact interface; and when the parameter $\delta \in (0,1)$, it indicates partial sliding. F, a constant with a stress unit, is added to ensure that the equation's dimensions are identical.

The displacement functions, together with the complete expression of χ, are subsequently employed in the application of boundary conditions in order to adequately capture heterogeneity effects. The application of the aforementioned boundary conditions yields a corresponding system of homogeneous equations. The matrix form that follows can be used to represent this system, which has three unknowns $(A_1, A_2, and\ A_4)$:

$$\begin{pmatrix} \lambda_1 & \lambda_2 & 0 \\ 0 & \lambda_3 & \lambda_4 \\ \lambda_5 & \lambda_6 & \lambda_7 \end{pmatrix} \begin{pmatrix} A_1 \\ A_2 \\ A_4 \end{pmatrix} = \begin{pmatrix} 0 \\ 0 \\ 0 \end{pmatrix} \tag{5.23}$$

where

$$\lambda_1 = \left(\xi_1 \mathrm{Sin}[H\xi_1]G_{44} - i\omega\left(\xi_1 \mathrm{Cos}[H\xi_1]\bar{G}_{44}\right)\right),\ \lambda_2 = \left(\xi_1 \mathrm{Cos}[H\xi_1]G_{44} - i\omega\left(\xi_1 \mathrm{Sin}[H\xi_1]\bar{G}_{44}\right)\right)$$

$$\lambda_3 = -(1-\delta)\left(\xi_1 G_{44} - i\omega\xi_1\bar{G}_{44}\right),\ \lambda_4 = G^*\left(-i\sqrt{\frac{\beta^2}{4} + k^2\xi_2^2} - \frac{\beta}{2}\right),\ \lambda_5 = (1-\delta)k\mathrm{F},$$

$$\lambda_6 = \delta\left(\xi_1 G_{44} - i\omega\xi_1\bar{G}_{44}\right),\ \lambda_7 = -(1-\delta)k\mathrm{F}$$

We now need $\left|(\lambda_i)_{3\times3}\right| = 0$, in order to get the nontrivial solution for $A_1, A_2\ and\ A_4$.

When the determinant indicated above is evaluated, the following equation is produced:

$$\tan[H\xi_1] = \frac{-FG^*k\left(\beta + i\sqrt{\beta^2 + 4k^2\xi_2^2}\right)(-1+\delta)}{\xi_1\left(2Fk(-1+\delta)^2 + G^*(\beta + i\sqrt{\beta^2 + 4k^2\xi_2^2})\delta\right)\left(G_{44} - i\omega\bar{G}_{44}\right)} \tag{5.24}$$

5.4 CASES

5.4.1 For partially sliding contact interface $(0 < \delta < 1)$

Case 5.4.1.1: If the upper medium is assumed to be free from initial stress and the viscoelastic properties, i.e., $P_1 = 0, \bar{G}_{44} = \bar{G}_{66} = 0$,

$$\tan\left[Hk\sqrt{\frac{G_{66}\left(\frac{c^2}{\alpha_1^2} - 1\right)}{G_{44}}}\right] = \frac{-FG^*(-1+\delta)\left(\beta + \sqrt{-\beta^2 + 4k^2\left(\frac{N^*}{G^*} - \eta_2 - \frac{c^2}{\alpha_2^2}\right)}\right)}{G_{44}\sqrt{\frac{G_{66}\left(\frac{c^2}{\alpha_1^2} - 1\right)}{G_{44}}}\left(2Fk(-1+\delta)^2 + G^*\delta\left(\beta + \sqrt{-\beta^2 + 4k^2\left(\frac{N^*}{G^*} - \eta_2 - \frac{c^2}{\alpha_2^2}\right)}\right)\right)} \tag{5.25}$$

The model transforms into an elastic orthotropic layer with a sliding contact over a pre-stressed, heterogeneous TI half-space.

Case 5.4.1.2: If the TI medium is assumed to be free from initial stress and the heterogeneity, i.e., $P_2^* = 0, \beta = 0$,

$$\tan\left[Hk\sqrt{\frac{G_{66}\left((-1+i\gamma_1)+\eta_1+\dfrac{c^2}{\alpha_1^2}\right)}{G_{44}(1-i\gamma_2)}}\right] = \frac{-2FG^*\sqrt{\left(\dfrac{N^*}{G^*}-\dfrac{c^2}{\alpha_2^2}\right)}(-1+\delta)}{\left(G_{44}-i\omega\bar{G}_{44}\right)\sqrt{\dfrac{G_{66}\left((-1+i\gamma_1)+\eta_1+\dfrac{c^2}{\alpha_1^2}\right)}{G_{44}(1-i\gamma_2)}}\left(2F(-1+\delta)^2+G^*2\delta\sqrt{\left(\dfrac{N^*}{G^*}-\dfrac{c^2}{\alpha_2^2}\right)}\right)} \tag{5.26}$$

The model transforms into a viscoelastic initially stressed orthotropic layer sliding over a homogenous TI half-space. In this case, the upper layer maintains all of its complexity, while the lower medium is converted to a homogeneous, stress-free elastic solid.

Case 5.4.1.3: When the initial stress from both the medium, viscoelastic properties from upper layer and heterogeneity from the lower half-space, is removed, i.e., $P_1 = 0, P_2^* = 0, \beta = 0, \bar{G}_{44} = \bar{G}_{66} = 0, G_{44} = G_{66} = \mu_1, G^* = N^* = \mu_2$;

$$\tan\left[Hk\sqrt{\left(\frac{c^2}{\alpha_1^2}-1\right)}\right] = \frac{-F\mu_2\sqrt{\left(1-\dfrac{c^2}{\alpha_2^2}\right)}(-1+\delta)}{\mu_1\sqrt{\left(\dfrac{c^2}{\alpha_1^2}-1\right)}\left(F(-1+\delta)^2+\mu_2\delta\sqrt{\left(1-\dfrac{c^2}{\alpha_2^2}\right)}\right)} \tag{5.27}$$

The model becomes the simplified sliding model.

5.4.2 For welded contact interface

Case 5.4.2.1: If the upper medium is assumed to be free from initial stress and the viscoelastic properties, i.e., $P_1 = 0, \bar{G}_{23} = \bar{G}_{12} = 0$,

$$\tan\left[Hk\sqrt{\frac{G_{66}\left(\dfrac{c^2}{\alpha_1^2}-1\right)}{G_{44}}}\right] = \frac{G^*\left(\beta+\sqrt{-\beta^2+4k^2\left(\dfrac{N^*}{G^*}-\eta_2-\dfrac{c^2}{\alpha_2^2}\right)}\right)}{2G_{44}\sqrt{\dfrac{k^2G_{66}\left(\dfrac{c^2}{\alpha_1^2}-1\right)}{G_{44}}}} \tag{5.28}$$

The system reduces to a pre-stressed heterogeneous TI half-space with an elastic orthotropic layer perfectly linked to it.

Case 5.4.2.2: If the TI medium is assumed to be free from initial stress and the heterogeneity, i.e., $P_2^* = 0, \beta = 0$,

$$\tan\left[Hk\sqrt{\frac{G_{66}\left((-1+i\gamma_1)+\eta_1+\dfrac{c^2}{\alpha_1^2}\right)}{G_{44}(1-i\gamma_2)}}\right] = \frac{G^*\left(\sqrt{\left(\dfrac{N^*}{G^*}-\dfrac{c^2}{\alpha_2^2}\right)}\right)}{\sqrt{\dfrac{G_{66}\left((-1+i\gamma_1)+\eta_1+\dfrac{c^2}{\alpha_1^2}\right)}{G_{44}(1-i\gamma_2)}}\left(G_{44}-i\omega\bar{G}_{44}\right)} \tag{5.29}$$

The model transforms into a homogenous TI half-space bonded to a visco-elastic initially stressed orthotropic layer. In this instance, the lower medium is reduced to a stress-free, uniform elastic solid, while the upper layer maintains all of its complexity.

Case 5.4.2.3: When the initial stress from both the medium, viscoelastic properties from upper layer and heterogeneity from the lower half-space is removed, i.e., $P_1 = 0, P_2^* = 0, \beta = 0, \bar{G}_{44} = \bar{G}_{66} = 0, G_{44} = G_{66} = \mu_1, G^* = N^* = \mu_2$;

$$\tan\left[Hk\sqrt{\left(\frac{c^2}{\alpha_1^2}-1\right)}\right] = \frac{\mu_2\sqrt{\left(1-\dfrac{c^2}{\alpha_2^2}\right)}}{\mu_1\sqrt{\left(\dfrac{c^2}{\alpha_1^2}-1\right)}} \tag{5.30}$$

The simplest idealized elastic waveguide model is achieved by removing all complications, including viscoelasticity, initial stress, and heterogeneity. The classical Love wave equation is validated by the equation [1].

5.5 NUMERICAL ILLUSTRATIONS

For the given system, the real part of Eq. (5.24) defines the dispersion equation, whereas its imaginary part characterizes wave damping. Mathematica software has been used for computation and graphical representation. Elastic constant values have been examined as follows:

For orthotropic viscoelastic [26]

$$G_{66}^* = 3.32\times10^9\ \text{N/m}^2,\ G_{44}^* = 6.15\times10^9\ \text{N/m}^2,\ \rho_1 = 1.5\times10^3\ \text{kg/m}^3.$$

For TI half-space [20]

$$G^* = 4\times10^{10}\ \text{N/m}^2,\ N^* = 6.3\times10^{10}\ \text{N/m}^2,\ \rho_2 = 7.1\times10^3\ \text{kg/m}^3.$$

5.6 GRAPHICAL INTERPRETATION

Figure 5.2(a) and (b) illustrates the coupled effects of interfacial sliding and heterogeneity on wave behaviour, taking into account the existence of a sliding interface between the TI heterogeneous half-space and the viscoelastic orthotropic upper layer, which is defined by a sliding contact parameter $\delta = 0.01$. As the wave number (kH) rises in Figure 5.2(a), the phase velocity shows a declining trend, which is characteristic of dispersive wave behaviour. Even with a compliant sliding interface, material heterogeneity in the half-space increases wave propagation speed, as seen by greater phase velocities over the wave number range when the heterogeneity parameter β increases. The damping velocity, on the other hand, decreases nearly linearly with increasing wave number and also further shows increase in damping velocity with increasing β, as seen in Figure 5.2(b).

In a layered system with a sliding interface ($\delta = 0.01$), the depiction of the figures shows how the initial stress parameter (η_1) of the first medium (viscoelastic orthotropic medium) affects the phase and the damping velocities. Dispersive wave behaviour is confirmed by Figure 5.3(a), which reflects a nonlinear reduction of phase velocity as the wave number grows. The phase velocity steadily drops as η_1 increases across the wave number range, suggesting that the higher η_1 in the upper layer softens the medium and slows wave propagation. Likewise, the damping velocity plot in Figure 5.3(b) indicates that it reduces as the wave number rises, and the damping velocity is further reduced by greater initial stress levels (η_1). Together, these findings show that the propagation speed and attenuation properties of waves are sensitive to the initial stress magnitude in the upper medium in a system with sliding interfaces, heterogeneous substrates, and pre-stressed upper layers.

The figures show how the initial stress parameter (η_2) in a TI medium affects the dispersive and dissipative properties of Love-type waves in the configuration with a sliding interface ($\delta = 0.01$). Dispersive wave behaviour is confirmed in Figure 5.4(a), exhibiting a nonlinear downward trend in phase velocity with respect to wave number. Across the wave number spectrum, the phase velocity increases noticeably with increasing η_2. This suggests that increased starting stress in the bottom layer stiffens the medium, increasing the speed at which waves propagate. Likewise, Figure 5.4(b) shows that the damping velocity exhibits a decreasing trend with rising wave number and an increasing trend with higher initial stress (η_2).

The changes in phase velocity and damping velocities in relation to wave number for various values of the viscoelastic parameters related to the orthotropic media are shown in Figure 5.5(a) and (b). Figure 5.5(a) shows that while the phase velocity drops nonlinearly as the wave number (kH) increases, it increases with greater viscoelastic parameters. Furthermore, Figure 5.5(b) shows that the damping velocity decreases with a rise in wave number (kH) and it increases with higher viscoelastic parameters.

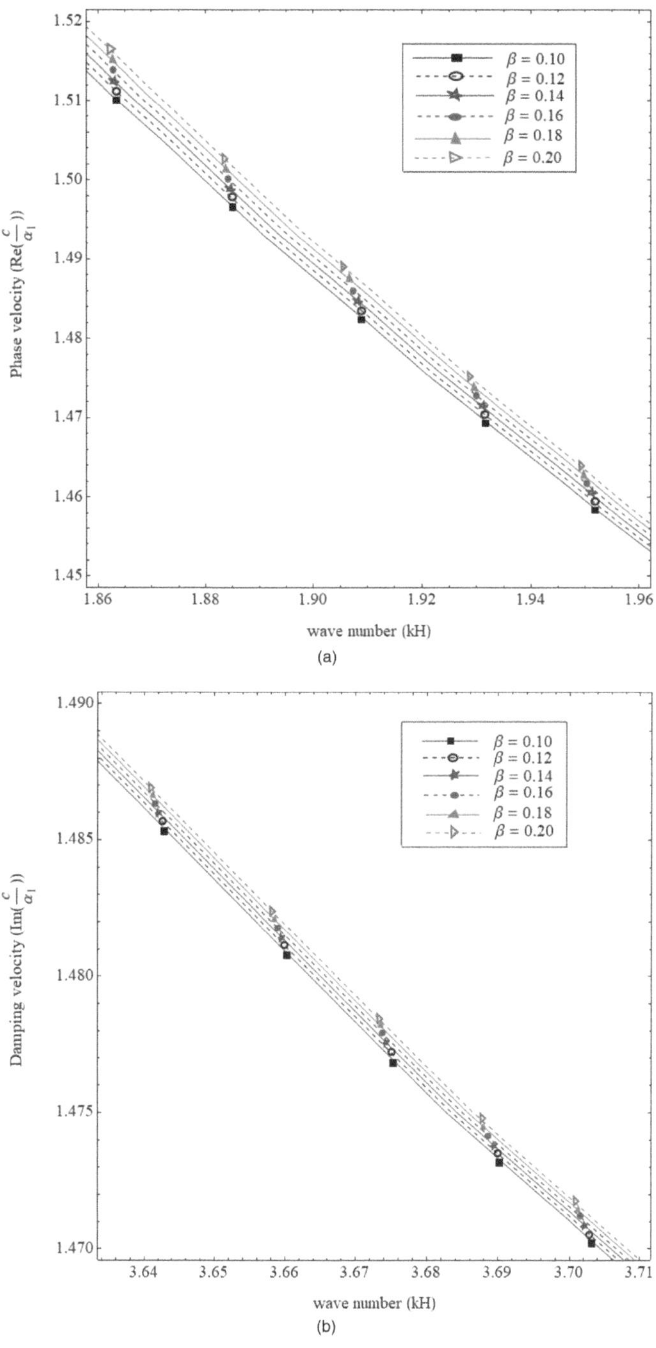

Figure 5.2 (a) Phase velocity $\left(\mathrm{Re}\left(\frac{c}{\alpha_1}\right)\right)$ and (b) damping velocity $\left(\mathrm{Im}\left(\frac{c}{\alpha_1}\right)\right)$ versus wave number (kH) for various values of the heterogeneity parameter of the half-space.

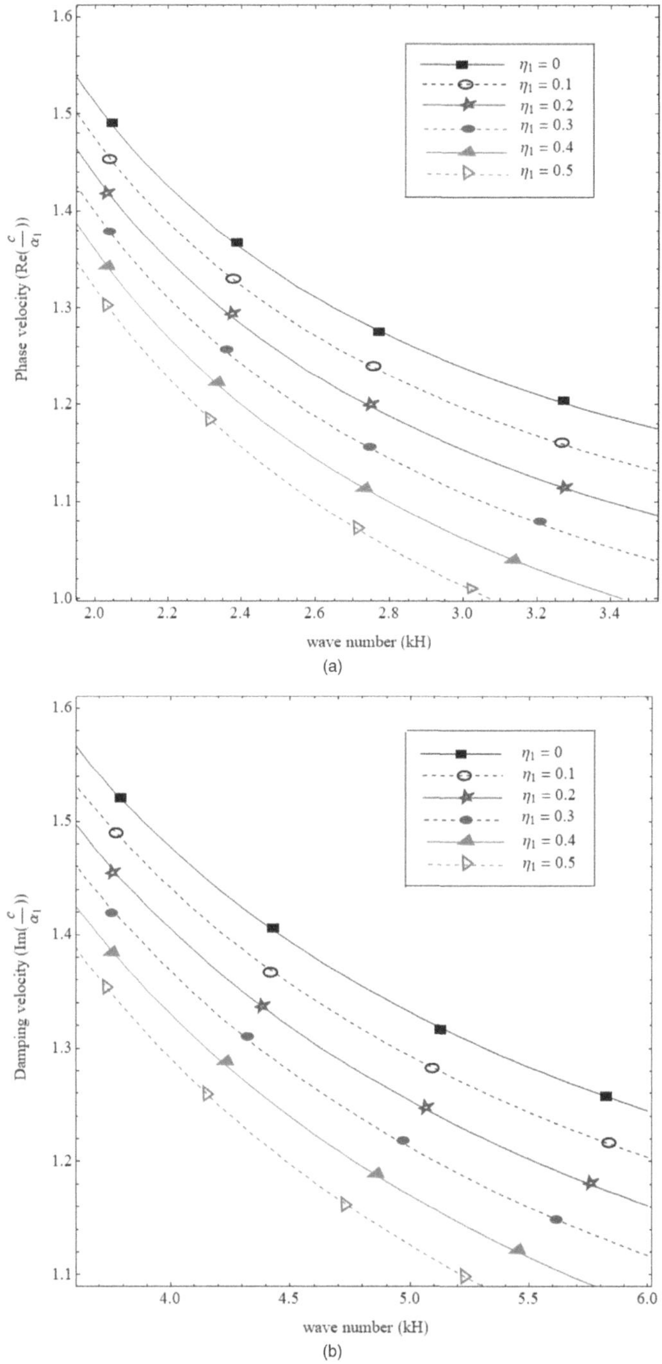

Figure 5.3 (a) Phase velocity $\left(\mathrm{Re}\left(\frac{c}{\alpha_1}\right)\right)$ and (b) damping velocity $\left(\mathrm{Im}\left(\frac{c}{\alpha_1}\right)\right)$ versus wave number (kH) for various values of the initial stress parameter of the upper medium.

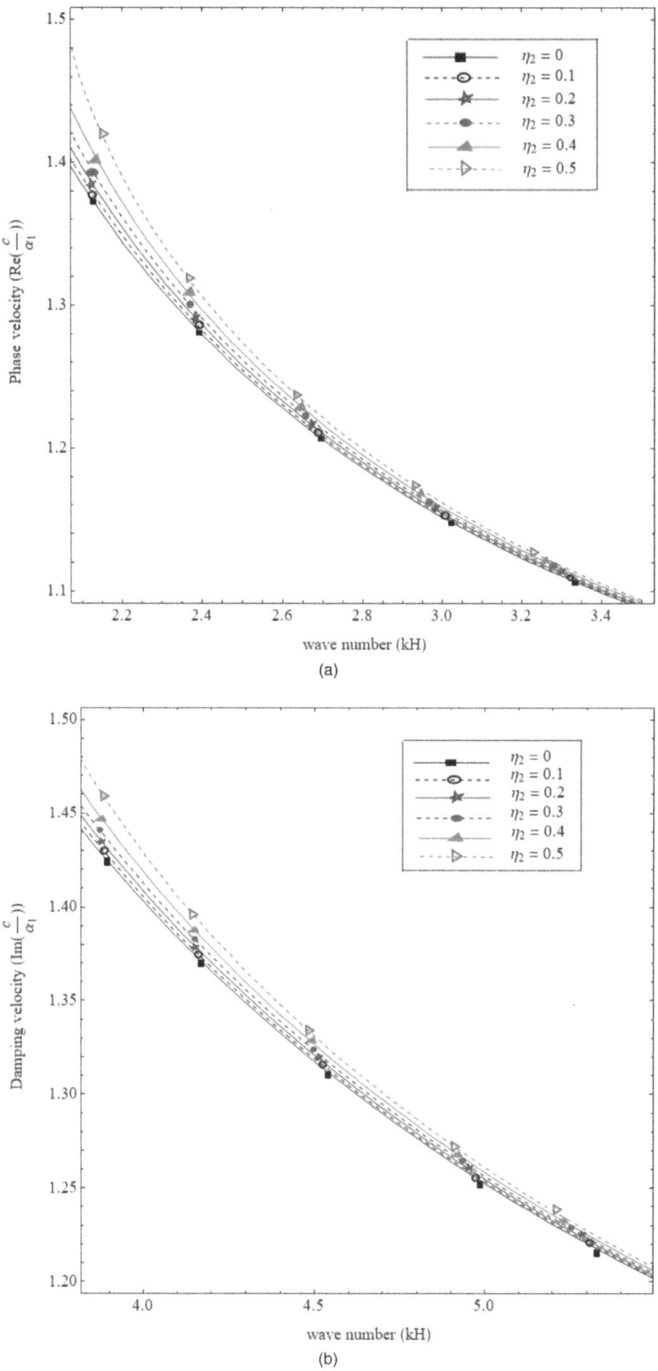

Figure 5.4 (a) Phase velocity $\left(\mathrm{Re}\left(\frac{c}{\alpha_1}\right)\right)$ and (b) damping velocity $\left(\mathrm{Im}\left(\frac{c}{\alpha_1}\right)\right)$ versus wave number (kH) for various values of the initial stress parameter of the lower medium.

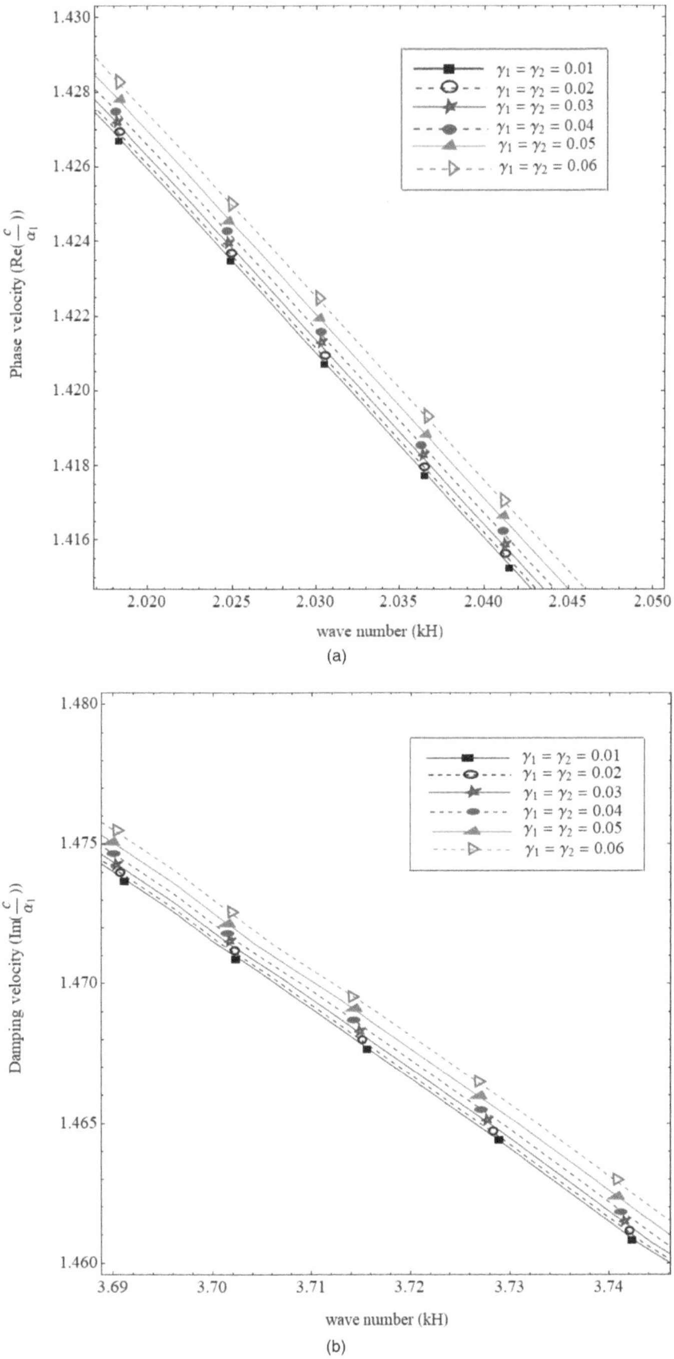

Figure 5.5 (a) Phase velocity $\left(\mathrm{Re}\left(\frac{c}{\alpha_1}\right)\right)$ and (b) damping velocity $\left(\mathrm{Im}\left(\frac{c}{\alpha_1}\right)\right)$ versus wave number (kH) for various values of the viscoelastic parameters of the upper medium.

The graphical analysis demonstrates the influence of the sliding parameter δ on both phase velocity and damping velocity in the presence of a sliding interface. As δ increases, the phase velocity exhibits a corresponding decrease for a specified wave number, indicating that greater interfacial slip weakens the wave propagation speed as shown in Figure 5.6(a). Concurrently, the damping velocity decreases with increasing δ, signifying a reduction in energy dissipation or attenuation of the wave in Figure 5.6(b).

Additionally, it can be seen by looking at the complex dispersion Eq. (5.24) that the interface acts as totally slipping with no resistance to relative motion between the adjacent layers as the sliding parameter δ approaches unity ($\delta \to 1$). In this scenario, the boundary requirements required for the existence of Love-type surface waves cannot be met since the mechanical coupling at the interface is low. This causes the Love-type wave to gradually change into regular shear waves that move through the bulk material. This transition

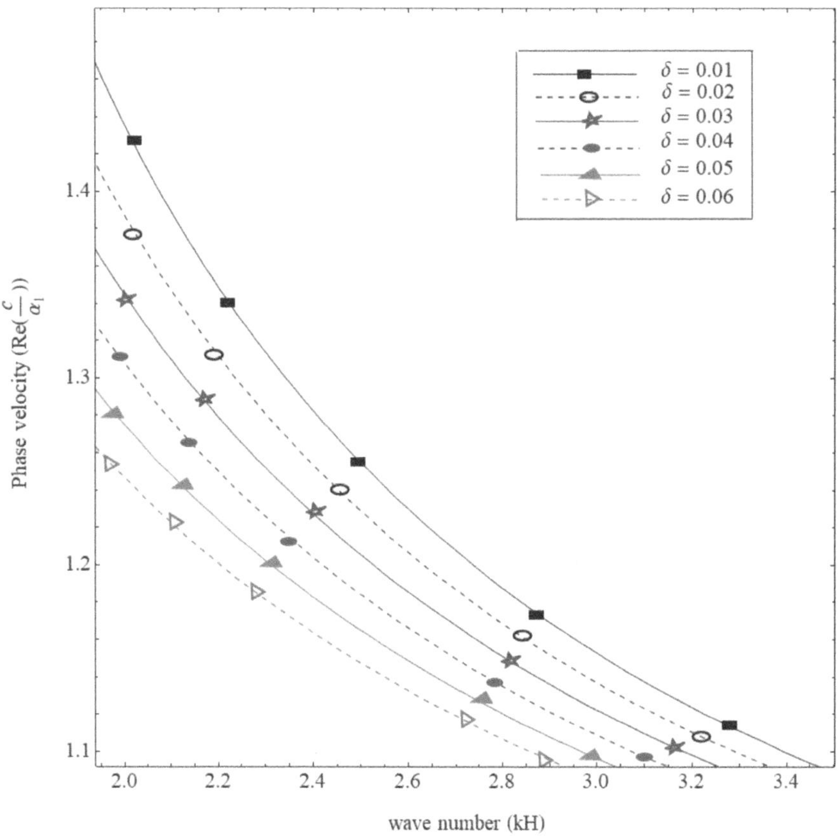

(a)

Figure 5.6 (a) Phase velocity $\left(\mathrm{Re}\left(\frac{c}{\alpha_1}\right)\right)$ and (b) damping velocity $\left(\mathrm{Im}\left(\frac{c}{\alpha_1}\right)\right)$ versus wave number (kH) for various values of the sliding parameter. *(Continued)*

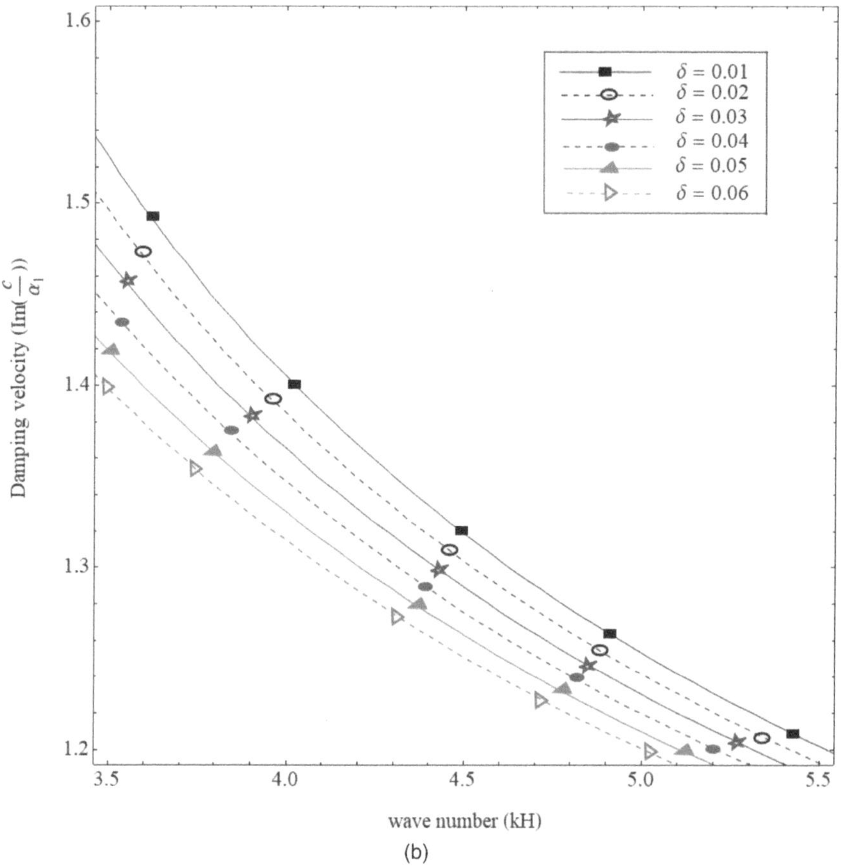

Figure 5.6 (Continued)

means that above a crucial sliding threshold, Love-type waves cease to exist due to the lack of sufficient interfacial restriction, leaving only shear wave propagation viable in the system.

5.7 CONCLUSION

The current work examines the intricate dispersion behaviour of Love waves in a multilayered medium, including an initially stressed viscoelastic orthotropic top medium and an initially stressed heterogeneous TI half-space with a sliding interface. The following is a summary of the main findings from the parametric analysis and their applications:

- The propagation of Love-type waves is substantially affected by the heterogeneity in the half-space. Due to heterogeneity, phase velocity stays higher even if it falls with wave number, indicating dispersion. A steady

decrease in damping velocity with increasing wave number, alongside its rise with greater heterogeneity, highlights the complex interplay of these parameters.
- The initial stress in the viscoelastic orthotropic upper medium significantly affects Love-type wave behaviour in a sliding interface system. Increased initial stress reduces both phase and damping velocities, indicating slower wave propagation and greater energy dissipation. This highlights the strong sensitivity of wave dynamics to pre-stress conditions in layered media.
- The initial stress in the TI bottom layer notably impacts Love-type wave behaviour in a sliding interface system. Increased initial stress enhances phase velocity, indicating medium stiffening and faster wave propagation. Meanwhile, damping velocity decreases with wave number but increases with higher initial stress, reflecting reduced energy loss in a stiffer medium.
- The results show that viscoelastic properties of the orthotropic layer significantly influence Love-type wave behaviour. Higher viscoelastic parameters increase both the phase velocity and damping velocity. As wave number increases, phase velocity decreases nonlinearly while damping velocity continues to drop, indicating dispersive and energy-dissipative wave characteristics.
- The analysis reveals that the sliding parameter significantly impacts Love-type wave behaviour. Increased interfacial slip weakens phase velocity and leads to slower wave propagation. Simultaneously, damping velocity decreases, indicating reduced energy dissipation across the sliding interface.

In geophysics, earthquake engineering, and subsurface exploration, the investigation of Love-type wave propagation in layered media with sliding surfaces, initial stress, and viscoelastic effects is crucial. It facilitates comprehension of surface wave behaviour in intricate geological environments, which is essential for fault zone analysis, geotechnical site characterization, and seismic hazard assessment.

REFERENCES

1. Love, A. E. H. (1911). *Some problems of geodynamics: Being an essay to which the Adams Prize in the University of Cambridge was adjudged in 1911*. Cambridge University Press.
2. Achenbach, J. (2012). *Wave propagation in elastic solids*. Elsevier.
3. Ewing, W. M., Jardetzky, W. S., Press, F., & Beiser, A. (1957). Elastic waves in layered media. *Physics Today*, *10*(12), 27–28.
4. Biot, M. A. (1965). *Mechanics of incremental deformations*. Wiley.
5. Gubbins, D. (1990). *Seismology and plate tectonics*. Cambridge University Press.

6. Abd-Alla, A. M., & Ahmed, S. M. (1999). Propagation of Love waves in a non-homogeneous orthotropic elastic layer under initial stress overlying semi-infinite medium. *Applied Mathematics and Computation*, *106*(2–3), 265–275.
7. Kundu, S., Gupta, S., & Manna, S. (2014). Propagation of Love wave in fiber-reinforced medium lying over an initially stressed orthotropic half-space. *International Journal for Numerical and Analytical Methods in Geomechanics*, *38*(11), 1172–1182.
8. Panja, S. K., & Mandal, S. C. (2022). Propagation of Love wave in multilayered viscoelastic orthotropic medium with initial stress. *Waves in Random and Complex Media*, *32*(2), 1000–1017.
9. Pandit, D. K., Kundu, S., & Gupta, S. (2017). Propagation of Love waves in a prestressed Voigt-type viscoelastic orthotropic functionally graded layer over a porous half-space. *Acta Mechanica*, *228*(3), 871–880.
10. Pramanik, D., & Manna, S. (2023). Love-like wave dispersion in a highly non-homogeneous viscoelastic orthotropic layer under the effect of non-local elasticity. *Mathematical Methods in the Applied Sciences*, *46*(14), 15048–15072.
11. Pranav, P., & Rahul, A. K. (2025). Love-type wave propagation in a heterogeneous medium between a liquid and a fractured porous half-space: A WKB and variable separation approach. *European Journal of Mechanics-A/Solids*, *114*, 105763.
12. Singh, M. K., & Alam, P. (2020). Surface wave analysis in orthotropic composite structure with irregular interfaces. *International Journal of Applied and Computational Mathematics*, *6*(1), 13.
13. Malik, N., & Sikka, J. S. (2024). Influence of initial stress on Love wave propagation in composite structures. *Journal of Vibration Engineering & Technologies*, *12*(3), 4163–4180.
14. Ahmad, F., & Khan, A. (2001). Effect of rotation on wave propagation in a transversely isotropic medium. *Mathematical Problems in Engineering*, *7*(2), 147–154.
15. Singh, A. K., Kumari, N., Chattopadhyay, A., & Sahu, S. A. (2016). Smooth moving punch in an initially stressed transversely isotropic magnetoelastic medium due to shear wave. *Mechanics of Advanced Materials and Structures*, *23*(7), 774–783.
16. Singh, B. (2016). Wave propagation in a rotating transversely isotropic two-temperature generalized thermoelastic medium without dissipation. *International Journal of Thermophysics*, *37*(1), 5.
17. Acharya, D. P., Roy, I., & Sengupta, S. (2009). Effect of magnetic field and initial stress on the propagation of interface waves in transversely isotropic perfectly conducting media. *Acta Mechanica*, *202*(1), 35–45.
18. Kundu, S., Gupta, S., & Manna, S. (2014). SH-type waves dispersion in an isotropic medium sandwiched between an initially stressed orthotropic and heterogeneous semi-infinite media. *Meccanica*, *49*(3), 749–758.
19. Bayones, F. S., Hussein, N. S., Abd-Alla, A. M., & Alharbi, A. M. (2021). The influence of heterogeneity and initial stress on the propagation of Love-type wave in a transversely isotropic layer subjected to rotation. *Science Progress*, *104*(3), 00368504211041496.
20. Alam, P., & Kundu, S. (2017). Influences of heterogeneities and initial stresses on the propagation of Love-type waves in a transversely isotropic layer over an inhomogeneous half-space. *Journal of Solid Mechanics*, *9*(5), 783–793.
21. Kakar, R. (2015). Dispersion of Love wave in an isotropic layer sandwiched between orthotropic and prestressed inhomogeneous half-spaces. *Latin American Journal of Solids and Structures*, *12*(10), 1934–1949.

22. Gajroiya, K., & Sikka, J. S. (2025). Dynamics of Love-type wave propagation in composite transversely isotropic porous structures. *Applied Mathematical Modelling*, *137*, 115723.
23. Zhu, H., Zhang, L., Han, J., & Zhang, Y. (2014). Love wave in an isotropic homogeneous elastic half-space with a functionally graded cap layer. *Applied Mathematics and Computation*, *231*, 93–99.
24. Pramanik, D., & Manna, S. (2024). Love-like wave fields at the interface of sliding contact with non-local elastic heterogeneous fluid-saturated fractured poro-viscoelastic layer. *European Journal of Mechanics-A/Solids*, *107*, 105350.
25. Sadab, M., & Kundu, S. (2025). Wave propagation in piezomagnetic-piezoelectric bilayer structure under sliding contact interface with external magnetic field. *Applied Physics A*, *131*(4), 274.
26. Yu, J. G., Ratolojanahary, F. E., & Lefebvre, J. E. (2011). Guided waves in functionally graded viscoelastic plates. *Composite Structures*, *93*(11), 2671–2677.

Chapter 6

On the characteristics of Love waves scattered due to the irregular surface of orthotropic poroelastic layer in an orthotropic composite structure

Anil Negi, Saurabh Kapoor, Pradeep Gautam, Manoj Kumar Singh, and Ram Prasad Yadav

6.1 INTRODUCTION

The study on the characteristics of Love wave propagation in a layered composite structure can become a topic of great interest among various researchers and scientists across the world. Ke et al. [1] examined the dynamic behavior of Love waves propagation in the in-homogeneous fluid-saturated poroelastic half-space having linearly varying characteristics. The characteristics of Love wave propagation in functionally graded piezoelectric material layer have been investigated by Du et al. [2].

Love wave propagation in certain kinds of structures exhibit some complicated characteristics that have to be explored more. These complicated propagation characteristics of Love waves in anisotropic structures can give some crucial information which are applicable in real life.

From the past recent years, there are growing engineering applications of anisotropic structures. Orthotropic materials are some special kind of structures having distinct material characteristics in two mutually perpendicular directions. The structures possessing orthotropic symmetry have various applications in geotechnical and civil engineering. There are various types of materials in the form of laminated composites, fibrous composite, and rolled sheets exist in real life which can be regarded as orthotropic materials. The structures of these kinds have high specific strength, thickness, and resistance to cracking [3, 4]. Therefore, these structures are widely used in the designing and manufacturing of several engineering constructions.

The knowledge regarding the orthotropic structure under the influence of surface acoustic wave propagation can be applicable for the structural health monitoring and non-destructive testing of engineering structures. As a result, several researchers have attracted toward the observation of the surface wave propagation phenomena in orthotropic structures.

The secular equation corresponds to the surface acoustic waves propagation was obtained by Destrade [5]. Destrade [6] further described the

 DOI: 10.1201/9781003470885-6

characteristics of Rayleigh waves propagation in an exponentially graded orthotropic semi-infinite structure. The mathematical model for the shock waves propagation in orthotropic structures has been provided by Mohd Nor et al. [7]. The analysis of the Rayleigh wave propagation in thermoelastic orthotropic structure has been accomplished by Biswas [8]. Yu and Zhang [9] analyzed the propagation behavior of guided waves in pre-stressed orthotropic functionally graded material plates.

The porosity of the orthotropic materials has substantial importance in the analysis and design of engineering structures. Seismic waves propagation is considerably influenced from the porosity of orthotropic material. Therefore, the consideration of porosity in orthotropic materials can be helpful to develop the earthquake-resistant materials effected by seismic waves. Keeping it in mind, several researchers studied the propagation of seismic waves in orthotropic poroelastic structure. Liu and Liu [10] explored the characteristics of Rayleigh waves propagation in fluid-saturated orthotropic poroelastic structure. Vinh et al. [11] obtained the exact equation for dispersion characteristics in explicit form for the propagation of Rayleigh waves in fluid-saturated orthotropic porous structure using polarization vector method.

Sometimes the surface defects caused by heavy loads, design process of structures, erosion, and during manufacturing process can be found in the orthotropic structures [12]. Due to the presence of surface defects, the dynamic performance of orthotropic material structures is substantially affected. In view of this, the effect of irregularity in upper surface on the dynamic characteristics of initially stressed orthotropic materials caused due to moving line load has been investigated by Singh et al. [13].

The interaction of the Love waves with the surface irregularity induces the phenomena of scattering of waves which further cause the reduction in the strength of composite structure. Therefore, it is imperative to investigate the scattering of Love waves in a layered composite structure. In view of this, several researchers made significant effort to understand the scattering characteristics of Love waves in layered composite structures. Slavin and Wolf [14] discussed Love waves scattering due to existence of irregularity in upper surface of an elastic-layere overlying a rigid half-space. Wolf [15] explored the characteristics of Love waves scattering in a layered isotropic elastic structure having irregular surface. The secular equation for the scattering of elastic waves in functionally graded isotropic plate was derived by Chen et al. [16]. Anagnostopoulos [17] studied the scattering of time harmonic elastic waves in an orthotropic structure. Chattopadhyay et al. [18] derived the relation for dispersion characteristics of shear waves propagating encountering the irregularity in upper surface of layered isotropic viscoelastic structure. The scattering characteristics of Love-type wave interacted by the surface defects in a layered piezoelectric composite structure have been explored by Negi and Singh [19].

However, no attempt has been made till now to examine the Love waves scattering in a layered orthotropic porous composite structure comprising an orthotropic porous layer lying over orthotropic half-space. These aforementioned observations stimulate the authors to investigate Love wave scattering (one type of guided waves) in a layered orthotropic porous composite structure containing irregular surface.

The present analysis provided the investigation on the scattering of Love waves caused by encounter of Love waves with the irregular surface in a layered orthotropic porous composite structure. In this study, the adopted analytical procedure includes the use of double-sided Fourier transformation along with the contour integral method. In the upper layer, irregularity of distinct shapes viz. rectangular, parabolic, and triangular notch has been assumed in the upper surface of said composite structure. The exact expressions for the dispersion characteristics and group velocity have been obtained analytically. The impacts of anisotropic and porosity parameters on the phase and group velocities are also considered. The deduction of the non-vanishing displacement component due to reflected wave field has also been done for all the considered scenarios of irregularity in upper surface. The influences of vertical irregularity parameter, porosity parameter, and anisotropic parameter of upper layer on reflected displacement have also been discussed for all said cases of surface irregularity.

6.2 PROBLEM FORMULATION

Let us assume the Love waves propagation in the orthotropic poroelastic layer of finite thickness H lying over an orthotropic half-space. The selection of cartesian coordinate system has been done in such a manner that y-axis denotes the wave propagation direction. On the other hand, x-axis is pointing vertically downward. In the aforesaid considered structure, thickness and span of the surface irregularity are assumed as b and S, respectively. The origin of the considered orthotropic porous composite structure is considered to be located at the interface $(x = 0)$ as exhibited by Figure 6.1.

The equation corresponds to the surface irregularity in the upper orthotropic poroelastic layer can be expressed as

$$x = x_B = -H + bh(y), \tag{6.1}$$

where

$$h(y) = \begin{cases} 0, & \text{for } y \le -\frac{S}{2} \text{ and } y \ge \frac{S}{2} \\ f(y), & \text{for } -\frac{S}{2} \le y \le \frac{S}{2} \end{cases} \tag{6.2}$$

where b $(\ll 1)$ represents the vertical irregularity parameter associated with the surface irregularity; function $f(y)$ is related with the irregularity's shape.

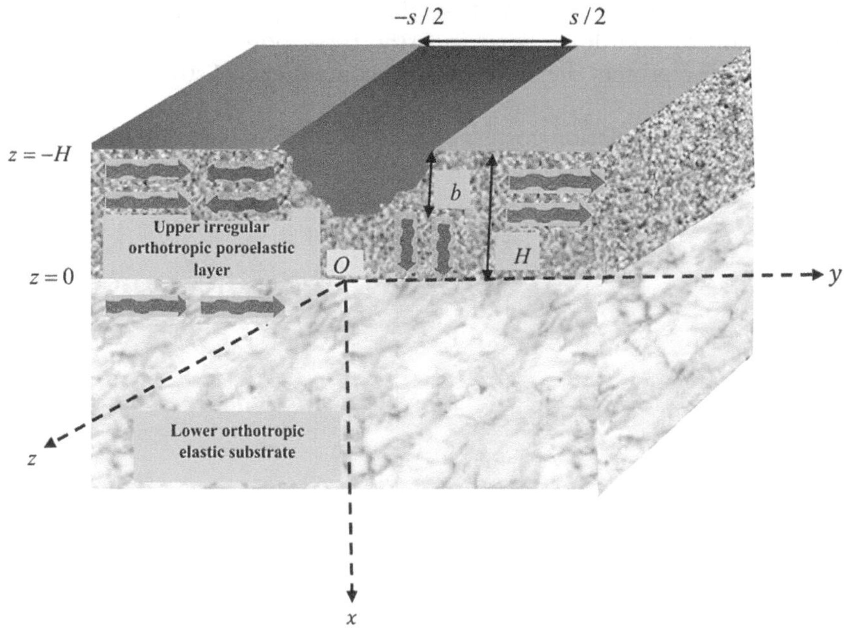

Figure 6.1 Geometry of the problem.

Let us consider (u_1, u_2, u_3) and (w_1, w_2, w_3) are the components for displacements associated with the solid skeleton in upper layer and lower half-space along the direction of x-, y-, and z-axis, respectively. Also, U_1^*, U_2^*, and U_3^* are considered to be the components for the fluid displacements in the direction of x-, y-, and z-axis, respectively. In view of the characteristics of Love waves propagation, the displacement components for the upper layer and lower half-space for the solid phase and for the fluid phase in upper layer are expressed in the following form

$$\left.\begin{array}{l} u_1 = u_2 = 0,\ u_3 = u_3(x,y,t),\ w_1 = w_2 = 0,\ w_3 = w_3(x,y,t), \\ U_1^* = 0,\ U_2^* = 0,\ \text{and}\ U_3^* = U_3^*(x,y,t) \end{array}\right\} \tag{6.3}$$

6.2.1 For the upper orthotropic poroelastic layer

The governing equations for the propagation of Love waves in the orthotropic poroelastic layer medium can be expressed as

$$\tau_{ij,j}^{l} = \rho_l \ddot{u}_i + \rho_f \ddot{U}_i \,(i, j = 1,2,3), \tag{6.4}$$

where ρ_l is the mass density of the solid skeleton part for the orthotropic poroelastic layer material; ρ_f is the density of fluid phase in the orthotropic poroelastic material; $u_i\ (i = 1,2,3)$ signifies the displacement components in the direction of coordinate axis in orthotropic poroelastic layer medium; $\tau_{ij}^l\ (i, j = 1,2,3)$ represent the non-vanishing components of stress tensor related to the upper layer.

The constitutive relation corresponds to the orthotropic poroelastic material medium can be written as

$$\left.\begin{aligned}
\tau_{11}^l &= c_{11}^l e_{11} + c_{12}^l e_{22} + c_{13}^l e_{33} + M_{11}\zeta,\\
\tau_{22}^l &= c_{12}^l e_{11} + c_{12}^l e_{22} + c_{33}^l e_{33} + M_{22}\zeta,\\
\tau_{33}^l &= c_{13}^l e_{11} + c_{23}^l e_{22} + c_{33}^l e_{33} + M_{33}\zeta,\\
\tau_{23}^l &= 2c_{44}^l e_{23},\\
\tau_{13}^l &= 2c_{55}^l \varepsilon_{13},\\
\tau_{12}^l &= 2c_{66}^l e_{12},\\
\tau^* &= M_{11}e_{xx} + M_{22}e_{yy} + M_{33}e_{zz} + M\zeta
\end{aligned}\right\} \tag{6.5}$$

where $c_{ij}^l\ (i,\ j = 1,2,\ldots6)$, $M_{ij}\ (i, j = 1,\ 2,\ 3)$, and M denote the constants associated with elastic properties and $e_{ij}\ (i, j = 1,\ 2,\ 3)$ are the components of strain.

The displacement and strain components can be related as

$$e_{11} = \frac{\partial u_1}{\partial x},\ e_{22} = \frac{\partial u_2}{\partial y},\ e_{33} = \frac{\partial u_3}{\partial z},\ e_{23} = \frac{1}{2}\left(\frac{\partial u_3}{\partial y} + \frac{\partial u_2}{\partial z}\right),\ e_{31} = \frac{1}{2}\left(\frac{\partial u_3}{\partial x} + \frac{\partial u_1}{\partial z}\right),$$

$$e_{12} = \frac{1}{2}\left(\frac{\partial u_1}{\partial y} + \frac{\partial u_2}{\partial x}\right),\ \zeta = \left(\frac{\partial U^*}{\partial x} + \frac{\partial V^*}{\partial y} + \frac{\partial W^*}{\partial z}\right). \tag{6.6}$$

In view of Eqs. (6.3) and (6.6), the constitutive Eq. (6.5) provides

$$\tau_{11}^l = 0,\ \tau_{22}^l = 0,\ \tau_{33}^l = 0,\ \tau_{12}^l = 0,\ \tau_{23}^l = c_{44}^l\frac{\partial u_3}{\partial y},\ \tau_{13}^l = c_{55}^l\frac{\partial u_3}{\partial x},\ \tau^* = 0. \tag{6.7}$$

Using Eq. (6.7) in Eq. (6.4), the equations for the Love wave propagation in orthotropic poroelastic layer can be written as

$$\frac{\partial^2 u_3}{\partial x^2} + \frac{1}{\gamma_l^2}\frac{\partial^2 u_3}{\partial y^2} = \frac{1}{\beta_l^2}\frac{\partial^2 u_3}{\partial t^2}, \tag{6.8}$$

where $\gamma_l = \sqrt{\frac{c_{44}^l}{c_{55}^l}}$ denotes the anisotropy parameter of the upper layer; u_3 represents the non-vanishing displacement components for the propagation of Love waves in upper layer; and $\beta_3 = \sqrt{\frac{c_{55}^l}{\rho_l - \rho_f}}$ signifies the shear wave's velocity in the upper orthotropic poroelastic layer.

In view of Eq. (6.8), the non-vanishing equation of motion for Love waves propagation in the upper orthotropic layer can be expressed as

$$\frac{\partial^2 u_3}{\partial x^2}+\frac{1}{\gamma_l^2}\frac{\partial^2 u_3}{\partial y^2}=\frac{1}{\beta_l^2}\frac{\partial^2 u_3}{\partial t^2}, \tag{6.9}$$

where $\gamma_l=\sqrt{\frac{c_{44}^l}{c_{55}^l}}$; and $\beta_l=\sqrt{\frac{c_{44}^l}{\rho_l}}$.

6.2.2 For the orthotropic half-space medium

The governing equations of motion for orthotropic elastic half-space medium can be expressed as

$$\tau_{ij,j}^h=\rho_h\ddot{w}_i,(i,j=1,2,3), \tag{6.10}$$

where ρ_h is the density of the orthotropic half-space; τ_{ij}^h are the components of stress tensor in orthotropic half-space; w_i is the component of displacement in orthotropic half-space medium.

The stress–strain relations associated with the orthotropic half-space medium can be expressed as

$$\left.\begin{aligned}
\tau_{11}^h&=c_{11}^h\varepsilon_{11}+c_{12}^h\varepsilon_{22}+c_{13}^h\varepsilon_{33},\\
\tau_{22}^h&=c_{12}^h\varepsilon_{11}+c_{12}^h\varepsilon_{22}+c_{33}^h\varepsilon_{33},\\
\tau_{33}^h&=c_{13}^h\varepsilon_{11}+c_{23}^h\varepsilon_{22}+c_{33}^h\varepsilon_{33},\\
\tau_{23}^h&=2c_{44}^h\varepsilon_{23},\\
\tau_{13}^h&=2c_{55}^h\varepsilon_{13},\\
\tau_{12}^h&=2c_{66}^h\varepsilon_{12},
\end{aligned}\right\} \tag{6.11}$$

where $c_{ij}^h(i,j=1,2,\ldots 6)$ represents the elastic constants and $\varepsilon_{ij}(i,j=1,2,3)$ related with the components of strain in orthotropic elastic half-space medium.

The relation between the strain and displacement components can be expressed as

$$\varepsilon_{11}=\frac{\partial w_1}{\partial x},\ \varepsilon_{22}=\frac{\partial w_2}{\partial y},\ \varepsilon_{33}=\frac{\partial w_3}{\partial z},\ \varepsilon_{23}=\frac{1}{2}\left(\frac{\partial w_3}{\partial y}+\frac{\partial w_2}{\partial z}\right),\ \varepsilon_{31}=\frac{1}{2}\left(\frac{\partial w_3}{\partial x}+\frac{\partial w_1}{\partial z}\right),$$

$$\varepsilon_{12}=\frac{1}{2}\left(\frac{\partial w_1}{\partial y}+\frac{\partial w_2}{\partial x}\right). \tag{6.12}$$

In view of Eqs. (6.3) and (6.12), the stress–strain relation (6.11) provides

$$\tau_{11}^{h}=0,\ \tau_{22}^{h}=0,\ \tau_{33}^{h}=0,\ \tau_{12}^{h}=0,\ \ \tau_{23}^{h}=c_{44}^{h}\frac{\partial w_3}{\partial y},\ \tau_{13}^{h}=c_{55}^{h}\frac{\partial w_3}{\partial x}. \tag{6.13}$$

Using Eq. (6.13) in Eq. (6.10), the equations for the Love wave propagation in orthotropic half-space can be written as

$$\frac{\partial^2 w_3}{\partial x^2}+\frac{1}{\gamma_h^{\ 2}}\frac{\partial^2 w_3}{\partial y^2}=\frac{1}{\beta_h^{\ 2}}\frac{\partial^2 w_3}{\partial t^2}, \tag{6.14}$$

where $\gamma_h=\sqrt{\frac{c_{44}^h}{c_{55}^h}}$ represents the anisotropy parameter of the lower orthotropic half-space; w_3 represents the non-vanishing components of displacement caused by the propagation of Love waves in lower orthotropic half-space; and $\beta_h=\sqrt{\frac{c_{55}^h}{\rho_h}}$ signifies the shear wave's velocity in the lower orthotropic half-space.

With the help of Eq. (6.14), the non-vanishing equations of motion for the propagation of Love waves in the lower orthotropic half-space can be expressed as

$$\frac{\partial^2 w_3}{\partial x^2}+\frac{1}{\gamma_h^{\ 2}}\frac{\partial^2 w_3}{\partial y^2}=\frac{1}{\beta_h^{\ 2}}\frac{\partial^2 w_3}{\partial t^2}, \tag{6.15}$$

where $\gamma_h=\sqrt{\frac{c_{44}^h}{c_{55}^h}}$ and $\beta_h=\sqrt{\frac{c_{44}^h}{\rho_h}}$.

Let us assume the harmonic nature of the waves, under this consideration, Eqs. (6.9) and (6.14) can be expressed as

$$u_3=U_3(x,y)e^{i\omega t},\ w_3=W_3(x,y)e^{i\omega t}, \tag{6.16}$$

where ω is the angular frequency.

Using Eq. (6.11) in Eqs. (6.9) and (6.14), we have

$$\frac{\partial^2 U_3}{\partial y^2}+\frac{1}{\gamma_l^2}\frac{\partial^2 U_3}{\partial x^2}+L_l^2U_3=0, \tag{6.17}$$

$$\frac{\partial^2 W_3}{\partial y^2}+\frac{1}{\gamma_h^2}\frac{\partial^2 W_3}{\partial x^2}+L_h^2W_3=0, \tag{6.18}$$

where $L_l^2=\frac{\omega^2}{\beta_l^2}$ and $L_h^2=\frac{\omega^2}{\beta_h^2}$.

6.2.3 Boundary conditions

i. The mechanical traction-free condition at the upper surface i.e. at $x = -H + bh(y)$ can be written as

$$\tau_{32}^{l} - \gamma_l^2 bh' \tau_{31}^{l} = 0. \tag{6.19}$$

ii. The continuity conditions at the interface $x = 0$ of the layer and half-space are
 a. $u_3 = w_3$ (6.20)
 b. $\tau_{13}^{l} = \tau_{13}^{h}$ (6.21)

6.3 SOLUTION OF THE PROBLEM

In view of Eqs. (6.17) and (6.18), the expression of the displacement components for the incident wave field can be written as [15, 18]

$$u_{3,in} = A\cos\gamma_l S_l (x + H) e^{-i\eta y}, \tag{6.22}$$

$$w_{3,in} = Be^{-\gamma_l S_l y} e^{-i\eta y}, \tag{6.23}$$

where $S_l^2 = L_l^2 - \eta^2$, $S_h^2 = \eta^2 - L_h^2$; $\eta = \frac{\omega}{c_{ph}}$ denotes the angular wave number; c_{ph} is the phase velocity of Love waves; L_l and L_h are wave numbers; A and B signify the arbitrary coefficients related with the incident wave field; $u_{3,in}$ and $w_{3,in}$ are non-vanishing displacement components associated with the incident wave fields related to Love wave's propagation in the considered orthotropic layer and half-space, respectively.

The solution corresponds to the scattered wave field in υ-plane can be considered [15, 18]

$$u_{3,scatt} = \int_c \left\{ C(\upsilon) e^{-i\gamma_l \xi_l x} + D(\upsilon) e^{i\gamma_l \xi_l x} \right\} e^{-i\upsilon y} d\upsilon, \tag{6.24}$$

$$w_{3,scatt} = \int_c E(\upsilon) e^{-\gamma_h \xi_h x} e^{-i\eta y} d\upsilon, \tag{6.25}$$

where $\xi_l^2 = L_l^2 - \upsilon^2$, $\xi_h^2 = L_h^2 - \upsilon^2$, $\upsilon = \frac{\omega}{c_{ph}}$ denotes the wave number; c_{ph} represents the phase velocity; L_l and L_h are the wave numbers for the scattered wave field; $C(\upsilon)$, $D(\upsilon)$, *and* $E(\upsilon)$ are associated with the arbitrary coefficients related to the scattered wave field. In the expressions mentioned in Eqs. (6.24) and (6.25), $u_{3,\,scatter}$ and $w_{3,\,scatter}$ are the non-vanishing displacement components associated with scattered wave field for the upper layer and lower half-space, respectively.

Keeping in view the expressions mentioned in Eqs. (6.22)–(6.25), the non-vanishing total mechanical displacement components related with the Love waves propagation in the considered layered orthotropic poroelastic structure can be expressed as

$$u_3 = u_{3,in} + u_{3,scatt} = A\cos\gamma_l S_l (x+H) e^{-i\eta y} + \int_c \left\{ C(\upsilon) e^{-i\gamma_l \xi_l x} + D(\upsilon) e^{i\gamma_l \xi_l x} \right\}, \quad (6.26)$$

$$w_3 = w_{3,in} + w_{3,scatt} = B e^{-\gamma_b S_b y} e^{-i\eta y} + \int_c E(\upsilon) e^{-\gamma_b \xi_b x} e^{-i\eta y} d\upsilon. \quad (6.27)$$

with the help of Eqs. (6.26) and (6.27), the boundary condition (6.19) provide

$$A\cos\gamma_l S_l H = B, \quad (6.28)$$

$$C(\upsilon) + D(\upsilon) = E(\upsilon). \quad (6.29)$$

Substituting the expression in Eq. (6.28) into Eq. (6.27), we get

$$w_3 = A\cos\gamma_l S_l H e^{-\gamma_b S_b y} e^{-i\eta y} + \int_c E(\upsilon) e^{-\gamma_b \xi_b x} e^{-i\eta y} d\upsilon. \quad (6.30)$$

With the use of expressions in Eqs. (6.26) and (6.30) in boundary condition (6.21), we get

$$C(\upsilon) - D(\upsilon) = \frac{E(\upsilon)}{\lambda}, \quad (6.31)$$

where $\lambda = \frac{i\xi_l \gamma_l c_{55}^l}{\xi_b \gamma_b c_{55}^b}$.

On solving Eqs. (6.29) and (6.31), we obtain

$$\frac{2\lambda}{\lambda+1} C(\upsilon) = E(\upsilon), \quad \frac{C(\upsilon)(\lambda-1)}{(\lambda+1)} = D(\upsilon). \quad (6.32)$$

Using Eq. (6.32) into Eq. (6.26) and Eq. (6.30), we get

$$u_3 = A\cos\gamma_l S_l (x+H) e^{-i\eta y} + \int_c \frac{2C}{(\lambda+1)} \left\{ \lambda\cos\gamma_l \xi_l x - i\sin\gamma_l \xi_l x \right\} e^{-i\upsilon y} d\upsilon, \quad (6.33)$$

$$w_3 = A\cos\gamma_l S_l H e^{-\gamma_b S_b x} e^{-i\eta y} + \int_c C \frac{2\lambda}{(\lambda+1)} e^{-\gamma_b \xi_b x} e^{-i\upsilon y} d\upsilon. \quad (6.34)$$

In the considered situation of surface irregularity, the irregularity of the surface is considered to be very small as $b \ll 1$, therefore, the arbitrary coefficient, sine, and cosine functions associated with the displacement components can be expressed using b as

$$C(\upsilon) = bC_1(\upsilon), \ \sin S_1 bh \cong S_1 bh, \ \cos S_1 bh \cong 1. \tag{6.35}$$

On using the expression of u_3 from Eq. (6.34) and concept from Eq. (6.35) in the boundary condition (6.19), we have

$$A\left[i\eta h' - S_l^2 h\right]\gamma_l e^{-i\eta y} + \int_c \frac{2C_1(\upsilon)\xi_l}{(\lambda+1)}(-i\cos\xi_l\gamma_l H + \lambda\sin\xi_l\gamma_l H)e^{-i\upsilon y}d\upsilon = 0. \tag{6.36}$$

The inversion of Eq. (6.36) yields

$$C_1(\upsilon) = \frac{A(\lambda+1)}{4\pi(-i\cos\xi_l\gamma_l H + \lambda\sin\xi_l\gamma_l H)\xi_l}\int_{-\infty}^{\infty}\left[i\eta h' - S_l^2 h\right]\gamma_l e^{ix(\upsilon-\eta)}dx. \tag{6.37}$$

On applying Eq. (6.37) into Eqs. (6.33) and (6.34), we get

$$u_3 = A\cos\gamma_l S_l(x+H)e^{-i\eta y} + \int_{-\infty}^{\infty}\frac{Ab}{2\pi}\left[i\eta h' - S_l^2 h\right]\gamma_l e^{-i\eta x}dx \times \int_c \Lambda_0(\upsilon)d\upsilon, \tag{6.38}$$

$$w_3 = A\cos\gamma_l S_l H e^{-\gamma_b S_b x}e^{-i\eta y} + \frac{Ab}{2\pi}\int_{-\infty}^{\infty}(i\eta h' - S_l^2 h)\gamma_l e^{-i\eta x}dx \times \int_c \Lambda_{00}(\upsilon)d\upsilon. \tag{6.39}$$

where

$$\Lambda_0(\upsilon) = \frac{\{\lambda\cos\gamma_l\xi_l x - i\sin\gamma_l\xi_l x\}}{(i\cos\xi_l\gamma_l H - \lambda\sin\xi_l\gamma_l H)\xi_l}e^{i\upsilon x - i\upsilon y}, \tag{6.40}$$

$$\Lambda_{00}(\upsilon) = \frac{\lambda e^{-\gamma_b\xi_b x}e^{-i\upsilon y + i\upsilon x}}{(i\cos\xi_b\gamma_b H - \lambda\sin\xi_b\gamma_b H)\xi_b}. \tag{6.41}$$

The singular points in the contour integrals of Eqs. (6.38) and (6.39) can be calculated as

$$(i\cos\xi_b\gamma_b H - \lambda\sin\xi_b\gamma_b H)\xi_b = 0. \tag{6.42}$$

On putting ξ_l and γ_l into Eq. (6.42), we obtain

$$\tan\left(\sqrt{\frac{c_{44}^l}{c_{55}^l}}\sqrt{(L_l^2 - \upsilon^2)}\right)H = \frac{c_{55}^b\sqrt{\frac{c_{44}^b}{c_{55}^b}}\sqrt{(\upsilon^2 - L_h^2)}}{c_{55}^l\sqrt{\frac{c_{44}^l}{c_{55}^l}}\sqrt{(L_l^2 - \upsilon^2)}}. \tag{6.43}$$

Eq. (6.43) denotes the generalized periodic equation associated with the scattered as well as incident wave field due to propagating Love waves in the studied orthotropic poroelastic composite structure. The roots of Eq. (6.43) in the real υ-axis can be represented by υ_m. From the graph of the function (in Figure 6.2), it is evident that there exist N_0 roots of Eq. (6.43) where N_0 signifies the integral part of the number $\left[\left\{(L_l^2 - L_b^2)\frac{H}{\pi}\right\}+1\right]$ [15].

Using the values of L_l^2, L_b^2 and ω, Eq. (6.43) yields

$$\tan\left(\sqrt{\frac{c_{44}^l}{c_{55}^l}}\sqrt{\left(\frac{c_{ph}^2}{\beta_1^2}-1\right)}\right)\upsilon H = \frac{c_{44}^b\sqrt{\frac{c_{44}^b}{c_{55}^b}}\sqrt{\left(1-\frac{c_{ph}^2}{\beta_2^2}\right)}}{c_{44}^l\sqrt{\frac{c_{44}^l}{c_{55}^l}}\sqrt{\left(\frac{c_{ph}^2}{\beta_1^2}-1\right)}}. \tag{6.44}$$

Eq. (6.44) denotes the relation associated with the dispersion characteristics of the Love waves scattered in the layered orthotropic poroelastic composite medium.

The phase velocity and group velocity of Love waves are related by the following expression [20–23]

$$c_g = c_{ph} + \upsilon\frac{dc_{ph}}{d\upsilon}, \tag{6.45}$$

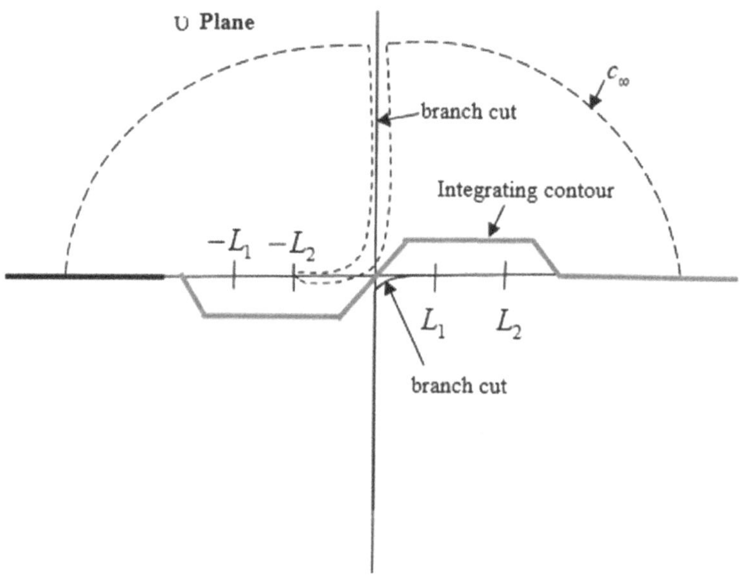

Figure 6.2 Contour of integration in υ-plane.

where c_g is the group velocity of Love waves.

In view of Eqs. (6.44) and (6.45), we get

$$\frac{c_g}{\beta_1} = \frac{K_h\left[\left(\frac{c_{ph}^2}{\beta_l^2}-1\right)\frac{c_{ph}^2}{\beta_h^2\upsilon}+\left(1-\frac{c_{ph}^2}{\beta_h^2}\right)\frac{c_{ph}^2}{\beta_l^2\upsilon}\right]+K_l\left[-\delta\left\{\gamma_l H\left(\frac{c_{ph}^2}{\beta_l^2}-1\right)-\frac{c_{ph}^2 H}{\beta_l^2}\right\}\right]}{K_h\left[\left(\frac{c_{ph}^2}{\beta_l^2}-1\right)\frac{c_{ph}\beta_l^2}{\beta_l\beta_h^2\upsilon}+\left(1-\frac{c_{ph}^2}{\beta_h^2}\right)\frac{c_{ph}}{\beta_l\upsilon}\right]+K_l\left[\frac{\delta\gamma_l\upsilon H c_{ph}}{\beta_l\upsilon}\right]}, \tag{6.46}$$

where $K_h = c_{55}^h\gamma_h$, $K_l = c_{55}^l\gamma_l$; $\delta = \left(\frac{c_{ph}^2}{\beta_l^2}-1\right)\sqrt{1-\frac{c_{ph}^2}{\beta_h^2}}\sec^2\left(\gamma_l\sqrt{\left(\frac{c_{ph}^2}{\beta_l^2}-1\right)}\upsilon H\right)$.

Eq. (6.46) represents the expression for the group velocity of scattered Love waves in the said orthotropic poroelastic composite structure.

After applying residue theorem, the expressions in $\Lambda_0(\upsilon)$ and $\Lambda_{00}(\upsilon)$ provided in Eqs. (6.38) and (6.39) and also using the concept defined in Figure 6.2, we get

$$\int_c \Lambda_0(\upsilon)d\upsilon = 2\pi i\sum \text{Res}\,\Lambda_0 - \int_{\substack{\text{Branch}\\ \text{line}}} \Lambda_0(\upsilon)d\upsilon - \int_{c_\infty} \Lambda_0(\upsilon)d\upsilon, \tag{6.47}$$

$$\int_c \Lambda_{00}(\upsilon)d\upsilon = 2\pi i\sum \text{Res}\,\Lambda_{00} - \int_{\substack{\text{Branch}\\ \text{line}}} \Lambda_{00}(\upsilon)d\upsilon - \int_{c_\infty} \Lambda_{00}(\upsilon)d\upsilon. \tag{6.48}$$

The residues of the integral Λ_0 and Λ_{00} at the poles υ_m are obtained as

$$\text{Res}\,\Lambda_0 = \frac{\cos\gamma_l\xi_{lm}(x+H)}{\gamma_l(\upsilon_m H)}e^{i\upsilon_m(x-y)}, \tag{6.49}$$

$$\text{Res}\,\Lambda_{00} = \frac{\cos\gamma_l\xi_{lm}He^{-\gamma_h\xi_{hm}x}}{\gamma_l(\upsilon_m H)}e^{i\upsilon_m(x-y)}. \tag{6.50}$$

The integrals presented in Eqs. (6.47) and (6.48) can be determined using the method developed by Sezawa [24]. It can be employed using a contour comprising the real axis and a huge-radius semicircle in the upper half-plane has been assumed. On owing to the characteristics of Love waves propagation, the branch cut points L_1 and L_2 can be taken along the real axis in contour of the integral. Due to having larger y and x values, the term involving $\frac{1}{x^{3/2}}$ in Eqs. (6.47) and (6.48) can be neglected in the field of branch line integral. Therefore, the integrals along the arcs of semicircle at infinity which contained in Eqs. (6.47) and (6.48) vanish.

Therefore, using Eqs. (6.49) and (6.50) and in view of the concept $z > x$ and $\upsilon_m < 0$, Eqs. (6.47) and (6.48) can be written as

$$\int_c \Lambda_0(\upsilon)d\upsilon = 2\pi i \sum_{m=1}^{N^o} \frac{\cos\gamma_l \xi_{lm}(x+H)}{\gamma_l(\upsilon_m H)} e^{i\upsilon_m(x-y)}, \tag{6.51}$$

$$\int_c \Lambda_{00}(\upsilon)d\upsilon = 2\pi i \sum_{m=1}^{N^0} \frac{\cos\gamma_l \xi_{lm} He^{-\gamma_b \xi_{bm} x}}{\gamma_l(\upsilon_m H)} e^{i\upsilon_m(x-y)}. \tag{6.52}$$

Similar, for $y < x$ and $\upsilon_m > 0$, we obtain

$$\int_c \Lambda_0(\upsilon)d\upsilon = -2\pi i \sum_{m=1}^{N^o} \frac{\cos\gamma_l \xi_{lm}(x+H)}{\gamma_l(\upsilon_m H)} e^{i\upsilon_m(x-y)}, \tag{6.53}$$

$$\int_c \Lambda_{00}(\upsilon)d\upsilon = -2\pi i \sum_{m=1}^{N^0} \frac{\cos\gamma_l \xi_{lm} He^{-\gamma_l \xi_{lm} x}}{\gamma_l(\upsilon_m H)} e^{i\upsilon_m(x-y)}, \tag{6.54}$$

where N^0 indicates the number of roots.

With the help of Eqs. (6.51)–(6.54) and $\upsilon_m > 0$, Eqs. (6.38) and (6.39) can be written as

$$\begin{aligned} u_3 &= A\cos\gamma_l S_l(x+H)e^{-i\eta y} - ibA\sum_{m=1}^{N^0} \frac{\cos\gamma_l \xi_{lm}(x+H)}{(\upsilon_m H)} \\ &\times \left\{ e^{-i\upsilon_m y} \int_{-\infty}^{x} (i\eta h' - S_l^2 h)e^{i(\upsilon_m - \eta)x} dx + e^{i\upsilon_m y} \int_{x}^{\infty} (i\eta h' - S_l^2 h)e^{-i(\upsilon_m + \eta)x} dx \right\}, \end{aligned} \tag{6.55}$$

$$\begin{aligned} w_3 &= A\cos\gamma_l S_l He^{-\gamma_b S_b x} e^{-i\eta y} - ibA\sum_{m=1}^{N^0} \frac{\cos\gamma_b \xi_{bm} He^{-\gamma_b \xi_{bm} x}}{(\upsilon_m H)} \\ &\times \left\{ e^{-i\upsilon_m y} \int_{-\infty}^{x} (i\eta h' - S_l^2 h)e^{i(\upsilon_m - \eta)x} dx + e^{i\upsilon_m y} \int_{x}^{\infty} (i\eta h' - S_l^2 h)e^{-i(\upsilon_m + \eta)x} dx \right\} \end{aligned} \tag{6.56}$$

In view of Eq. (6.2), Eqs. (6.55) and (6.56) provide

$$\begin{aligned} u_3 &= A\cos\gamma_l S_l(x+H)e^{-i\eta y} + ibA\sum_{m=1}^{N^0} \frac{\cos\gamma_l \xi_{lm}(x+H)}{(\upsilon_m H)} \\ &\times \left\{ e^{i\upsilon_m y} (\eta\upsilon_m + \eta^2 + S_l^2) \int_{s/2}^{s/2} he^{-i(\upsilon_m + \eta)x} dx \right\}, \end{aligned} \tag{6.57}$$

$$w_3 = A\cos\gamma_l S_l H e^{-\gamma_b S_b x} e^{-i\eta y} + ibA\sum_{m=1}^{N^0} \frac{\cos\gamma_b \xi_{2m} H e^{-\gamma_b \xi_{bm} x}}{(\upsilon_m H)} \times \left\{ e^{i\upsilon_m y}(\eta\upsilon_m + \eta^2 + S_1^2) \int_{-s/2}^{s/2} h e^{-i(\upsilon_m+\eta)x} dx \right\}. \tag{6.58}$$

where the expressions in Eqs. (6.57) and (6.58) represent the non-vanishing components correspond to displacement caused by Love waves propagation in the structure containing layer over the half-space in the said orthotropic composite poroelastic structure.

6.4 SPECIAL CASES

Let us consider three distinct physical scenarios involving irregularities in the orthotropic structure: (I) a parabolic-shaped irregularity, (II) a triangular notch-shaped irregularity, and (III) a rectangular-shaped irregularity.

6.4.1 Case I: Parabolic shape surface irregularity in the upper layer

The equation describing the parabolic-shaped irregularity in upper layer in the surface is defined as (as displayed in Figure 6.3)

$$h(y) = \begin{cases} 0, & y > \dfrac{s}{2}, \text{ and } y < \dfrac{-s}{2} \\ \left(1 - \dfrac{4y^2}{s^2}\right), & \dfrac{-s}{2} \le y \le \dfrac{s}{2}. \end{cases} \tag{6.59}$$

The upper layer in the surface of the considered orthotropic poroelastic material is considered to support only the primary wave propagation. This wave satisfies the dispersion relation mentioned in Eq. (6.44).

With the help of Eq. (6.59) and the substitution $\eta = \upsilon_l$, $S_l = \xi_l$; $m = 1$ (for first mode) in Eq. (6.57), we obtain

$$\frac{u^P_{3,ref}}{A} = ib\frac{\cos\gamma_l \xi_{l1}(x+H)}{(\upsilon_l H)} e^{i\upsilon_l y}(2\upsilon_l^2 + \xi_{l1}^2)\left(\frac{2\sin\upsilon_l s}{s^2\upsilon_l^3} - \frac{2\cos\upsilon_l s}{s\upsilon_l^2}\right). \tag{6.60}$$

Eq. (6.60) provides the non-vanishing component of displacement associated with the reflected wave field induced in the upper orthotropic poroelastic layer for the scenario of parabolic-shaped surface irregularity in the said orthotropic composite structure influenced by Love wave propagation.

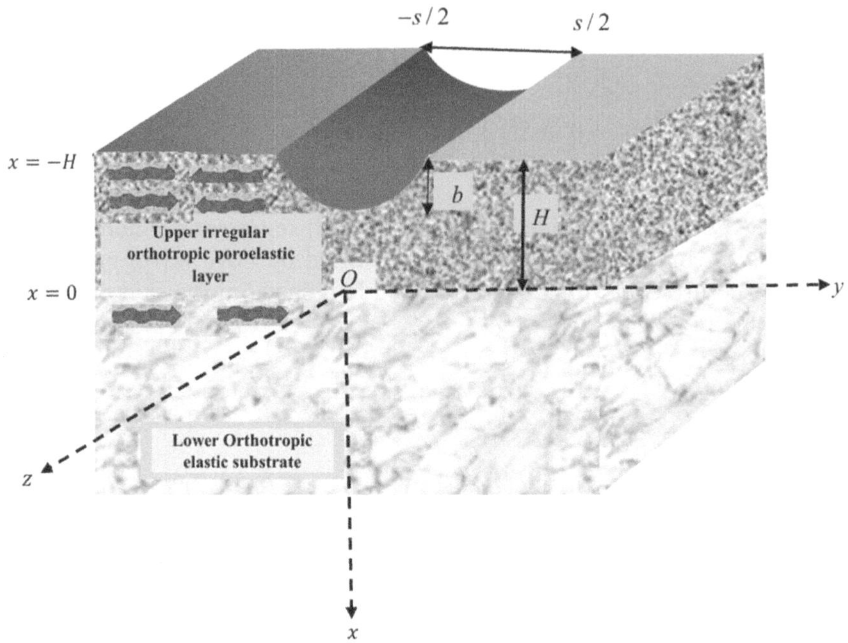

Figure 6.3 Geometry of the structure containing parabolic-shaped irregularity in the upper surface.

6.4.2 Case II: Surface irregularity in the form of a triangular notch in the upper layer

The equation corresponds to the irregularity of triangular notch-shaped in upper layer is represented as (as depicted in Figure 6.4)

$$h(y)=\begin{cases} 0, & y>\frac{s}{2}, y<\frac{-s}{2} \\ 1+\frac{2y}{s}, & \frac{-s}{2}\le y\le 0 \\ 1-\frac{2y}{s}, & 0\le y\le \frac{s}{2}. \end{cases} \tag{6.61}$$

Using the similar analytical procedure as employed for the situation of parabolic-shaped surface irregularity and inserting Eq. (6.61) in Eq. (6.57), we get

$$\frac{u_{3,ref}^{T}}{A}=ib\frac{\cos\gamma_l\xi_{l1}(x+H)}{(\upsilon_l H)}\times e^{i\upsilon_l y}(2\upsilon_l^{2}+\xi_{l1}^{2})\left(\frac{2\sin^{2}\frac{\upsilon_l s}{2}}{s\upsilon_l^{2}}\right). \tag{6.62}$$

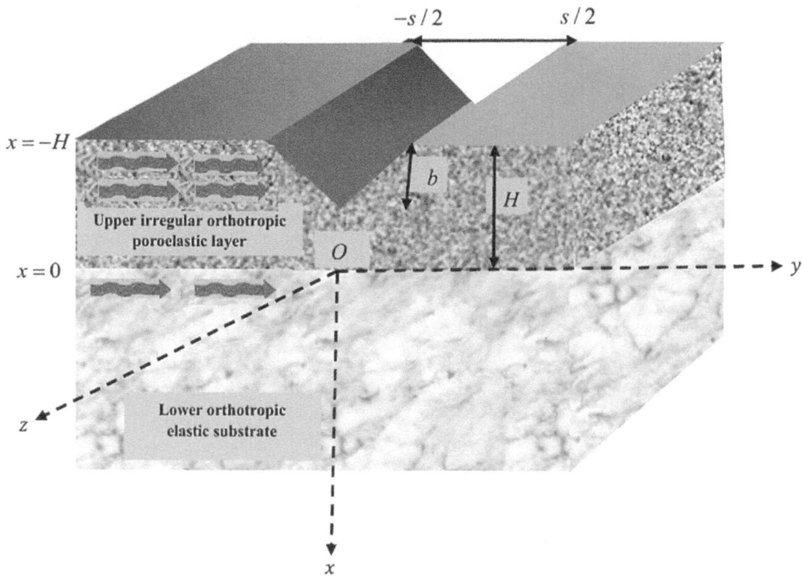

Figure 6.4 Geometry of triangular notch-shaped surface irregularity.

Eq. (6.62) provides the non-vanishing component associated with the displacement for reflected wave field induced in the upper orthotropic poroelastic layer for the situation of triangular notch type surface irregularity in the said orthotropic composite structure affected by Love wave propagation.

6.4.3 Case III: Rectangular-shaped irregularity in the upper surface layer

The equation containing the expression of the surface irregularity of rectangular shape in upper irregular surface can be expressed as (as depicted in Figure 6.5)

$$h(y) = \begin{cases} 0, & y > \dfrac{s}{2} \text{ and } y < \dfrac{-s}{2}, \\ 1, & \dfrac{-s}{2} \le y \le \dfrac{s}{2}. \end{cases} \tag{6.63}$$

The application of similar analytical procedure which described for the situation of irregularity with parabolic shape in upper surface and inserting the expression of triangular notch-shaped irregularity Eq. (6.63) into Eq. (6.57), we get

$$\frac{u_{3,\,ref}^{R}}{A} = ib\frac{\cos\gamma_l\xi_{l1}(x+H)}{(\upsilon_l H)} \times e^{i\upsilon_l y}(2\upsilon_l^{\,2} + \xi_{l1}^2)\left(\frac{2\sin\upsilon_l s}{\upsilon_l}\right). \tag{6.64}$$

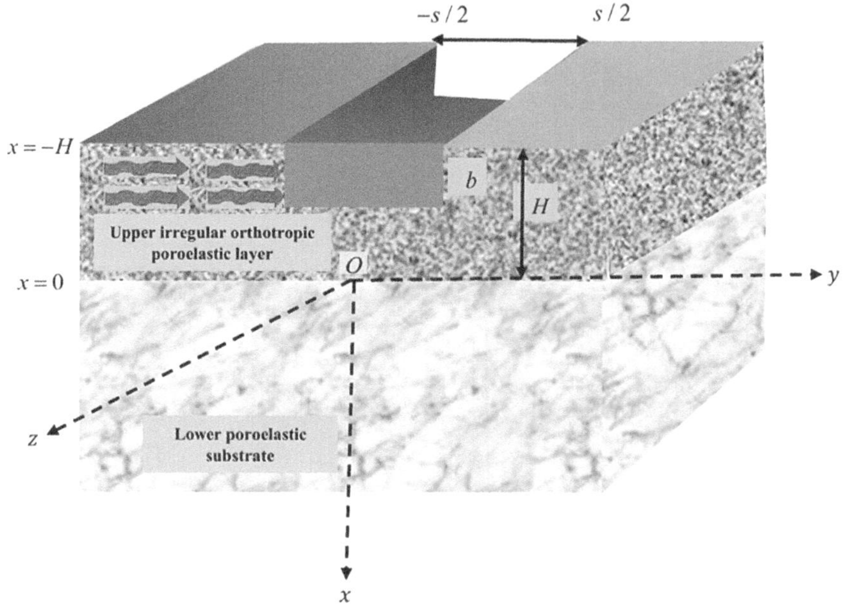

Figure 6.5 Geometry of rectangular-shaped surface irregularity.

Eq. (6.64) delves into the expression of non-vanishing component of induced displacemenet caused by reflected wave field in the upper layer having the irregularity of rectangular shape in the studied orthotropic composite structure.

6.5 RESULTS VALIDATION

6.5.1 Validation for dispersion relation

If $c_{44}^{l}=\mu^{l}$, $c_{55}^{b}=\mu^{b}$ and $\gamma_{l}=\sqrt{\frac{c_{44}^{l}}{c_{55}^{l}}}=1$, $\gamma_{b}=\sqrt{\frac{c_{44}^{b}}{c_{55}^{b}}}=1$, the dispersion relation mentioned in Eq. (6.44) reduces into the following equation

$$\tan\sqrt{\left(\frac{c_{pb}^{2}}{\beta_{1}^{2}}-1\right)}\upsilon H=\frac{\mu^{b}\sqrt{\left(1-\frac{c_{pb}^{2}}{\beta_{b}^{2}}\right)}}{\mu^{l}\sqrt{\left(\frac{c_{pb}^{2}}{\beta_{l}^{2}}-1\right)}}, \tag{6.65}$$

where μ^{l} and μ^{b} signify the shear modulus related with the isotropic layer and isotropic substrate, respectively. Eq. (6.65) provides the dispersion relation associated with the scattered wave field due to Love wave propagation in the layered and irregular isotropic elastic structure. The expression derived in Eq. (6.65) exactly coincides with the classical Love wave equation proposed by Love [25].

6.5.2 Validation of the non-vanishing component of displacement due to reflected wave field under the presence of parabolic-shaped surface irregularity

For the case of layered isotropic elastic structure i.e. $\gamma_l = \sqrt{\frac{c_{44}^l}{c_{55}^l}} = 1$, the expression in Eq. (6.60) converts into

$$\frac{u_{3,\,ref}^P}{A} = ib\frac{\cos\xi_{l1}(x+H)}{(\upsilon_l H)}e^{i\upsilon_l y}(2\upsilon_l^{\,2}+\xi_{l1}^2)\left(\frac{2\sin\upsilon_l s}{s^2\upsilon_l^{\,3}}-\frac{2\cos\upsilon_l s}{s\upsilon_l^{\,2}}\right). \tag{6.66}$$

The expression in Eq. (6.66) signifies the non-vanishing displacement component for reflected wave field due to Love waves propagation in a layered isotropic elastic structure. Eq. (6.66) yields the results that coincide perfectly with the findings of Wolf [15] and Chattopadhyay et al. [18] in the context of a layered isotropic elastic structure.

6.5.3 Validation for the displacement due to reflected wave field associated with a triangular notch-shaped irregularity in upper surface layer

Using similar approach, if $\gamma_l = \sqrt{\frac{c_{44}^l}{c_{55}^l}} = 1$, the expression in Eq. (6.66) gives

$$\frac{u_{3,\,ref}^T}{A} = ib\frac{\cos\xi_{l1}(x+H)}{(\upsilon_l H)}\times e^{i\upsilon_l y}(2\upsilon_l^{\,2}+\xi_{l1}^2)\left(\frac{2\sin^2\frac{\upsilon_l s}{2}}{s\upsilon_l^{\,2}}\right). \tag{6.67}$$

Eq. (6.67) represents the expression of non-vanishing displacement component for reflected wave field caused by the interaction of irregularity in upper surface with the Love waves in upper orthotropic poroelastic layer in the considered orthotropic material structure containing triangular notch-shaped surface irregularity. Eq. (6.67) presents the outcomes that aligns perfectly with the similar results reported by Wolf [15] and Chattopadhyay et al. [18] for the scenario of a layered isotropic elastic structure.

6.6 ANALYSIS OF FINDINGS

To facilitate numerical computation as well as graphical depictions of the dimensionless phase velocity, group velocity, and reflected displacement due

Table 6.1 Materials characteristic of orthotropic poroelastic layer and orthotropic material half-space [10, 16]

	For solid skeleton			For pore fluids		
Properties	c_{44}	c_{55}	c_{66}	ρ_s	ρ_f	K_f
Upper orthotropic poroelastic layer	3	3.2	3.8	2.6	1	2
Lower orthotropic substrate	4	4.2	4.6	1.5	–	–

to Love wave scattering in an irregular orthoporous-layered composite, the physio-mechanical parameters provided in Table 6.1 have been employed.

Unit: $c_{ij}\left(\times 10^9 N/m^2\right)$, $\rho_s\left(\times 10^3 kg/m^3\right)$, $\rho_f\left(\times 10^3 kg/m^3\right)$.

Unless otherwise stated,

$$(y+H)/H = 10,\ \ x/H = 10,\ \ s/H = 0.01,\ \ b/H = 0.001.$$

The effectuality of porosity parameter on phase and group velocities of scattering of Love waves in an orthotropic porous composite structure has been graphically illustrated by Figure 6.6(a) and (b). The upsurge of the porosity parameter represents the decline of the porosity in orthotropic porous composite structure. It is examined from Figure 6.6(a) that the increase in porosity parameter, i.e. decrease in porosity, causes the reduction of the phase velocity of scattered Love waves. Figure 6.6(b) depicts the efficacy of porosity parameter on the group velocity of scattered Love waves. It is observed from Figure 6.6(b) that the decrease in the porosity causes the upsurge in the group velocity of scattered Love waves.

The influence of the anisotropy parameter on the phase and group velocities of scattered Love waves in an orthotropic porous composite rock structure is depicted in Figure 6.7(a) and (b), respectively. It is manifested from the analysis of Figure 6.7(a) and (b) that with the increase of anisotropy parameter (anisotropy prevails) in orthotropic porous composite structure, scattered Love waves propagate with reduced phase and group velocities of scattered Love waves in the considered orthotropic porous composite structure.

Figure 6.8(a) and (b) graphically demonstrate the effects of the vertical irregularity parameter and the anisotropy parameter, respectively, on the reflected displacement in an orthotropic porous composite structure with a parabolic-shaped surface irregularity. As observed from Figure 6.8(a), increasing the vertical irregularity parameter, i.e. the maximum depth of the surface irregularity, results in a corresponding rise in the reflected displacement in the orthotropic porous composite structure with a parabolic surface irregularity. Figure 6.8(b) demonstrates the variation of the anisotropy parameter and its impact on the reflected displacement within the orthotropic porous composite structure possessing a parabolic-shaped surface irregularity.

Figure 6.9(a) and (b) represent the effect of vertical irregularity parameter and anisotropy parameter, respectively, on the reflected displacement induced in orthotropic porous composite structure having rectangular-shaped surface

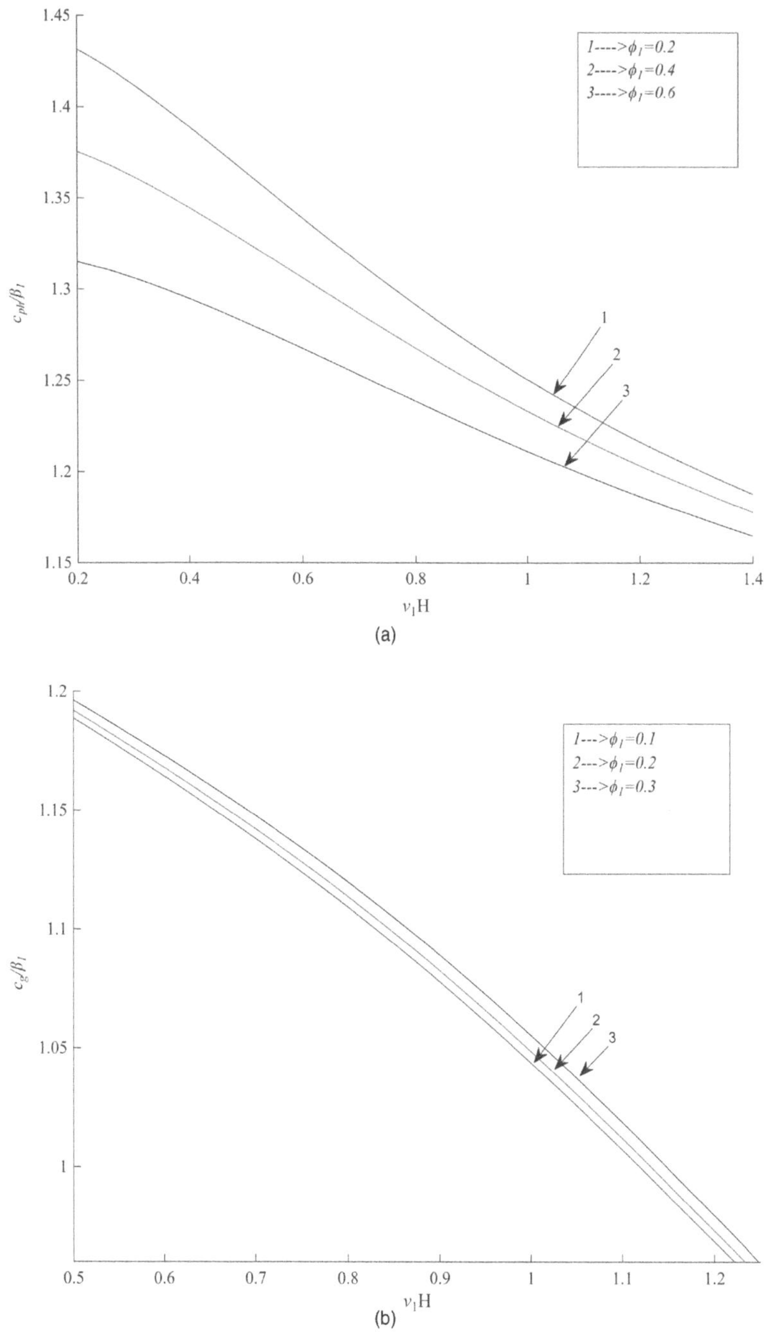

Figure 6.6 Non-dimensional (a) phase velocity $\left(\frac{c_{ph}}{\beta_1}\right)$ and (b) group velocity $\left(\frac{c_g}{\beta_1}\right)$ against non-dimensional wave number $\left(\upsilon_1 H\right)$ for distinct values of porosity parameter (ϕ_1) of the upper layer in the orthotropic composite material structure.

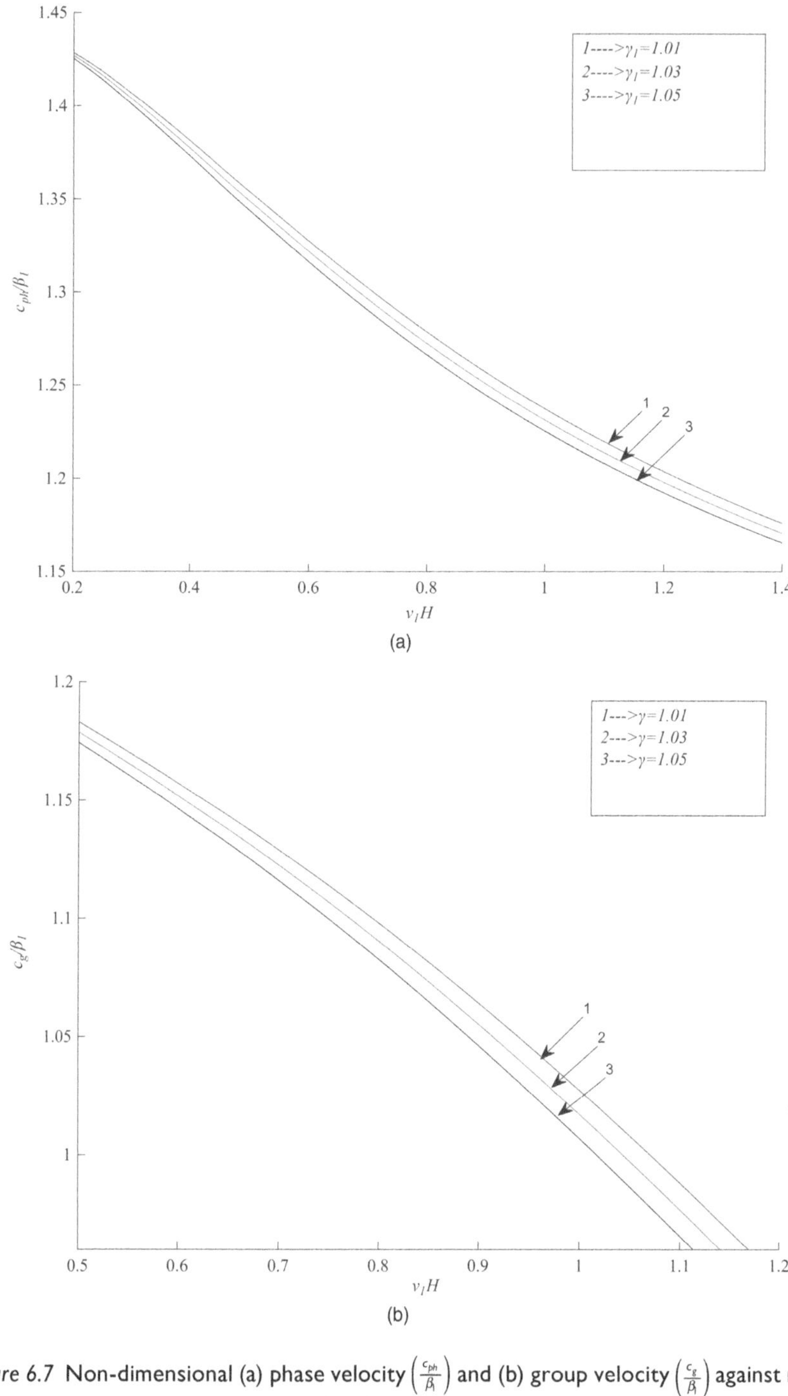

Figure 6.7 Non-dimensional (a) phase velocity $\left(\frac{c_{ph}}{\beta_1}\right)$ and (b) group velocity $\left(\frac{c_g}{\beta_1}\right)$ against non-dimensional wave number $\left(\upsilon_1 H\right)$ for diverse values of anisotropy parameter (γ_1) of the upper layer in the orthotropic porous composite material structure.

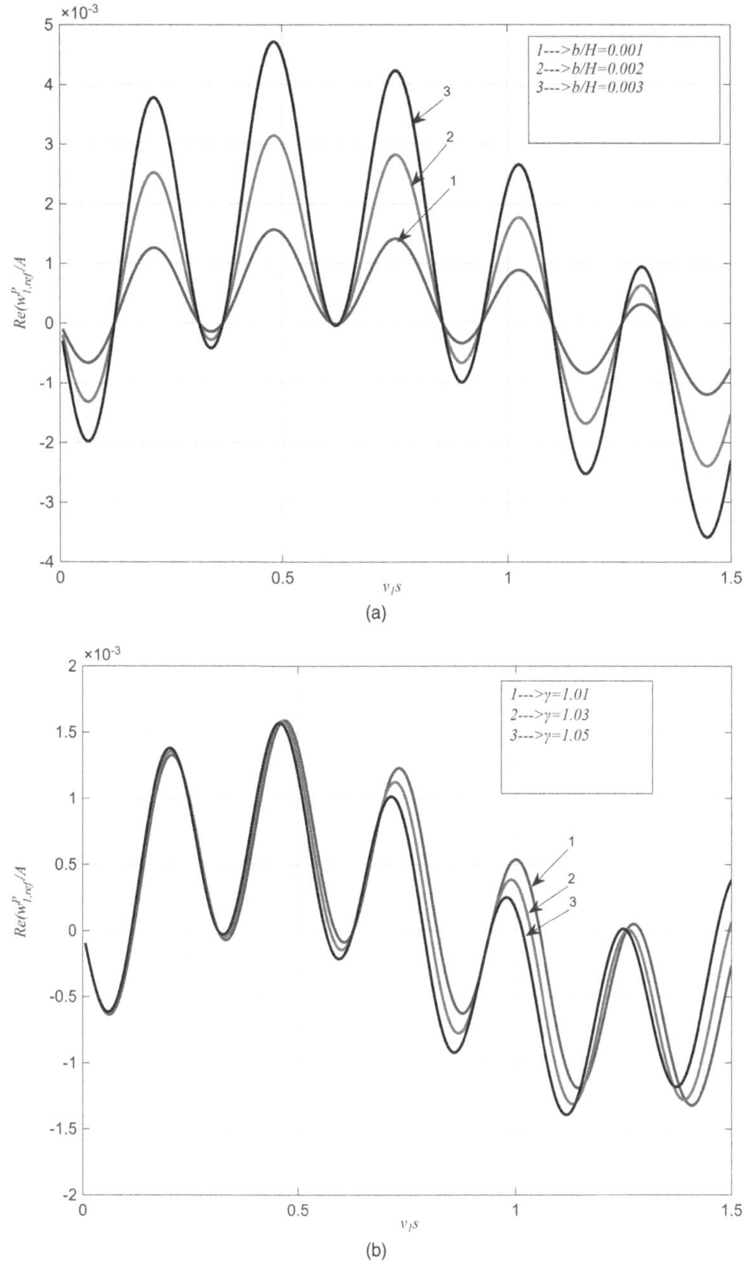

Figure 6.8 Non-dimensional reflected displacement $\mathrm{Re}\left(w_{1,ref}^{P} / A\right)$ against non-dimensional width $\left(\upsilon_1 s\right)$ for diverse values related with (a) vertical irregularity parameter (b / H) and (b) anisotropy parameter (γ) associated with surface irregularity of the parabolic shape in the upper orthotropic poroelastic layer of the considered orthotropic poroelastic composite structure.

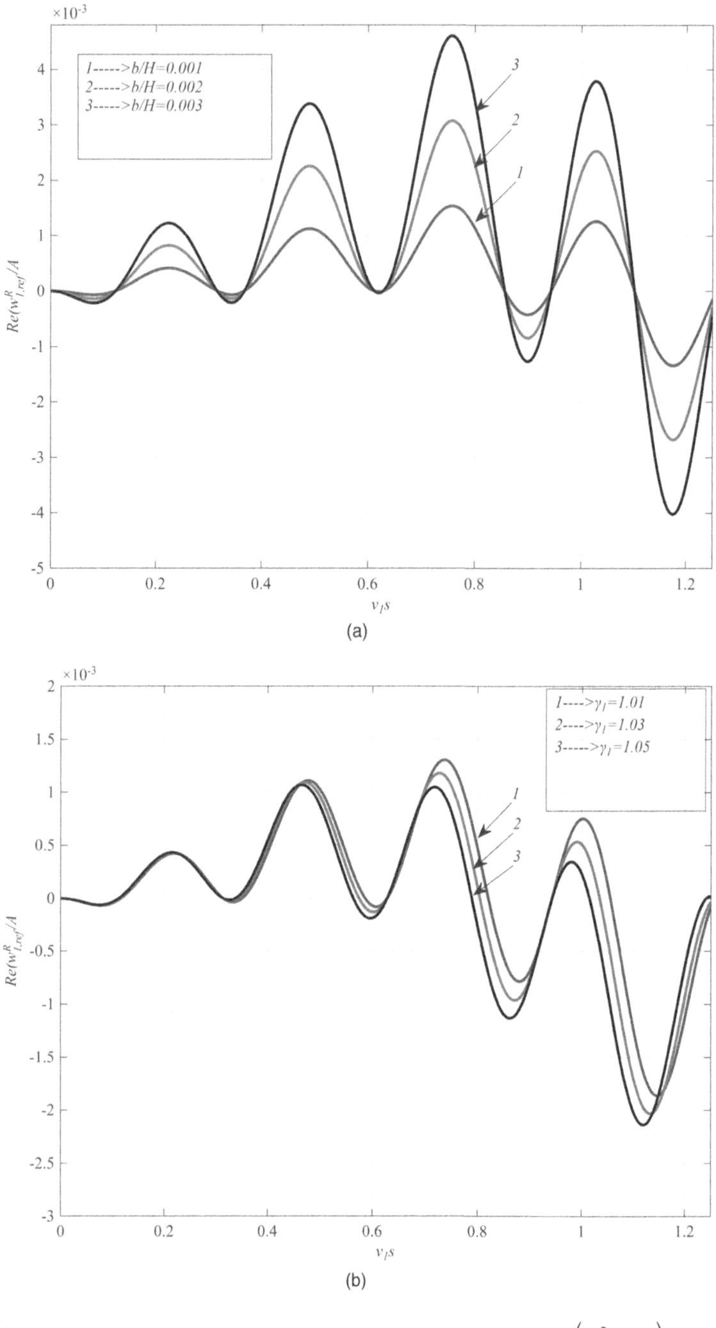

Figure 6.9 Non-dimensional reflected displacement component $\text{Re}\left(w_{1,ref}^{R} / A\right)$ against non-dimensional width $\left(\upsilon_1 s\right)$ for several values of (a) vertical irregularity parameter $\left(b / H\right)$ and (b) anisotropy parameter (γ_1) of the upper orthotropic poroelastic layer having surface irregularity of the rectangular shape in the said orthotropic poroelastic composite structure.

irregularity. It has been observed from Figure 6.9(a) that the enhancement of vertical irregularity parameter leads to the escalation in the vertical irregularity parameter induced in the orthotropic porous composite structure containing rectangular-shaped surface irregularity. The examination of Figure (6.9) provides that as the anisotropy prevails in the medium, the reflected displacement causes drop in the reflected displacement in the orthotropic porous composite structure having rectangular-shaped surface irregularity.

The behavior of reflected displacement under the impacts of vertical irregularity parameter and anisotropy parameter in orthotropic poroelastic composite structure containing triangular notch surface irregularity have been demonstrated by Figure 6.10(a) and (b), respectively. From Figure 6.10(a), it is evident that the vertical irregularity parameter contributes to an increase in the reflected displacement in the orthotropic porous composite structure with a triangular notch surface irregularity. It is analyzed from Figure 6.10(b) that a rise in the degree of anisotropy in the considered orthotropic porous composite structure causes the drop in the value of reflected displacement

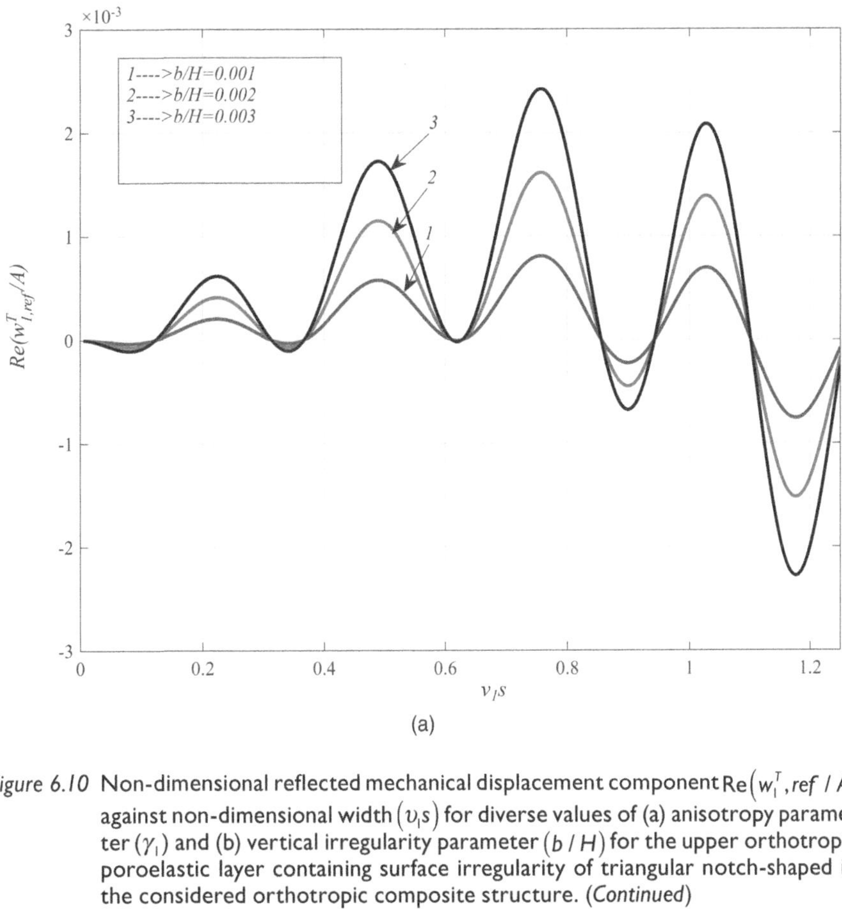

(a)

Figure 6.10 Non-dimensional reflected mechanical displacement component $\mathrm{Re}\left(w_1^T, ref / A\right)$ against non-dimensional width $(\upsilon_1 s)$ for diverse values of (a) anisotropy parameter (γ_1) and (b) vertical irregularity parameter (b / H) for the upper orthotropic poroelastic layer containing surface irregularity of triangular notch-shaped in the considered orthotropic composite structure. *(Continued)*

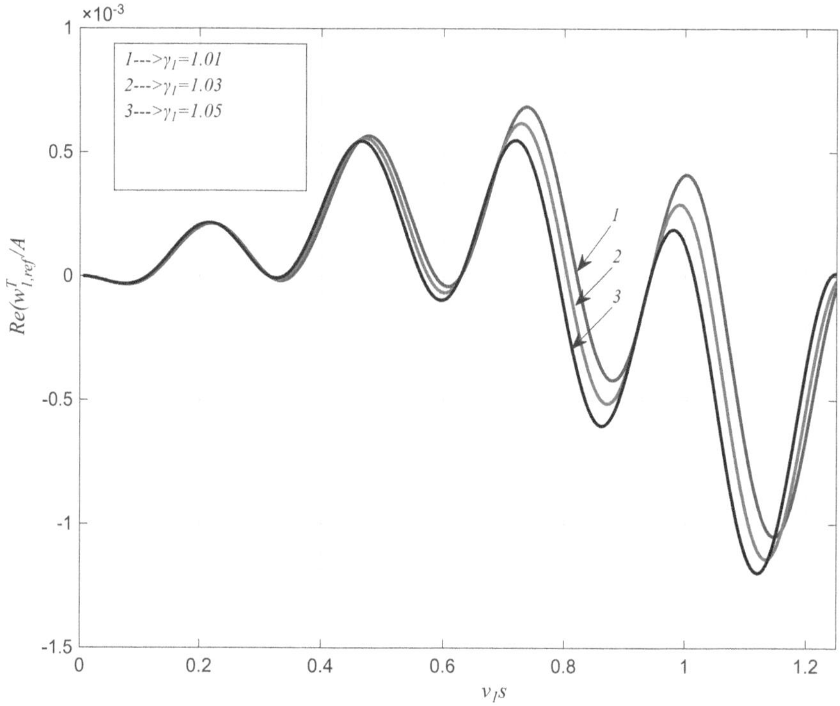

(b)

Figure 6.10 (Continued)

induced in orthotropic porous composite structure containing the triangular notch-shaped surface irregularity.

The minute observation of Figures 6.7–6.10 also provides that the induced reflected displacement is maximum for the case of parabolic-shaped surface irregularity while it is minimum for the scenario of triangular notch-shaped surface irregularity. Furthermore, the reflected displacement resulting from the rectangular-shaped surface irregularity is found to be between the levels observed for the other two surface irregularity configurations.

6.7 CONCLUSIONS

This research presents an analysis of Love wave scattering and reflection behavior in a layered orthotropic porous composite structure featuring irregularity in upper surface layer. Different types of surface irregularities specifically parabolic, triangular notch, and rectangular have been assumed in the orthotropic porous composite configuration. The irregularity in the upper layer has been dealt mathematically using perturbation technique. The expressions correspond to the dispersion relation and group velocity have now been derived analytically. The effects of porosity (in terms of porosity parameter) and anisotropy (in terms of anisotropy parameter) have been examined. The exact expressions of induced displacement component for reflected wave field caused by different shaped of surface irregularity such as parabolic-shaped, rectangular-shaped, and triangular notch-shaped have also been derived. In addition, the major conclusions drawn from this study are outlined below:

1. As the porosity of the upper layer decreases (represented by an increase in the porosity parameter), the phase velocity of Love waves due to scattered wave field in the orthotropic poroelastic composite structure reduces.
2. The group velocity of scattered Love waves increases with the rise in the porosity parameter (corresponding to a reduction in porosity) of the upper layer in the orthotropic poroelastic composite structure.
3. If the anisotropy prevails in the upper poroelastic layer of orthotropic poroelastic composite structure, there is a decline in the phase and group velocities of scattered Love waves.
4. The augmentation of vertical irregularity parameter (related with the amplitude of vertical irregularity) results in the increase of induced reflected displacement for all said shapes viz. parabolic, rectangular, triangular notch of irregularity in upper surface.
5. Enhancing the anisotropy of the upper poroelastic layer in the orthotropic composite structure causes the decrease in reflected displacement across all considered surface irregularity geometries, including parabolic, rectangular, and triangular notch shapes.
6. Among the three types of surface irregularities, the induced reflected displacement is highest for the parabolic-shaped irregularity in the upper poroelastic layer and lowest for the triangular notch-shaped irregularity.

6.8 APPLICATIONS IN ENGINEERING

The results obtained from the current investigation may find its possible applications to construct the safer design of bridge, pavements, and other engineering structures made by composite materials. The observations of obtained results provide the information about the physical parameters (properties) through the velocity of scattered Love waves and resulted reflected displacement can be reduced. The present analysis reveals that increased porosity and anisotropy in orthotropic poroelastic composite materials lead to a reduction of scattered Love waves velocity. Also, the lower amplitude of irregularity and higher anisotropy can cause the minimum reflected displacement for all considered shapes viz. parabolic, rectangular, and triangular notch surface irregularity. It is also evident that among all the considered surface irregularities, the triangular notch shape results in the minimum induced reflected displacement. Therefore, a composite material with higher porosity, greater anisotropy, and a triangular notch-shaped surface irregularity can be considered effective in minimizing the damaging effects of seismic activity on engineering structures.

ACKNOWLEDGMENT

The authors sincerely grateful to UPES, Dehradun for providing the research facilities that supported the accomplishment of this work.

DECLARATION OF INTEREST STATEMENT

There are no conflicts of interest to declare, as confirmed by the corresponding author on behalf of all authors.

DATA AVAILABILITY

The data used in this study are fully provided and referenced in the manuscript.

REFERENCES

1. Ke, L. L., Wang, Y. S., & Zhang, Z. M. (2006). Love waves in an inhomogeneous fluid saturated porous layered half-space with linearly varying properties. *Soil Dynamics and Earthquake Engineering*, *26*(6–7), 574–581.
2. Du, J., Jin, X., Wang, J., & Xian, K. (2007). Love wave propagation in functionally graded piezoelectric material layer. *Ultrasonics*, *46*(1), 13–22.
3. Zhu, X. Q., & Law, S. S. (2003). Dynamic behavior of orthotropic rectangular plates under moving loads. *Journal of Engineering Mechanics*, *129*(1), 79–87.
4. Maljaars, J., van Dooren, F., & Kolstein, H. (2012). Fatigue assessment for deck plates in orthotropic bridge decks. *Steel Construction*, *5*(2), 93–100.
5. Destrade, M. (2001). Surface waves in orthotropic incompressible materials. *The Journal of the Acoustical Society of America*, *110*(2), 837–840.
6. Destrade, M. (2007). Seismic Rayleigh waves on an exponentially graded, orthotropic half-space. *Proceedings of the Royal Society A: Mathematical, Physical and Engineering Sciences*, *463*(2078), 495–502.
7. Mohd Nor, M. K., Ho, C. S., Ma'at, N., & Kamarulzaman, M. F. (2020). Modelling shock waves in composite materials using generalised orthotropic pressure. *Continuum Mechanics and Thermodynamics*, *32*, 1217–1229.
8. Biswas, S. (2020). Surface waves in porous nonlocal thermoelastic orthotropic medium. *Acta Mechanica*, *231*, 2741–2760.
9. Yu, J., & Zhang, C. (2014). Effects of initial stress on guided waves in orthotropic functionally graded plates. *Applied Mathematical Modelling*, *38*(2), 464–478.
10. Liu, K., & Liu, Y. (2004). Propagation characteristic of Rayleigh waves in orthotropic fluid-saturated porous media. *Journal of Sound and Vibration*, *271*(1–2), 1–13.

11. Vinh, P. C., Aoudia, A., & Giang, P. T. H. (2016). Rayleigh waves in orthotropic fluid-saturated porous media. *Wave Motion*, *61*, 73–82.
12. Singh, A. K., Pal, M. K., Negi, A., & Mistri, K. C. (2019). Analytical study on dynamic response due to a moving load on distinctly characterized orthotropic half-spaces under different physical conditions with comparative approach. *Arabian Journal for Science and Engineering*, *44*(5), 4863–4883.
13. Singh, A. K., Lakshman, A., & Chattopadhyay, A. (2016). Effect of irregularity and anisotropy on the dynamic response due to a shear load moving on an irregular orthotropic half-space under influence of gravity. *Multidiscipline Modeling in Materials and Structures, 12*(1), 194–214.
14. Slavin, L. M., & Wolf, B. (1970). Scattering of Love waves in a surface layer with an irregular boundary for the case of a rigid underlying half-space. *Bulletin of the Seismological Society of America*, *60*(3), 859–877.
15. Wolf, B. (1970). Propagation of Love waves in layers with irregular boundaries. *Pure and Applied Geophysics*, *78*(1), 48–57.
16. Chen, W. Q., Wang, H. M., & Bao, R. H. (2007). On calculating dispersion curves of waves in a functionally graded elastic plate. *Composite Structures*, *81*(2), 233–242.
17. Anagnostopoulos, K. A., Charalambopoulos, A., & Massalas, C. V. (2005). On the investigation of elasticity equation for orthotropic materials and the solution of the associated scattering problem. *International Journal of Solids and Structures*, *42*(24–25), 6376–6408.
18. Chattopadhyay, A., Gupta, S., Sharma, V. K., & Kumari, P. (2010). Propagation of shear waves in viscoelastic medium at irregular boundaries. *Acta Geophysica*, *58*(2), 195–214.
19. Negi, A., & Kumar Singh, A. (2019). Analysis on scattering characteristics of Love-type wave due to surface irregularity in a piezoelectric structure. *The Journal of the Acoustical Society of America*, *145*(6), 3756–3783.
20. Graff, K. F. (2012). *Wave motion in elastic solids*. Courier Corporation.
21. Miklowitz, J. (2012). *The theory of elastic waves and waveguides*. Elsevier.
22. Li, X. Y., Wang, Z. K., & Huang, S. H. (2004). Love waves in functionally graded piezoelectric materials. *International Journal of Solids and Structures*, *41*(26), 7309–7328.
23. Novotny, O. (1999). Seismic surface waves. *Bahia, Salvador: Instituto de Geociencias, 61*.
24. Sezawa, K. (1935). Love-waves generated from a source of a certain depth. *Bulletin of the Earthquake Research Institute, Tokyo Imperial University*, *13*(1), 1–17.
25. Love, A. E. H. (1944). A treatise on the mathematical theory of elasticity, Dover Publications. *New York*, *1*.

Chapter 7

Dynamics of gravity affected reflected waves generated due to incident qP wave at the impedance boundary of a micro-mechanically modeled PFRC half-space

Sayantan Guha, Suman Nain, Ajeet Kumar Singh, and Sonam Singh

7.1 INTRODUCTION

The application of wave reflection/transmission analysis in key areas, such as seismic analysis, geophysics, and earthquake engineering, makes it a significant field. There have been a number of notable studies done in the past on the wave reflection/refraction phenomenon in various media by [1–3], etc.

The phenomenon of piezoelectricity was put forth by [4]. Piezoelectricity property allows non-centrosymmetric crystals to generate electricity influenced by external stress (direct effect) and causes deformation of the crystal in an electric field (inverse effect). Piezoelectricity property is widely used in applications involving sensors, ultrasound scans, SONAR, actuators, motors, Non Destructive Testing (NDT)/Non Destructive Evaluation (NDE), etc. Studies on waves traversing in smart (and other) structures were conducted by [5–14], etc.

An increasing number of scientific and engineering fields are using smart composites, which is driving rapid development in this field. Piezoelectric fiber-reinforced composites (PFRCs), among other composites, are of remarkable significance due to several advantages like less weight and more strength. The authors of Refs. [15–24] did some works on designing composite materials.

Impedance boundary conditions are applied in different problems of acoustics, electromagnetism, physics, material science, etc. However, they are generally not employed in wave reflection problems. [25] discovered the impedance boundary condition while analyzing wave propagation in an isotropic solid encased in a minute film of different constituents overlying the half-space. This boundary relates stresses with mechanical displacements. Some analyses have been conducted in the past involving impedance boundary conditions such as [26–31]. Some studies have also been conducted undertaking the influence of gravitational effects in wave reflection problems, as seen from [32–34], etc.

 DOI: 10.1201/9781003470885-7

However, a thorough survey of the literature shows that no studies are performed analyzing the reflection of waves affected by gravity at a PFRC half-space's impedance boundary. This is the main source of motivation driving this analysis.

7.2 MICRO-MECHANICAL MODELING OF PFRC

Some assumptions are made to compose a homogeneous PZT-5A–epoxy PFRC. These can be found in the work done by [5]. The PFRC is perceived as a congress of rectangular RVEs consisting of parts of fiber and matrix [7]. Using strength of materials (SM) and rule of mixtures (RM) techniques (which can be applied regardless of the fiber's cross-sectional structure), the PFRC's mechanical and electrical terms are obtained.

The basic constitutive equations for piezoelectric fibers (Eq. (7.1)) and piezoelectrically inactive matrix (Eq. (7.2)) are as follows:

$$\begin{aligned}\{\tau^f\} &= [c^f]\{s^f\} - [e^f]^T\{E^f\}\\ \{D^f\} &= [e^f]\{s^f\} + [\epsilon^f]\{E^f\}\end{aligned} \tag{7.1}$$

$$\begin{aligned}\{\tau^m\} &= [c^m]\{s^m\}\\ \{D^m\} &= [\epsilon^m]\{E^m\}\end{aligned} \tag{7.2}$$

where

$$\{\tau^\gamma\} = \begin{bmatrix}\tau_{11}^\gamma\\ \tau_{22}^\gamma\\ \tau_{33}^\gamma\\ \tau_{23}^\gamma\\ \tau_{13}^\gamma\\ \tau_{12}^\gamma\end{bmatrix},\ [c^\gamma] = \begin{bmatrix} c_{11}^\gamma & c_{12}^\gamma & c_{13}^\gamma & 0 & 0 & 0\\ c_{12}^\gamma & c_{22}^\gamma & c_{23}^\gamma & 0 & 0 & 0\\ c_{13}^\gamma & c_{23}^\gamma & c_{33}^\gamma & 0 & 0 & 0\\ 0 & 0 & 0 & c_{44}^\gamma & 0 & 0\\ 0 & 0 & 0 & 0 & c_{55}^\gamma & 0\\ 0 & 0 & 0 & 0 & 0 & c_{66}^\gamma\end{bmatrix},\ \{s^\gamma\} = \begin{bmatrix} s_{11}^\gamma\\ s_{22}^\gamma\\ s_{33}^\gamma\\ 2s_{23}^\gamma\\ 2s_{13}^\gamma\\ 2s_{12}^\gamma\end{bmatrix},\ \{E^\gamma\} = \begin{bmatrix}E_1^\gamma\\ E_2^\gamma\\ E_3^\gamma\end{bmatrix},$$

$$\{D^\gamma\} = \begin{bmatrix}D_1^\gamma\\ D_2^\gamma\\ D_3^\gamma\end{bmatrix},\ [e^f] = \begin{bmatrix}0 & 0 & e_{31}^f\\ 0 & 0 & e_{32}^f\\ 0 & 0 & e_{33}^f\\ 0 & e_{24}^f & 0\\ e_{15}^f & 0 & 0\\ 0 & 0 & 0\end{bmatrix}^T,\ [\epsilon^\gamma] = \begin{bmatrix}\epsilon_{11}^\gamma & 0 & 0\\ 0 & \epsilon_{22}^\gamma & 0\\ 0 & 0 & \epsilon_{33}^\gamma\end{bmatrix}.$$

In the above terms, γ can either be f signifying fiber part or m signifying matrix part.

Due to rigid bonding and x-axis orientation, $s_{11}^{f} = s_{11}^{m} = s_{11}$. Employing RM, the strains are obtained as [17]

$$\begin{aligned} s_{22} &= V^{F} s_{22}^{f} + V^{M} s_{22}^{m} \\ s_{33} &= V^{F} s_{33}^{f} + V^{M} s_{33}^{m} \\ s_{23} &= V^{F} s_{23}^{f} + V^{M} s_{23}^{m} \cdot \\ s_{13} &= V^{F} s_{13}^{f} + V^{M} s_{13}^{m} \\ s_{12} &= V^{F} s_{12}^{f} + V^{M} s_{12}^{m} \end{aligned} \tag{7.3}$$

where $V^{F}\left(V^{M}\right) \rightarrow$ fiber (matrix) volume fraction and $V^{F} + V^{M} = 1$.

Thus, $V^{F} = 1\left(V^{M} = 1\right)$ implies only piezoelectric fibers (epoxy matrix) are present.

In order to attain axial equilibrium as [35], the stresses in the composite are expressed as

$$\begin{aligned} \tau_{11} &= V^{F} \tau_{11}^{f} + V^{M} \tau_{11}^{m} \\ \tau_{22} &= \tau_{22}^{f} = \tau_{22}^{m} \\ \tau_{33} &= \tau_{33}^{f} = \tau_{33}^{m} \\ \tau_{23} &= \tau_{23}^{f} = \tau_{23}^{m} \\ \tau_{13} &= \tau_{13}^{f} = \tau_{13}^{m} \\ \tau_{12} &= \tau_{12}^{f} = \tau_{12}^{m} \end{aligned} \quad . \tag{7.4}$$

The electric displacements are expressed as

$$\begin{aligned} D_1 &= V^{F} D_1^{f} + V^{M} D_1^{m} \\ D_2 &= V^{F} D_2^{f} + V^{M} D_2^{m} \; . \\ D_3 &= V^{F} D_3^{f} + V^{M} D_3^{m} \end{aligned} \tag{7.5}$$

The PFRC's density ρ is given as

$$\rho = V^{F} \rho^{f} + V^{M} \rho^{m}. \tag{7.6}$$

The expressions of all obtained PFRC constants can be found from the work of [5].

Figure 7.1 represents some electro-mechanical properties of the PFRC. To scale the terms, we use the notations: $\overline{e_{31}} = e_{31} / e_{31}^{f}$, $\overline{\varepsilon_{11}} = \varepsilon_{11} / \varepsilon_{11}^{f}$, $\overline{\varepsilon_{33}} = \varepsilon_{33} / \varepsilon_{33}^{f}$, $\overline{c_{ij}} = c_{ij} / c_{ij}^{f}$, and $\bar{\rho} = \rho / \rho^{f}$.

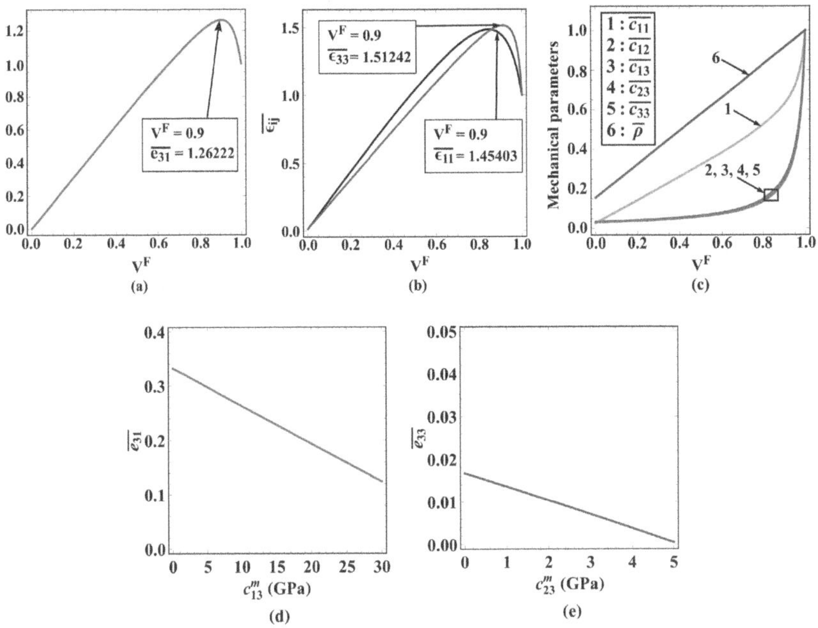

Figure 7.1 Variations of scaled (a) piezoelectric constant $\overline{e_{31}}$ against V^F, (b) dielectric permittivities $(\overline{\varepsilon_{11}}, \overline{\varepsilon_{33}})$ against V^F, (c) stiffness constants $(\overline{c_{ij}})$ & density $(\overline{\rho})$ against V^F, (d) $\overline{e_{31}}$ against c_{13}^m, and (e) $\overline{e_{33}}$ against c_{23}^m of the considered PFRC.

The values of V^F and V^M are chosen to be 0.9 and 0.1, because the following observations are made from Figure 7.1:

- The PFRC generates greater electric response (greater magnitudes of piezoelectric constant and dielectric permittivities) compared to monolithic PZT-5A. Piezoelectric constants regulate functional capability of piezoelectric devices like sensors and actuators. Moreover, dielectric permittivities are essential components that control voltage in applications using piezoelectric sensors and capacitors.
- The magnitudes of the PFRC's density and stiffness are lesser than PZT-5A. This ensures that the PFRC is mechanically flexible and has low density which is required for electroelastic transducers.
- As the matrix material's stiffness constants increase, $\overline{e_{31}}$ and $\overline{e_{33}}$ drop. Thus, softer matrix materials—like the epoxy material we used—can improve the PFRC's effective piezoelectric characteristics. Using a micromechanical model based on the theory of multiple-scattering, Nan and Jin [36] also proved a comparable effect of stiffness of matrix on the coefficients of PE composites.

7.3 GOVERNING EQUATIONS

The governing equations disregarding electric/body forces are as follows:

- Stress, strain, electric field: $\tau_{ij} = c_{ijkl}s_{kl} - e_{kij}E_k$. (7.7)
- Strain, displacement: $s_{ij} = \left(u_{i,j} + u_{j,i}\right)/2$. (7.8)
- General motion equation: $\tau_{ji,j} = \rho \ddot{u}_i$, (7.9)
  which, considering gravity following [37] and the undertaken geometrical model, takes the form of Eqs. (12) and (13) provided later on.
- Electric field: $D_i = e_{ijk}s_{jk} + \varepsilon_{ij}E_j$. (7.10)
  where $E_i = -\phi_{,i}$ considering quasi-static electric field.
- The Gauss equation: $D_{i,i} = 0$. (7.11)
  In the above equations, $i, j, k, l = 1, 2, 3$.

7.4 PROBLEM FORMULATION

The impedance boundary of the undisturbed PFRC half-space lies along x-axis, and z-axis is vertically down. “O” represents the area of incident quasi-longitudinal (qP) (IqP) wave incidence (shown in Figure 7.2) with an inclination of θ along z-axis. Owing to the IqP wave at the impedance boundary, three reflected waves are generated, viz., reflected qP (RqP), reflected quasi-transverse (qSV) (RqSV), and reflected electroacoustic (REA) waves.

A plane strain condition is assumed, which gives us $\frac{\partial(.)}{\partial y} = 0$. Thus, all functions are dependent on the coordinates x and z and time. Let $\{u = u(x,z,t),\ v = 0,\ w = w(x,z,t)\}$ and $\phi = \phi(x,z,t)$ denote the mechanical displacement components and electric potential, respectively.

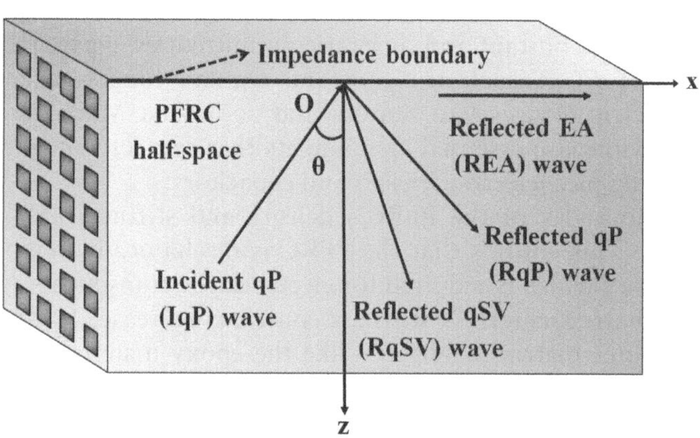

Figure 7.2 Geometry of the problem showing the incident and reflected waves.

Using Eqs. (7.7)–(7.11), we derive the following governing equations:

$$c_{11}u_{,11} + c_{44}u_{,33} + (c_{13} + c_{44})w_{,13} + (e_{31} + e_{15})\phi_{,13} + \rho g w_{,1} = \rho \ddot{u}, \tag{7.12}$$

$$(c_{13} + c_{44})u_{,13} + c_{44}w_{,11} + c_{33}w_{,33} + e_{15}\phi_{,11} + e_{33}\phi_{,33} - \rho g u_{,1} = \rho \ddot{w}, \tag{7.13}$$

$$(e_{31} + e_{15})u_{,13} + e_{15}w_{,11} + e_{33}w_{,33} - \varepsilon_{11}\phi_{,11} - \varepsilon_{33}\phi_{,33} = 0. \tag{7.14}$$

The natures of all waves can be understood when the above highly coupled system of PDEs is solved.

7.5 CONDITIONS AT THE IMPEDANCE BOUNDARY ($z = 0$)

Mechanical-type conditions:

$$\tau_{13} + \omega Z1u = 0, \tag{7.15}$$

$$\tau_{33} + \omega Z3w = 0. \tag{7.16}$$

Electrical-type condition:

$$D_3 = 0. \tag{7.17}$$

The conditions at the impedance boundary are structured in a way that encompasses the conditions of both extreme boundaries (stress-free/rigid surfaces). When $|Z_i| \to 0$, we get the case when the surface is stress-free, and when $|Z_i| \to \infty$, we get the case when the surface is rigid. As a result, the impedance boundary provides us an in-depth assessment of the changes that take place at each stage of the transition from stress-free to rigid surfaces.

7.6 SOLUTION

We assume solutions to Eqs. (12)–(14) assuming the generalized Snell's law as follows:

$$(u, w, \phi) = (1, B, C)A exp\left[ik(x + qz - ct)\right] \tag{7.18}$$

where $k \to$ wave number, $c \to$ phase velocity, Real (q) = cot θ.

Using Eq. (7.18) in Eqs. (7.12)–(7.14), we get

$$\begin{aligned} c_{11} + c_{44}q^2 - \rho c^2 + \left[(c_{13} + c_{44})q + \rho g / ik\right]B + (e_{31} + e_{15})qC = 0 \\ \left[(c_{13} + c_{44})q - \rho g / ik\right] + (c_{44} + c_{33}q^2 - \rho c^2)B + (e_{15} + e_{33}q^2)C = 0 \\ (e_{15} + e_{31})q + (e_{15} + e_{33}q^2)B - (\varepsilon_{11} + \varepsilon_{33}q^2)C = 0 \end{aligned} \tag{7.19}$$

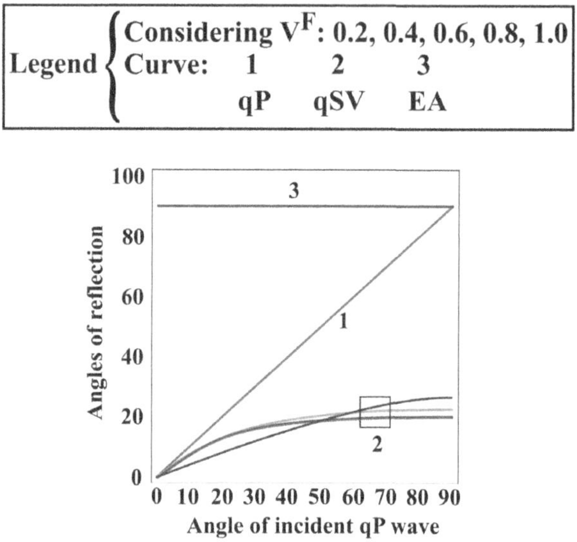

Figure 7.3 Angles of all reflected waves varying against incident angle considering different fiber volume fractions.

On solving the determinant of coefficients of 1, B, C, we get the expression of c of IqP wave, as well as the reflected waves' directions. The propagation angles of the RqP, RqSV, and REA waves are displayed in Figure 7.3.

We make the following observations from Figure 7.3 for all considered values of V^F:

- Angles of IqP and RqP wave are equal.
- The angles of RqSV are lesser compared to RqP wave as qSV wave is relatively slow and, thus, propagate closer to the z-axis.
- REA wave moves along the surface at a constant angle of $\theta = 90^\circ$, being an evanescent surface wave.

The mechanical displacement and electric potential expressions are written as follows:

$$
\begin{aligned}
u &= A_2 \exp\left[\Lambda_2\right] + \sum_{n=1,3,5} A_n \exp\left[\Lambda_n\right] \\
w &= B_2 A_2 \exp\left[\Lambda_2\right] + \sum_{n=1,3,5} B_n A_n \exp\left[\Lambda_n\right] \qquad (7.20) \\
\phi &= C_2 A_2 \exp\left[\Lambda_2\right] + \sum_{n=1,3,5} C_n A_n \exp\left[\Lambda_n\right].
\end{aligned}
$$

The mechanical stress and electric displacement expressions are written as follows:

$$\begin{aligned}
\tau_{33} &= J_{12}A_2\exp[\Lambda_2]+\sum_{n=1,3,5} J_{1n}A_n\exp[\Lambda_n]\\
\tau_{13} &= J_{22}A_2\exp[\Lambda_2]+\sum_{n=1,3,5} J_{2n}A_n\exp[\Lambda_n]\\
D_3 &= J_{32}A_2\exp[\Lambda_2]+\sum_{n=1,3,5} J_{3n}A_n\exp[\Lambda_n].
\end{aligned} \tag{7.21}$$

where $J_{1k} = ik(c_{13}+c_{33}q_kB_k+e_{33}q_kC_k)$, $J_{2k} = ik(c_{44}(q_k+B_k)+e_{15}C_k)$, and $J_{3k} = ik(e_{31}+e_{33}q_kB_k-\epsilon_{33}q_kC_k)$, $\Lambda_k = ik(x+q_kz-ct)$ (subscript $k=2, n$).

The first terms in Eqs. (7.20) and (7.21) refer to the IqP wave component, and the suffixes 1, 3, and 5 in the summation refer to the RqP, RqSV, and REA wave components.

Applying the boundary conditions Eqs. (7.15)–(7.17), Eqs. (7.20) and (7.21) result in

$$\begin{bmatrix} J_{21}-icZ1 & J_{23}-icZ1 & J_{25}-icZ1\\ J_{11}-icZ3B_1 & J_{13}-icZ3B_3 & J_{15}-icZ3B_5\\ J_{31} & J_{33} & J_{35}\end{bmatrix}\begin{bmatrix}A_1\\A_3\\A_5\end{bmatrix}=\begin{bmatrix}-J_{22}+icZ1\\-J_{12}+icZ3B_2\\-J_{32}\end{bmatrix}A_2. \tag{7.22}$$

With the aid of Cramer's rule, Eq. (7.22) gives the amplitude ratios, which demonstrate that they rely on incident angle, impedance boundary, gravity, and PFRC properties. Since amplitude ratios are not sufficient to shed some light on whether energy remains conserved or not in this system, we have derived the energy ratio expressions in the next section.

7.7 ENERGY RATIOS

The shared energy between IqP wave and RqP/RqSV/REA waves is considered at an infinitely thin surface element with unit area lying on $z=0$. In accordance with [38], the mean energy flux over the surface element is given as follows:

$$P = -Re\left[\tau_{13}\bar{\dot{u}}+\tau_{33}\bar{\dot{w}}-\overline{\dot{D}_3\phi}\right]/2. \tag{7.23}$$

Using Eqs. (7.20) and (7.21), Eq. (7.23) becomes

$$P_{\alpha\beta} = k^2c\left[J_{1\alpha}\overline{B_\beta}A_\alpha\overline{A_\beta}+J_{2\alpha}A_\alpha\overline{A_\beta}+\overline{J_{3\alpha}}C_\beta\overline{A_\alpha}A_\beta\right]/2. \tag{7.24}$$

The energy ratio expression is $E_{\alpha\beta} = P_{\alpha\beta}/P_{\zeta\zeta}$. (7.25)

The interaction energy expression is

$$E_{\text{int}} = \sum_{\alpha=1,3,5} \left(E_{\zeta\alpha} + E_{\alpha\zeta}\right) + \sum_{\alpha=1,3,5} \left(\sum_{\beta=1,3,5} E_{\alpha\beta} - E_{\alpha\alpha} \right), \tag{7.26}$$

and as per the Energy Conservation Law, we should obtain

$$\sum_{\alpha=1,3,5} E_{\alpha\alpha} + E_{\text{int}} = 1. \tag{7.27}$$

The ratios of energy of RqP wave |ER1|, RqSV wave |ER2|, interaction energy |Eint|, and overall energy of the system |E| are obtained and graphically demonstrated. The evanescent REA wave traverses along the surface and carries almost nil energy. Thus, we have plotted the amplitude ratio of reflected EA wave, denoted by |AR3|. Numerical simulations and graphical demonstrations reveal that the net energy is ≈1 for IqP wave considering all variations of the chosen parameters.

7.8 RESULTS AND DISCUSSION

The considered PFRC comprises PZT-5A fiber [39] and epoxy matrix [17]. The data are

PZT-5A data: $c_{11} = c_{22} \left(Nm^{-2}\right) = 139 \times 10^9$, $c_{12} \left(Nm^{-2}\right) = 77.8 \times 10^9$, $c_{13} = c_{23} \left(Nm^{-2}\right) = 75.4 \times 10^9$, $c_{33} \left(Nm^{-2}\right) = 113 \times 10^9$, $c_{44} \left(Nm^{-2}\right) = 25.6 \times 10^9$, $e_{15} = e_{24} \left(Cm^{-2}\right) = 13.4$, $e_{31} = e_{32} \left(Cm^{-2}\right) = -6.98$, $e_{33} \left(Cm^{-2}\right) = 13.8$, $\rho \left(Kg\ m^{-3}\right) = 7750$, $\epsilon_{11} = \epsilon_{22} \left(C^2N^{-1}m^{-2}\right) = 60 \times 10^{-10}$, $\epsilon_{33} \left(C^2N^{-1}m^{-2}\right) = 54.7 \times 10^{-10}$.

Epoxy data: $c_{11} = c_{22} \left(Nm^{-2}\right) = 3.86 \times 10^9$, $c_{12} \left(Nm^{-2}\right) = 2.57 \times 10^9$, $c_{13} = c_{23} \left(Nm^{-2}\right) = 2.57 \times 10^9$, $c_{33} \left(Nm^{-2}\right) = 3.86 \times 10^9$, $c_{44} \left(Nm^{-2}\right) = 2.57 \times 10^9$, $e_{15} = e_{24}\left(Cm^{-2}\right) = 0$, $e_{31} = e_{32}\left(Cm^{-2}\right) = 0$, $e_{33} \left(Cm^{-2}\right) = 0$, $\epsilon_{11} = \epsilon_{22} \left(C^2N^{-1}m^{-2}\right) = 7.9 \times 10^{-11}$, $\epsilon_{33} \left(C^2N^{-1}m^{-2}\right) = 7.9 \times 10^{-11}$, $\rho \left(Kg\ m^{-3}\right) = 1210$.

We chose particular values of fiber volume fraction as $V^F = 0.2, 0.4, 0.6, 0.8$, and 1; particular values of impedance boundary parameters as $Z1 = Z3 = -10, -5, 0, 5$, and 10; $\omega = 2\ s^{-1}$; and gravitational acceleration parameter as $g = 9.8\ ms^{-2}$.

Here, graphical representations (using Mathematica 7.0, MS Excel, and Inkscape) of performed simulations are provided for IqP wave.

7.8.1 Energy ratios showing impacts of V^F

In Figure 7.4, the impacts of varying magnitudes of V^F on energy ratios of waves considering fixed impedance boundary parameters ($Z1 = 10$, $Z3 = -5$) and gravitational effect are illustrated.

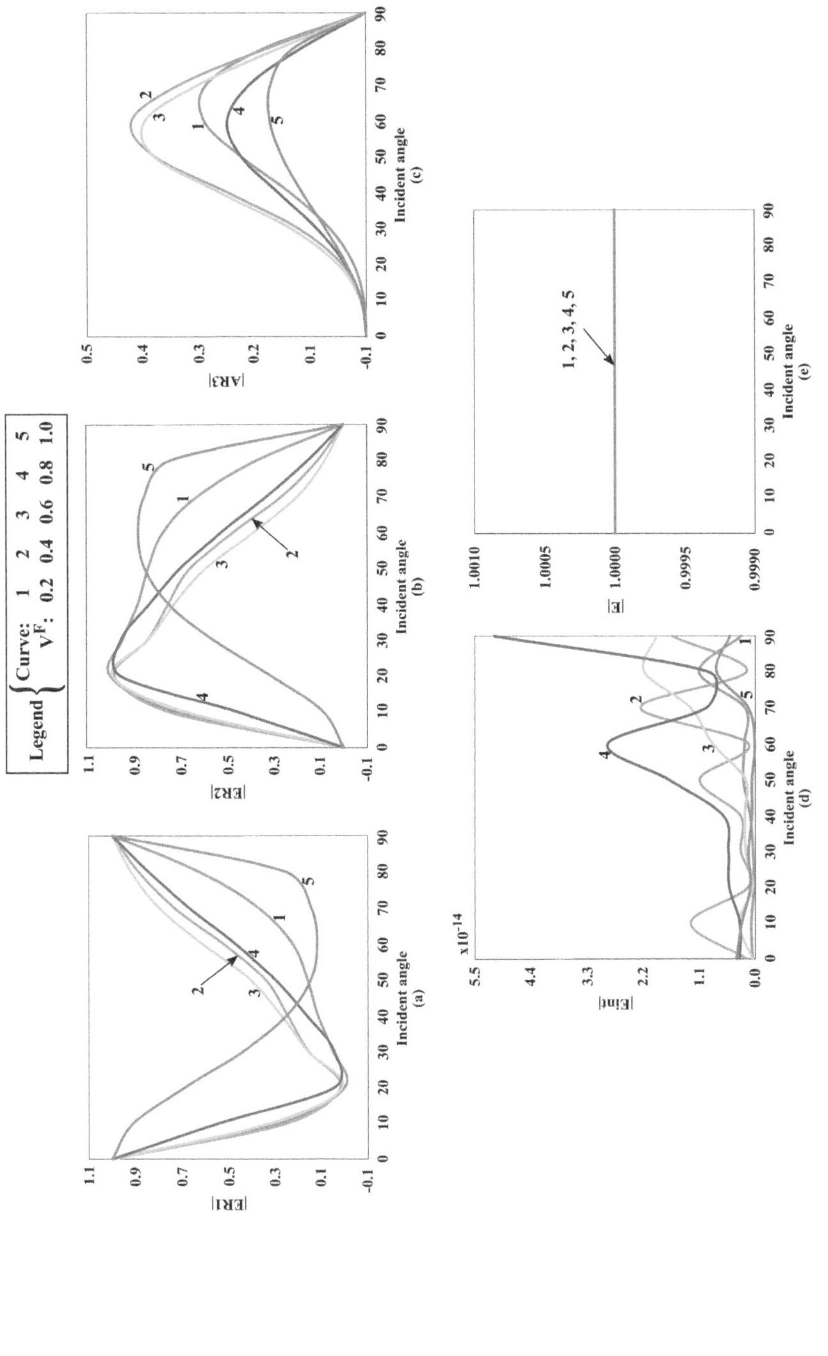

Figure 7.4 Effect of fiber volume fraction on the reflected waves.

When $\theta = 0^\circ$, IqP wave generates RqP wave which propagates in the z-direction with maximum energy, for which we observe |ER1|=1 $\forall V^F$ from Figure 7.4(a). This case is called normal incidence. As θ keeps increasing, RqP wave keeps losing energy at a drastic rate in a low range of θ. Thereafter, for other values of θ, |ER1| increases steadily and finally; when $\theta = 90^\circ$, IqP wave generates RqP wave which propagates in the x-direction with maximum energy, for which we observe |ER1|=1. This case is called grazing incidence. For low to mid-range values of θ, |ER1| generally decreases with V^F. However, for moderate to high values of θ, |ER1| increases with V^F, up to certain values, and then starts diminishing. Since RqSV wave gains the energy that RqP wave loses, the exact opposite natures of Figure 7.4(a) are seen in Figure 7.4(b). Figure 7.4(c) shows that |AR3| increases with V^F (for low values) and then starts diminishing as V^F increases. The order of magnitude of |Eint|, as per Figure 7.4(d), is also quite negligible, and thus, it has no overall impact on the system's energy. |E|≈1, showing that energy remains conserved.

7.8.2 Energy ratios showing impacts of g

In Figure 7.5, the effects of gravitational parameter g on energy ratios of waves considering fixed impedance boundary parameters ($Z1 = 10$, $Z3 = -5$) and fiber volume fraction $\left(V^F = 0.8\right)$ are illustrated.

Figure 7.5(a) shows that the magnitude of |ER1| without gravity is greater than the case when g is considered in the analysis $\forall \theta \in [0°, 90°]$. The opposite happens for |ER2| and |AR3|, i.e. $\forall \theta \in [0°, 90°]$; their magnitudes with gravity under consideration are greater than when g is disregarded. Figure 7.5(d) shows that |Eint| does not contribute much to the system's overall energy because it has order of magnitude of −14. Moreover, for most values of θ, |Eint| with gravity is greater than the case when g is ignored in the analysis. Last, Figure 7.5(e) shows that |E|≈1, which establishes energy conservation.

7.8.3 Energy ratios showing impacts of Z1 and Z3

In Figure 7.6, the effects of impedance boundary parameters (Z1, Z3) on energy ratios of waves considering fixed fiber volume fraction $\left(V^F = 0.6\right)$ and gravitational effect are illustrated. When the effects of Z1 are studied, we consider a fixed value of Z3 = 10, and when the influences of Z3 are studied, we consider a fixed value of Z1 = 10.

Figure 7.6(a) shows that |ER1| rises with rising values of Z1. However, |ER1| diminishes for increasing values of Z3. Since the energy lost by RqP wave is carried forward by RqSV wave, the exact opposite influences of Z1 and Z3 on |ER2| are displayed in Figure 7.6(b). It is inferred from Figure 7.6(c) that |AR3| diminishes with increasing values of Z1, whereas it increases for increasing values of Z3. Figure 7.6(d) shows that the natures of |Eint| are somewhat erratic for the chosen values of V^F. However, since the overall order of magnitude is −14, its overall influence on the system's net

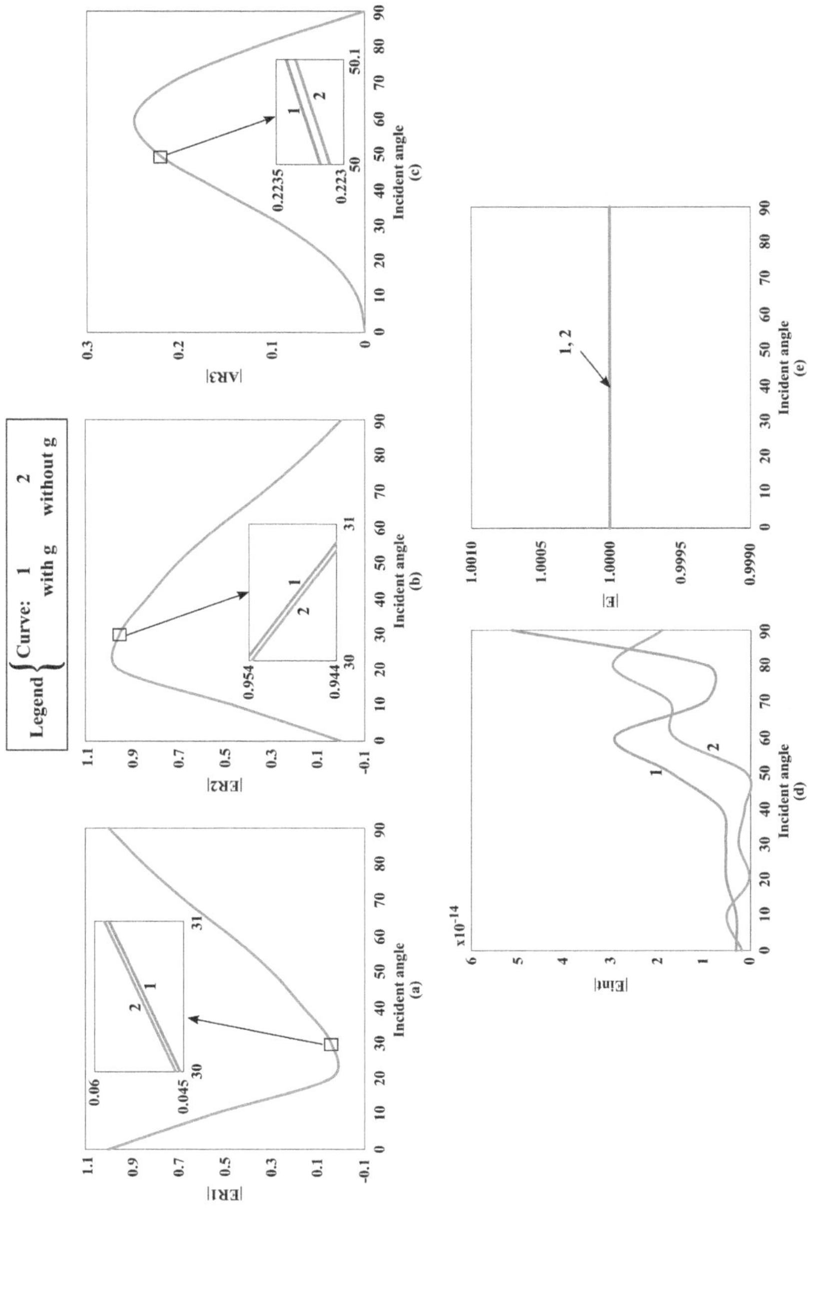

Figure 7.5 Effect of gravitation on the reflected waves.

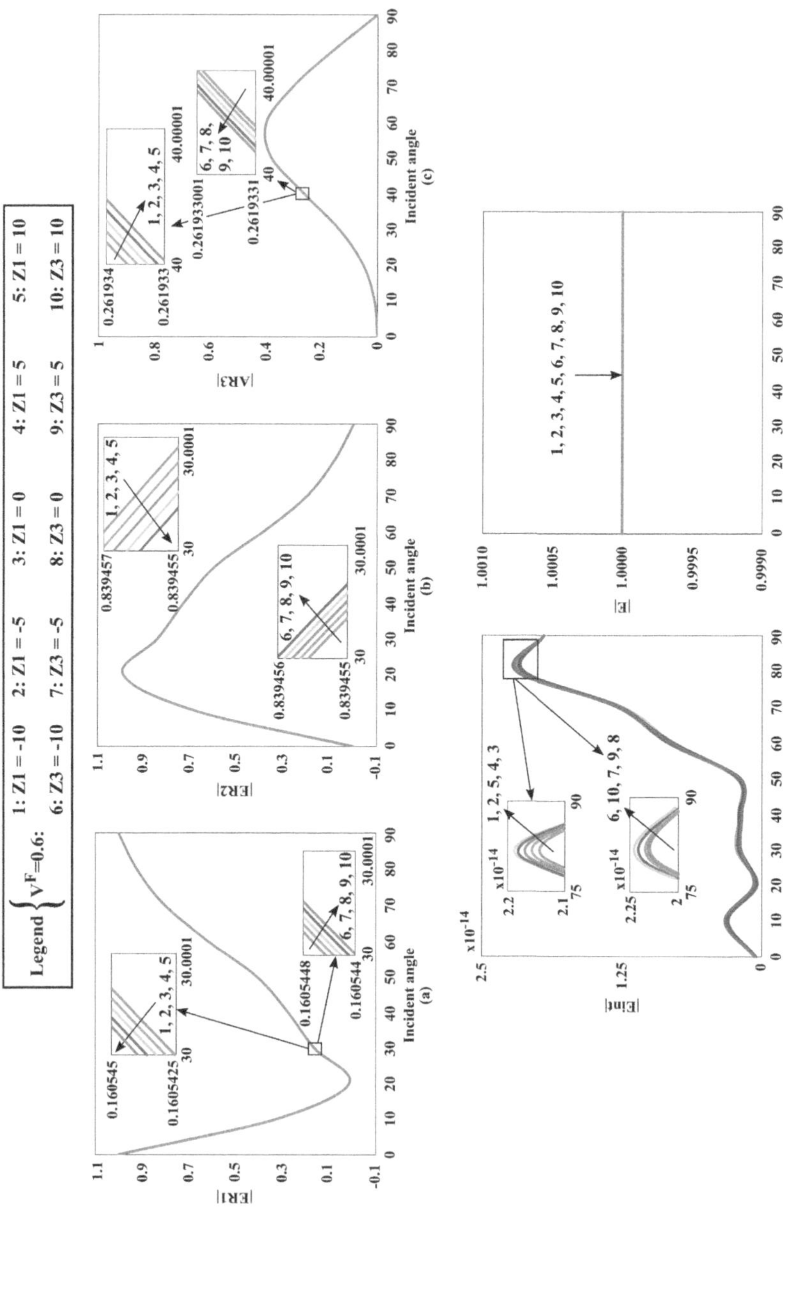

Figure 7.6 Effect of impedance boundary on the reflected waves.

energy is negligible. Finally, Figure 7.6(e) reveals |E| to be ≈1, which establishes energy conservation.

7.9 CONCLUSIONS

- After deriving the material data for all values of $V^F \in [0,1]$ using SM and RM, it is found that for high values of V^F, the chosen PFRC obtains desirable electro-mechanical properties compared to PZT-5A. For a more robust analysis, we have chosen different values of V^F in this work to observe their particular influences.
- For all considered values of V^F, the angles of IqP and RqP waves are equal. The REA wave traverses along the x-axis constantly with $\theta = 90°$. Since the qSV wave is slower compared to qP wave, it propagates closer to the z-axis, for which the propagation angles of RqSV wave are lesser.
- When $\theta = 0°$ for IqP wave, the RqP wave carries maximum energy and others are meager. When $\theta = 90°$ for IqP wave, the RqP wave carries maximum energy, while others are meager. These are special cases also referred to as normal incidence and grazing incidence, respectively.
- For low- to mid-range values of θ, |ER1| mostly diminishes with V^F, and when θ takes moderate to high values, |ER1| increases for certain values V^F, followed by a diminishing pattern. The exact opposite characteristic is exhibited by |ER2|. |AR3| rises with increasing (but low) V^F and then diminishes for increasing V^F.
- |ER1| without gravity is comparatively greater than when g is assumed $\forall \theta \in [0°, 90°]$. The opposite nature is displayed for |ER2| and |AR3|.
- |ER1| rises with Z1. However, |ER1| diminishes as Z3 increases. The exact opposite influences of Z1 and Z3 on |ER2| and |AR3| are observed.
- For all variations of the chosen parameters, |Eint| does not contribute significantly to the system's net energy because it has order of magnitude of −14.
- The magnitude of the overall energy is 1 for all variations of the chosen parameters. This ensures no energy loss, and thereby, fulfillment of the Energy Conservation Law.

ACKNOWLEDGMENTS

The authors are thankful to their respective universities for providing support and facilities for the research work.

REFERENCES

1. Yadav, A. K. (2021). Reflection of plane waves in a fraction-order generalized magneto-thermoelasticity in a rotating triclinic solid half-space. *Mechanics of Advanced Materials and Structures*, 29(25), 4273–4290. https://doi.org/10.1080/15376494.2021.1926017

2. Singh, P., Singh, A. K., Chattopadhyay, A., & Guha, S. (2020). Mathematical study on the reflection and refraction phenomena of three-dimensional plane waves in a structure with floating frozen layer. *Applied Mathematics and Computation*, *386*, 125488. https://doi.org/10.1016/j.amc.2020.125488
3. Nain, S., & Guha, S. (2023). Influence of fiber orientation on reflection and attenuation phenomenon in fiber-reinforced viscoelastic medium. *Archive of Applied Mechanics*, *93*(7), 2993–3005. https://doi.org/10.1007/s00419-023-02422-3
4. Curie, J., & Curie, P. (1880). Développement par compression de l'électricité polaire dans les cristaux hémièdres à faces inclinées. *Bulletin de La Société Minéralogique de France*, *3*(4), 90–93. https://doi.org/10.3406/bulmi.1880.1564
5. Guha, S., & Singh, A. K. (2022). Influence of varying fiber volume fractions on plane waves reflecting from the stress-free/rigid surface of a piezoelectric fiber-reinforced composite half-space. *Mechanics of Advanced Materials and Structures*, *29*(27), 5758–5772. https://doi.org/10.1080/15376494.2021.1964046
6. Guha, S., & Singh, A. K. (2021). Plane wave reflection/transmission in imperfectly bonded initially stressed rotating piezothermoelastic fiber-reinforced composite half-spaces. *European Journal of Mechanics – A/Solids*, *88*, 104242. https://doi.org/10.1016/j.euromechsol.2021.104242
7. Singh, S., Singh, A. K., & Guha, S. (2021-a). Impact of interfacial imperfections on the reflection and transmission phenomenon of plane waves in a porous-piezoelectric model. *Applied Mathematical Modelling*, *100*, 656–675. https://doi.org/10.1016/j.apm.2021.08.022
8. Guha, S., & Singh, A. K. (2020). Effects of initial stresses on reflection phenomenon of plane waves at the free surface of a rotating piezothermoelastic fiber-reinforced composite half-space. *International Journal of Mechanical Sciences*, *181*, 105766. https://doi.org/10.1016/j.ijmecsci.2020.105766
9. Singh, A. K., Rajput, P., Guha, S., & Singh, S. (2022). Propagation characteristics of Love-type wave at the electro-mechanical imperfect interface of a piezoelectric fiber-reinforced composite layer overlying a piezoelectric half-space. *European Journal of Mechanics-A/Solids*, *93*, 104527. https://doi.org/10.1016/j.euromechsol.2022.104527
10. Guha, S., & Singh, A. K. (2022). Transference of SH waves in a piezoelectric fiber-reinforced composite layered structure employing perfectly matched layer and infinite element techniques coupled with finite elements. *Finite Elements in Analysis and Design*, *209*, 103814. https://doi.org/10.1016/j.finel.2022.103814
11. Biswas, M., & Guha, S. (2025). Influence of dynamic fiber volume fraction on Love wave velocity in PFRC plate imperfectly bonded with piezoelectric-viscoelastic substrate. *Acta Mechanica*, *236*(1), 321–341. https://doi.org/10.1007/s00707-024-04147-y
12 Singh, S., & Guha, S. (2023). Modeling of SH waves in a functionally graded piezo-poroelastic structure with sensitive coating in presence of point source of disturbance. *Corrosion Mitigation Coatings: Functionalized Thin Film Fundamentals and Applications*, 369. https://doi.org/10.1515/9783111016160-015
13. Dholey, B., Alneamy, A., Mistri, K., Guha, S., & Tharwan, M. (2025). The influence of SH-wave propagation in a tri-layered composite structure with interfacial imperfections. *Journal of Vibration Engineering & Technologies*, *13*(2), 197. https://doi.org/10.1007/s42417-025-01776-y
14. Dev, A., & Guha, S. (Eds.). (2025). *Applied engineering mathematics: Fluid and solid mechanics in life science* (Vol. 11). Walter de Gruyter GmbH & Co KG. https://doi.org/10.1515/9783111638782
15. Singh, A. K., Mahto, S., & Guha, S. (2024). Analysis of plane wave reflection phenomenon from the surface of a micro-mechanically modeled piezomagnetic fiber-reinforced composite half-space. *Waves in Random and Complex Media*, *34*(6), 5382–5403. https://doi.org/10.1080/17455030.2021.2006822

16. Singh, A. K., Mahto, S., & Guha, S. (2022). Analysis of plane wave reflection and transmission phenomenon at the interface of two distinct micro-mechanically modeled rotating initially stressed piezomagnetic fiber-reinforced half-spaces. *Mechanics of Advanced Materials and Structures*, *29*(28), 7623–7639. https://doi.org/10.1080/15376494.2021.2003490
17. Kumar, A., & Chakraborty, D. (2008). Effective properties of thermo-electro-mechanically coupled piezoelectric fiber reinforced composites. *Materials & Design*, *30*(4), 1216–1222. https://doi.org/10.1016/j.matdes.2008.06.009
18. Guha, S., & Singh, A. K. (2022). Frequency shifts and thermoelastic damping in distinct micro-/nano-scale piezothermoelastic fiber-reinforced composite beams under three heat conduction models. *Journal of Ocean Engineering and Science*. https://doi.org/10.1016/j.joes.2022.06.015
19. Guha, S., Singh, A. K., & Singh, S. (2023). Thermoelastic damping and frequency shift of different micro-scale piezoelectro-magneto-thermoelastic beams. *Physica Scripta*, *99*(1), 015203. https://iopscience.iop.org/article/10.1088/1402-4896/ad0bbd/meta
20. Mallik, N., & Ray, M. C. (2001). Effective coefficients of piezoelectric fiber-reinforced composites. *AIAA Journal*, *41*(4), 704–710. https://doi.org/10.2514/2.2001
21. Singh, A. K., Guha, S., & Maji, A. (2025). Analysis of thermoelastic damping and frequency shift of nano-scale piezoelectric fiber-reinforced thermoelastic composite beam under single, dual, and three phase-lag models: A comparative approach. *Thin-Walled Structures*, *211*, 113134. https://doi.org/10.1016/j.tws.2025.113134
22. Alneamy, A. M., Guha, S., & Tharwan, M. Y. (2024). Modeling and analysis of thermoelastic damping in a piezoelectro-magneto-thermoelastic imperfect flexible beam. *Mathematics*, *12*(24). https://doi.org/10.3390/math12244011
23. Singh, S., Singh, A. K., & Guha, S. (2022). Shear waves in a piezo-fiber-reinforced-poroelastic composite structure with sandwiched functionally graded buffer layer: Power series approach. *European Journal of Mechanics-A/Solids*, *92*, 104470. https://doi.org/10.1016/j.euromechsol.2021.104470
24. Guha, S., & Alneamy, A. (2025). The influence of flexible support on the material properties of piezothermoelastic fiber-reinforced composite beam. *Journal of Vibration Engineering & Technologies*, *13*(2), 191. https://doi.org/10.1007/s42417-025-01770-4
25. Tiersten, H. F. (2013). *Linear piezoelectric plate vibrations: Elements of the linear theory of piezoelectricity and the vibrations piezoelectric plates.* Springer. https://doi.org/10.1007/978-1-4899-6453-3
26. Yadav, A. K., Carrera, E., Schnack, E., & Marin, M. (2023). Effects of memory response and impedance barrier on reflection of plane waves in a nonlocal micropolar porous thermo-diffusive medium. *Mechanics of Advanced Materials and Structures*, 1–17. https://doi.org/10.1080/15376494.2023.2217556
27. Jaensch, S., Sovardi, C., & Polifke, W. (2016). On the robust, flexible and consistent implementation of time domain impedance boundary conditions for compressible flow simulations. *Journal of Computational Physics*, *314*, 145–159. https://doi.org/10.1016/j.jcp.2016.03.010
28. Godoy, E., Durán, M., & Nédélec, J.-C. (2012). On the existence of surface waves in an elastic half-space with impedance boundary conditions. *Wave Motion*, *49*(6), 585–594. https://doi.org/10.1016/j.wavemoti.2012.03.005
29. Guha, S., Gupta, V., Biswas, M., Mahanty, M., Sarangi, B. P., Haty, A., & Nain, S. (2025). Acoustic wave behavior in rotating functionally graded piezomagnetic media with impedance boundaries. *European Journal of Mechanics-A/Solids*, 105786. https://doi.org/10.1016/j.euromechsol.2025.105786

30. Hoppe, D. J. (2018). *Impedance boundary conditions in electromagnetics*. CRC Press.
31. Guha, S., & Singh, A. K. (2023). On-plane waves reflecting at the impedance boundary of an initially stressed micromechanically modeled piezomagnetic fiber-reinforced composite half-space. *Mechanics of Advanced Materials and Structures*, 1–18. https://doi.org/10.1080/15376494.2023.2251194
32. Othman, M. I. A., & Ahmed, E. A. A. (2016). Effect of gravity field on piezo-thermoelastic medium with three theories. *Journal of Thermal Stresses*, *39*(4), 474–486. https://doi.org/10.1080/01495739.2016.1152136
33. Abo-Dahab, S. M., Jahangir, A., & Abd-alla, A. N. (2018). Reflection of plane waves in thermoelastic microstructured materials under the influence of gravitation. *Continuum Mechanics and Thermodynamics*, *32*(3), 803–815. https://doi.org/10.1007/s00161-018-0739-2
34. Guha, S., Gupta, V., Biswas, M., Singh, A. K., & Barak, M. S. (2025). Complex wave interactions in graded piezomagnetic materials: A gravity and flexomagnetic perspective. *International Journal of Numerical Methods for Heat & Fluid Flow*, *35*(4), 1459–1483. https://doi.org/10.1108/HFF-10-2024-0804
35. Benveniste, Y., & Dvorak, G. J. (1992). Uniform fields and universal relations in piezoelectric composites. *Journal of the Mechanics and Physics of Solids*, *40*(6), 1295–1312. https://doi.org/10.1016/0022-5096(92)90016-u
36. Nan, C.-W., & Jin, F.-S. (1993). Multiple-scattering approach to effective properties of piezoelectric composites. *Physical Review B*, *48*(12), 8578–8582. https://doi.org/10.1103/physrevb.48.8578
37. Biot, M. A. (1965). *Mechanics of incremental deformations: Theory of elasticity and viscoelasticity of initially stressed solids and fluids, including thermodynamic foundations and applications to finite strain*. John Wiley & Sons.
38. Guo, X., & Wei, P. (2014). Effects of initial stress on the reflection and transmission waves at the interface between two piezoelectric half spaces. *International Journal of Solids and Structures*, *51*(21–22), 3735–3751. https://doi.org/10.1016/j.ijsolstr.2014.07.008
39. Alshaikh, F. A. (2012). The mathematical modelling for studying the influence of the initial stresses and relaxation times on reflection and refraction waves in piezothermoelastic half-space. *Applied Mathematics*, *03*(08), 819–832. https://doi.org/10.4236/am.2012.38123

Chapter 8

Crack propagation and interaction in piezoelectric materials

A mathematical approach and numerical simulation

Sarita Pippal and Pooja Raj Verma

8.1 INTRODUCTION

Piezoelectric materials generate an electric charge under mechanical stress, and they can be classified as isotropic or anisotropic based on their directional properties. Isotropic piezoelectric materials exhibit uniform properties in all directions, which is rare since piezoelectricity is inherently direction-dependent; such materials are usually engineered, like certain polymers. In contrast, anisotropic piezoelectric materials have properties that vary with direction, making them more common in nature and applications. Examples include quartz, barium titanate ($BaTiO_3$), and lead zirconate titanate, which exhibit strong directional piezoelectric responses. Isotropic materials are useful in applications requiring uniform stress distribution, while anisotropic materials are widely used in sensors, actuators, and transducers due to their enhanced piezoelectric efficiency. Due to intrinsic coupling effects that take place between electric fields and mechanical deformation, Piezoelectric materials are highly useful in applications like sensors, actuators, energy harvesting devices, medical implants, and aircraft components because they can transform mechanical stress into electrical charge and vice versa. Piezoelectric materials are vital to modern engineering applications and intelligent structural systems due to their unique electromechanical coupling properties. Because they can convert mechanical energy into electrical energy and vice versa, these materials are crucial to many cutting-edge technologies, such as sensors, actuators, energy harvesting devices, and adaptable buildings. From medical equipment to aeronautical engineering, their ability to generate an electrical charge in reaction to mechanical inputs has led to significant advancements in a variety of scientific and industrial fields. Their structural integrity and electrical performance, however, may be significantly impacted by interface fractures [1–2]. These defects, which can result from manufacturing errors, environmental stressors, or operating wear and tear, jeopardize the reliability and durability of piezoelectric components. Interface cracks can alter the stress distribution, impede electrical responsiveness, and result in catastrophic failure. Understanding and fixing these cracks is crucial to expanding the applications of piezoelectric materials [3–7].

DOI: 10.1201/9781003470885-8

The theoretical and analytical methods for investigating interface cracks in piezoelectric materials are examined in this chapter. It discusses the basic mathematical formulas that control how they behave under different loading scenarios. This chapter examines models for describing and forecasting crack propagation, such as stress intensity factors, energy release rates, and crack-tip fields, using the concepts of fracture mechanics. The interplay between the electrical and mechanical fields at the fracture contact is given particular attention.

This chapter also discusses current developments in computational approaches that improve crack prediction and mitigation tactics, including phase-field modeling, boundary element methods (BEM), and finite element analysis (FEA). Real-time monitoring and predictive maintenance of piezoelectric systems have been further enhanced by the incorporation of artificial intelligence and machine learning into computational fracture mechanics. In order to present a comprehensive viewpoint, the chapter examines important research papers on cutting-edge experimental methods, creative material designs, and new developments that enhance the robustness and effectiveness of piezoelectric components. This chapter lays the groundwork for more robust smart structures by providing researchers, engineers, and industry experts with a thorough understanding of interface crack processes in piezoelectric materials through the integration of theoretical, analytical, and computational insights. This chapter also discusses potential future directions and challenges in ensuring the long-term reliability of piezoelectric components, emphasizing the importance of interdisciplinary research in this evolving field.

8.2 FUNDAMENTALS OF PIEZOELECTRICITY

In crystalline materials, the linear electromechanical connection between the mechanical and electrical states is referred to as the piezoelectric effect. This section will go over its basic formulas and guiding ideas.

8.2.1 Linear piezoelectric materials

Linear piezoelectric materials obey a direct proportionality between the mechanical strain and the electric field, which can be regulated by linear constitutive equations can be written as:

8.2.1.1 Deformation of a piezoelectric medium

$$S_{ij} = s_{ijkl}\,\sigma_{kl} + d_{kij}\,E_k \tag{8.1}$$

The above expression represents the deformation of a piezoelectric medium under mechanical stress and an electrical field, which describes how strain in a piezoelectric material results from both mechanical stress and an applied electric field.

The first term $s_{ijkl}\, \sigma_{kl}$ in Equation (8.1) represents standard elastic deformation due to stress, following Hooke's Law. The second term $d_{kij}\, E_k$ accounts for the piezoelectric effect, where an electric field can induce strain. This coupling between mechanical and electrical properties is fundamental to piezoelectric applications like sensors, actuators, and energy harvesters.

where:

- S_{ij}: Strain tensor components, representing the deformation in the material. Deformation refers to the change in shape or size of a material under an applied force. The indices i and j denote the directions of strain within the material, corresponding to the Cartesian coordinate system.
- s_{ijkl}: Elastic compliance tensor components, describing the material's mechanical response to stress. The indices i, j, k, l indicate that this is a fourth-order tensor, providing a generalized relation between stress and strain in three-dimensional space.
- d_{kij}: Piezoelectric coefficient tensor, coupling mechanical strain to the electric field. The index k represents the electric field direction, while i and j denote the strain components influenced by this field.
- E_k: Electric field components, representing the applied or induced electric field, with the index k indicating its direction.
- σ_{kl}: Stress tensor components, representing internal forces per unit area within the material. The indices k and l specify the directions of the applied stress.

8.2.1.2 Electric displacement in a piezoelectric medium

The electric displacement in a piezoelectric material is given by:

$$D_i = d_{ikl}\, \sigma_{kl} + e_{ik}\, E_k \qquad (8.2)$$

- D_i: Electric displacement vector, indicating the electric flux density in the material, with the index i showing the direction of displacement.
- e_{ik}: Permittivity tensor components, describing the dielectric properties of the material. The indices i and k define how the electric field affects electric displacement.

Equation (8.2) explains how an applied electric field and mechanical stress combine to produce the electric displacement in a piezoelectric material. The piezoelectric effect, in which mechanical stress produces an electric reaction, is represented by the first term, $d_{ikl}\, \sigma_{kl}$. The second term, $e_{ik}\, E_k$, represents the material's typical dielectric response, in which polarization is induced by an applied electric field. In devices where mechanical and electrical characteristics interact, such as piezoelectric sensors, capacitors, and transducers, this relationship is crucial.

8.2.2 Nonlinear piezoelectric materials

Nonlinear piezoelectric materials, however, exhibit more complex behavior where higher-order terms and material dependencies come into play. These materials require nonlinear constitutive equations:

$$S_{ij} = s_{ijkl}\,\sigma_{kl} + d_{kij}\,E_k + \alpha_{ijkl}\,\sigma^2_{kl} + \beta_{ijk}\,E^2_k \tag{8.3}$$

$$D_i = d_{ikl}\,\sigma_{kl} + e_{ik}\,E_k + \gamma_{ikl}\,\sigma^2_{kl}\,E_k \tag{8.4}$$

where

- α_{ijkl} represents the second-order nonlinear elastic effects in the material, describing how stress quadratically influences strain.
- β_{ijk} describes the nonlinear coupling between the electric field and strain, capturing the quadratic dependence of strain on the electric field.
- γ_{ikl} represents the electromechanical coupling term in the nonlinear regime, indicating how stress and electric field interact beyond the linear approximation.

The indices i, j, k, l represent coordinate directions in a Cartesian system and are used in tensor notation to generalize the equations across three dimensions.

8.2.3 Derivation of constitutive equations

The linear constitutive equations are derived from fundamental thermodynamic principles, particularly by considering:

- The strain–stress relationship:
  $S_{ij} = s_{ijkl}\,\sigma_{kl}$ (Hooke's law for linear elasticity)
- The electric displacement relation:

  $$D_i = e_{ik}\,E_k$$

- The coupling between mechanical and electrical fields through the piezoelectric coefficients d_{kij}.

In nonlinear cases, higher-order terms are introduced to account for material responses beyond the linear regime, leading to additional stress–strain and electric field interactions.

8.2.4 Strain and electric field computation

In piezoelectric materials, the mechanical strain and the electric field are fundamental quantities that determine their behavior under external forces and electric potentials.

8.2.4.1 Strain tensor computation

The strain tensor S_{ij} is given by:

$$s_{ij} = \frac{1}{2}\left(u_{i,j} + u_{j,i}\right) \tag{8.5}$$

where:

- u_i represents the displacement in the ith direction.
- The notation $u_{i,j}$ denotes the partial derivative of displacement u_i with respect to coordinate x_j

The equation ensures that the strain tensor accounts for both normal and shear deformations in the material. The factor $\frac{1}{2}$ symmetrizes the strain tensor, ensuring correct representation of relative displacements.

8.2.4.2 Electric field computation

The electric field components E_i are related to the electric potential φ by:

$$E_i = -\varphi_{,i}\,. \tag{8.6}$$

The negative sign follows from the definition of the electric field as the gradient of the potential, ensuring that electric field lines point in the direction of decreasing potential.

8.2.5 Governing equations of piezoelectric materials

The fundamental governing equations for piezoelectric materials are expressed as:

$$\frac{\partial S_{ij}}{\partial x_i} + f_i = 0. \tag{8.7}$$

$$\frac{\partial D_i}{\partial x_i} - q_i = 0. \tag{8.8}$$

where:

- f_i represents the body force acting within the material
- q_i represents the charge density within the material

The balance of mechanical forces in the piezoelectric material is shown in Equation (8.7). It asserts that in order to maintain equilibrium, the external body force f_i and the divergence of the stress tensor S_{ij} must both be zero. The equilibrium of the external and internal forces acting on the material is guaranteed by this equation. The material's charge conservation is described by Equation (8.8). It suggests that the charge density q_i and the divergence of the electric displacement vector D_i must match. This formula guarantees that Gauss's law for electricity is followed by the electric field and charge distribution inside the piezoelectric material. These formulas collectively provide the basic framework for studying piezoelectric materials, which combine electric fields and mechanical stress.

Note: Body forces being neglected and free charges being absent, equations of static equilibrium and Maxwellian equations are of the form:

$$\frac{\partial S_{ij}}{\partial x_i} = 0. \tag{8.9}$$

$$\frac{\partial D_i}{\partial x_i} = 0. \tag{8.10}$$

8.2.6 Governing equations with substitutions

Substituting the constitutive equations into the governing equations for a linear piezoelectric medium under mechanical stress:

$$\frac{\partial}{\partial x_i}\left(s_{ijkl}\,\sigma_{kl} + d_{kij}\,E_k\right) + f_i = 0, \tag{8.11}$$

$$\frac{\partial}{\partial x_i}\left(d_{ikl}\,\sigma_{kl} + e_{ik}\,E_k\right) - q_i = 0. \tag{8.12}$$

8.2.7 Boundary conditions for piezoelectric medium

Certain boundary conditions that must be met at the material interfaces or surfaces control the behavior of piezoelectric materials.

8.2.7.1 Mechanical boundary conditions

The behavior of stress or displacement at the material borders is determined by the mechanical boundary conditions

$$S_{ij}\,n_j = T_j \tag{8.13}$$

where:

- n_j represents the outward unit normal vector to the surface.
- T_j represents the externally applied traction force.

8.2.7.2 *Displacement boundary condition*

$$u_i = u_i^* \tag{8.14}$$

where u_i^* is the specified displacement at the boundary.

8.2.7.3 *Electrical boundary conditions*

The electrical boundary conditions describe how the electric displacement and potential behave at material interfaces:

8.2.7.3.1 *Electric displacement continuity*

At an interface between two different materials, the normal component of the electric displacement vector must be continuous

$$D_i^1 n_i = D_i^2 n_i + S_f \tag{8.15}$$

where:

- D_i^1 and D_i^2 are the electric displacement vectors in the two adjacent materials.
- S_f is the free surface charge density at the interface.

8.2.7.3.2 *Electric potential boundary condition*

If an electric potential is prescribed on a boundary, it must satisfy

$$\varphi_i = \varphi_i^* \tag{8.16}$$

where φ_i^* is the given electric potential at the boundary.

8.2.7.3.3 *Perfect electric conductor (PEC) condition*

On a perfectly conducting surface, the tangential component of the electric field must vanish, i.e.,

$$E_t = 0. \tag{8.17}$$

These boundary conditions ensure that both mechanical and electrical behaviors are properly constrained in piezoelectric materials.

8.3 CRACKS IN PIEZOELECTRIC MATERIALS

Cracks in piezoelectric materials can seriously reduce their performance, leading to altered electromechanical response, reduced efficiency, and structural failure. In contrast to conventional materials, where crack behavior is primarily controlled by mechanical stress and strain, piezoelectric materials exhibit significant complexity due to the interaction between mechanical and electrical fields. This relationship makes the study of fracture propagation and interaction essential, requiring the application of particular mathematical models and analytical techniques to assess the effects of different loading scenarios. Fracture frequently takes the form of debonding and cracks at interfaces, which impairs mechanical and electrical performance.

8.3.1 Type of cracks in piezoelectric materials

Cracks can be classified into different types based on their location and propagation characteristics.

8.3.1.1 Interface cracks

Interface cracks occur at the junction between different material layers, leading to stress concentration and electrical discontinuity. These cracks often result from thermal mismatches, mechanical stresses, or manufacturing defects. Due to the practical and fundamental importance of this problem, interface cracks in piezoelectric materials have been actively studied over the last few decades [1, 2]. Figure 8.1 represents an interface crack in dissimilar piezoelectric materials, illustrating its impact on mechanical and electrical properties.

8.3.1.2 Surface cracks

Surface cracks appear on the outer surface of the piezoelectric material and can propagate inward. They are primarily caused by external loading, environmental degradation, or fatigue.

An **interface crack** occurs at the interface between two dissimilar piezoelectric materials due to weak bonding, stress mismatch, or thermal expansion differences, leading to potential delamination or separation of layers.

In contrast, a **surface crack** forms on the upper surface of a single material (see Figure 8.2) or the topmost layer of a composite structure due to external mechanical or electrical stress, initiating material fatigue and propagating inward. While an interface crack weakens the bond between layers, a surface

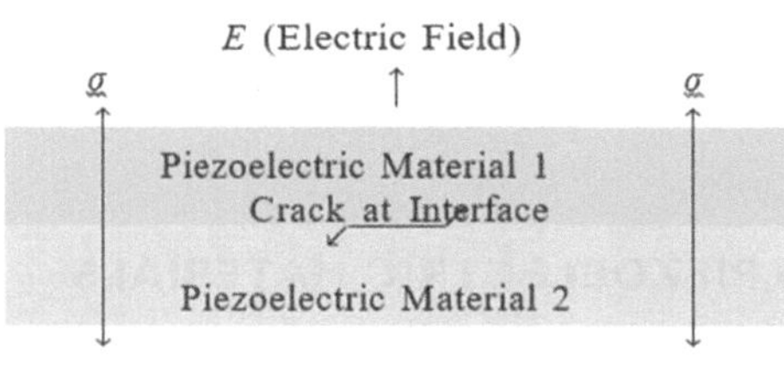

Figure 8.1 Interface crack in dissimilar piezoelectric materials.

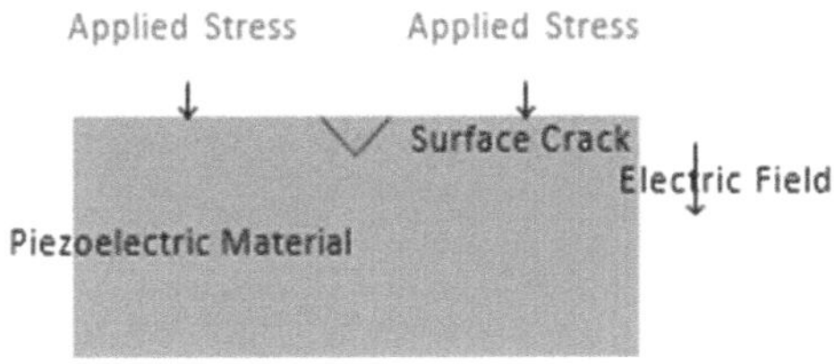

Figure 8.2 Surface crack in piezoelectric materials.

crack primarily affects the top layer, potentially compromising the structural integrity of the material itself without affecting the interface.

8.3.1.3 Transverse cracks

One kind of defect that appears in a piezoelectric material perpendicular to the polarization direction is called a transverse crack. Environmental variables, cyclic loading, or external mechanical stress can all contribute to these fissures. Transverse cracks are especially troublesome because they impair the material's piezoelectric efficacy in addition to reducing its mechanical strength. The presence of a transverse crack reduces the ability of the material to withstand external forces, leading to a higher risk of fracture and failure under stress. Piezoelectric materials generate an electric charge when subjected to mechanical deformation. Transverse cracks in piezoelectric materials are a major concern as they affect both structural integrity and electrical performance. Understanding their causes and effects helps in designing more reliable piezoelectric devices with improved durability.

A transverse crack interrupts the continuity of the material, leading to:

- Reduced charge transfer across the crack.
- Distortion of the internal electric field.
- Decreased energy conversion efficiency.

Several factors contribute to the formation of transverse crack (Figure 8.3):

- **Repeated Mechanical Loading**: Stress cycles cause fatigue, leading to crack formation.
- **High Electric Fields**: Excessive electric fields can generate stress within the material.
- **Temperature Variations:** Thermal expansion and contraction can initiate cracks over time.

8.3.1.4 Internal micro-cracks

Piezoelectric materials find extensive utility in energy harvesting, sensing, and actuators. Nevertheless, internal damage in the form of micro-cracks may

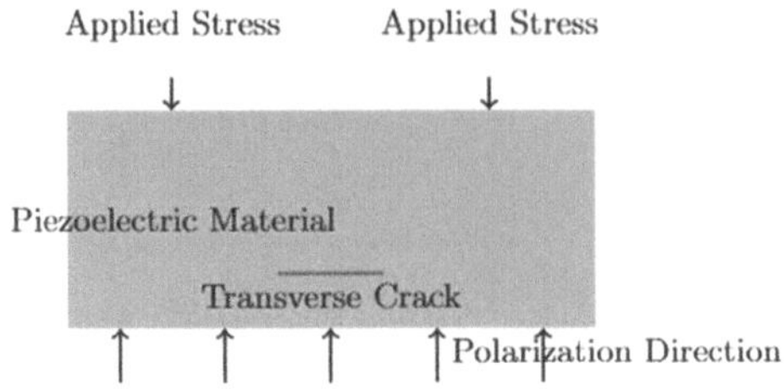

Figure 8.3 Transverse crack in piezoelectric materials.

result from their frequent exposure to mechanical or electrical loading cycles. These internal micro-cracks steadily increase under stress, finally leading to catastrophic failure. Internal micro-cracks form inside the bulk of the material rather than on the surface (see Figure 8.4). The presence of micro-cracks weakens the material, reducing its load-bearing capacity and making it susceptible to sudden fracture.

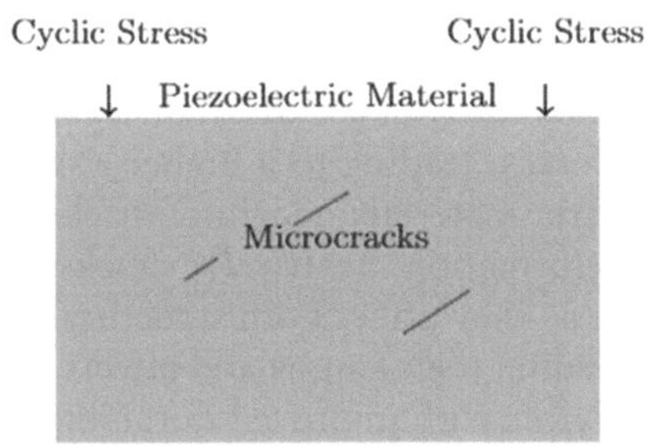

Figure 8.4 Internal micro-crack in piezoelectric materials.

These cracks develop due to:

- **Repeated mechanical loading:** Fatigue from cyclic stress leads to crack initiation.
- **Electrical degradation:** High electric fields cause localized stress accumulation.
- **Thermal fluctuations:** Expansion and contraction create micro-cracks over time.
- Environmental Impact

8.3.1.5 Permeable crack and impermeable crack

In a piezoelectric material, an impermeable crack prevents the electric displacement field from passing through the crack interface, causing charge accumulation at the crack surfaces. In contrast, a permeable crack permits the electric displacement field to pass through the crack interface.

8.4 ARRHENIUS-BASED DEFINITIONS IN PIEZOELECTRIC CRACK ANALYSIS

8.4.1 Arrhenius law in piezlectric material degradation

The long-term reliability of piezoelectric materials follows an Arrhenius-type dependence, where the degradation rate due to thermal and environmental factors is given by:

$$k = Ae^{\frac{-E_a}{RT}} \tag{8.18}$$

where k is the degradation rate, A, is the pre-exponential factor, E_a is the activation energy, R is the universal gas constant, and T is the absolute temperature.

8.4.2 Arrhenius-based crack growth in piezoelectric systems

Crack propagation in piezoelectric materials can be modelled using an Arrhenius function to describe temperature-dependent fracture kinetics:

$$v = v_0 e^{\frac{-E_a}{RT}} \tag{8.19}$$

where v is the crack growth rate and v_0 is material-dependent factor.

8.4.3 Thermally activated fatigue in piezoelectric crystals

The structural fatigue of piezoelectric materials is influenced by temperature and stress cycling, with failure probabilities often modelled using Arrhenius equations in reliability assessments.

8.4.4 Temperature-driven interface crack propagation

The effect of thermal variations on interface cracks in piezoelectric structures follows Arrhenius kinetics, impacting the stress intensity factor and material fatigue life.

8.5 CRACK INTERACTION MODELS

In fracture mechanics, cracks can be classified based on their interaction with surrounding cracks or interfaces. The two primary types are single-interaction cracks and multiple-interaction cracks.

8.5.1 Single interaction crack (SIC)

A single-interaction crack is an isolated crack that interacts with the surrounding medium without additional crack branches or multiple interacting crack tips. The crack may experience loading, but there is no interaction with other fractures.

Example of such cracks are: A **semi-infinite crack** in a piezoelectric medium and a **center crack** in a finite plate under tension (see Figure 8.5).

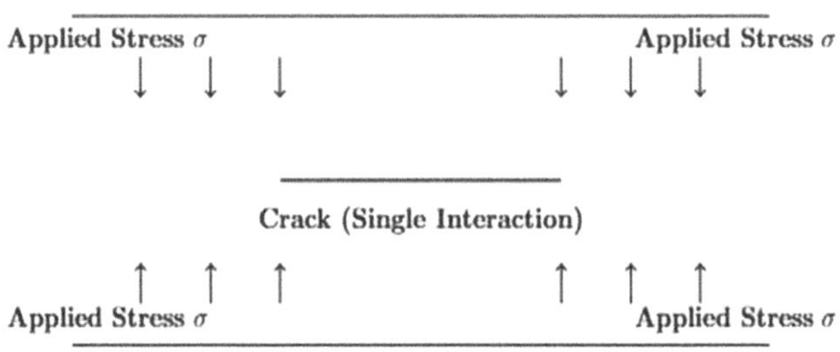

Figure 8.5 Single interaction crack.

8.5.1.1 Characteristics of SIC

- Only **one** crack exists in the medium.
- The stress and displacement fields are affected only by applied loads and material properties.
- No interference from other cracks.
- Governing equations involve a single stress intensity factor (SIF) at the crack tip.

8.5.1.2 Example problem on single-interaction crack in a transversely isotropic piezoelectric medium

Consider a two-dimensional piezoelectric material containing a single crack of length $2a$ along the x-axis, with its center at the origin. Figure 8.6, the diagram illustrates the single crack of length $2a$, located along the x-axis in a piezoelectric material, subjected to applied mechanical stress σ_∞ and electric displacement D_∞.

The mechanical equilibrium equations are given by

$$\sigma_{ij,j} = 0, \; i, j = 1, 2 \tag{8.20}$$

The electro-elastic equilibrium equations satisfy

$$D_{i,i} = 0, \; i = 1, 2 \tag{8.21}$$

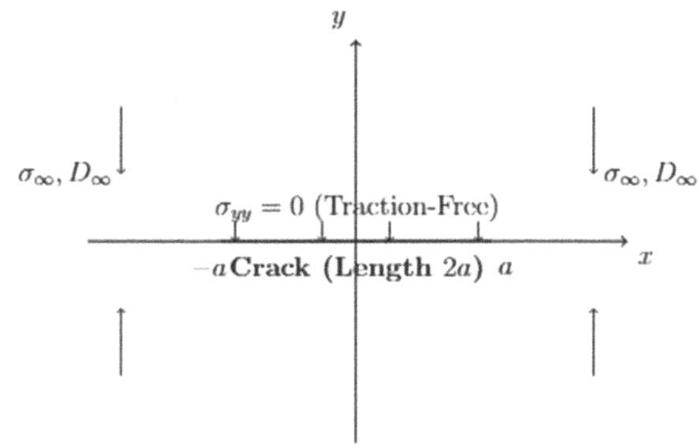

Figure 8.6 Schematic representation of a two-dimensional piezoelectric material containing a single crack of length $2a$ along the x-axis. The crack is subjected to applied mechanical stress σ_∞ and electric displacement D_∞. The boundary conditions at the crack faces determine the electromechanical field distribution around the crack tip.

- BOUNDARY CONDITIONS ACROSS THE CRACK

  The crack is assumed to be traction-free, and the electric boundary conditions depend on its charge state

  1. **Mechanically Free Crack Faces**

$$\sigma_{yy}(x, 0) = 0,\ \sigma_{xy}(x, 0) = 0,\ |x| < a. \tag{8.22}$$

  2. **Electrically Permeable Crack (No Charge Accumulation)**

$$D_y^+ - D_y^- = 0,\ E_x^+ = E_x^-,\ |x| < a \tag{8.23}$$

  3. **Electrically Impermeable Crack (Charge Discontinuity)**

$$D_y^+ = D_y^-,\ |x| < a. \tag{8.24}$$

- SOLUTION OF CRACK INTERACTION STUDY IN PIEZOELECTRIC MATERIALS USING COMPLEX POTENTIALS FUNCTIONS

  The governing equations in terms of complex potentials $\Phi(z)$ and $\Psi(z)$ are:

$$\sigma_{xx} + \sigma_{yy} = 2\mathrm{Re}\left[\Phi'(z)\right]. \tag{8.25}$$

$$\sigma_{yy} - \sigma_{xx} + 2i\sigma_{xy} = 2\mathrm{Re}\left[\Psi'(z)\right]. \tag{8.26}$$

$$D_x + iD_y = -\Phi'(z). \tag{8.27}$$

where $\Phi(z)$ and $\Psi(z)$ represent the mechanical and electrical fields, respectively.

- **Complex Potential Functions for a single crack:** For a single crack under remote stress σ_∞ and electric displacement D_∞, the complex potential are given as

$$\Phi(z) = \frac{\sigma_\infty}{2}\left(z + \frac{a^2}{z}\right) - i\,\frac{D_\infty}{2}\left(z - \frac{a^2}{z}\right). \tag{8.28}$$

$$\Psi(z) = \frac{\sigma_\infty}{2}\left(z - \frac{a^2}{z}\right) + i\,\frac{D_\infty}{2}\left(z + \frac{a^2}{z}\right). \tag{8.29}$$

- **Stress Intensity Factor (SIF) and Electric Displacement Intensity Factor (EDIF)**

  The stress intensity factor and electric displacement factor at the crack tip are given by

$$K_I = \sigma_\infty\sqrt{\pi\, a},\ K_D = D_\infty\sqrt{\pi\, a} \tag{8.30}$$

Figure 8.7 illustrates the normal stress distribution σ_{xx} around a single crack in a piezoelectric material for varying crack lengths (a) and applied remote stresses (σ_∞). The results indicate that as the crack length increases,

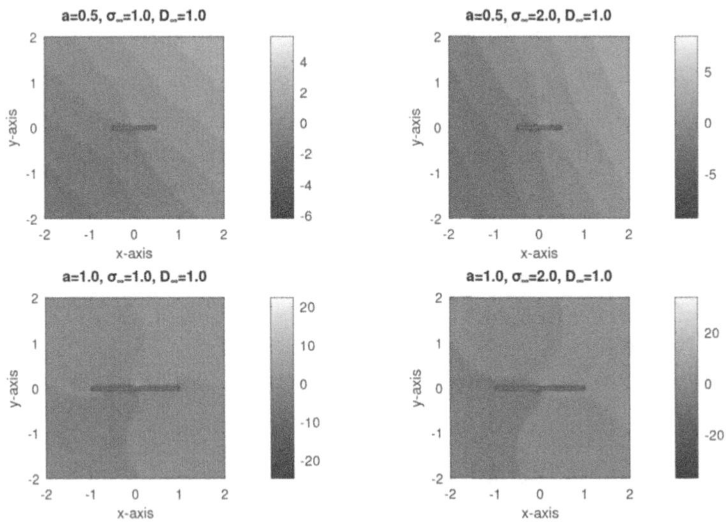

Figure 8.7 Stress distribution σ_{xx} around a crack in a piezoelectric material for different crack lengths (a) and applied stresses (σ_∞). The crack is represented by the red line along the x-axis. The contour plots visualize the variation of stress fields due to different loading conditions.

stress concentration near the crack tips becomes more pronounced, aligning with fracture mechanics principles that suggest larger cracks experience higher local stresses. Additionally, increasing the applied stress from $\sigma_\infty = 1$ to $\sigma_\infty = 2$ results in a proportional rise in stress magnitudes, as evidenced by the contour intensity variations. The stress field is symmetric along the y-axis, and the highest stress concentrations appear at the crack tips, highlighting critical points for potential crack propagation. The applied electric displacement field.

8.5.2 Multiple-interaction cracks

A multiple-interaction crack scenario occurs when two or more cracks exist in close proximity, leading to interactions between their stress fields. This can significantly change the crack propagation behavior. Parallel cracks in a piezoelectric material where electrical and mechanical stress fields overlap. Intersecting cracks where two cracks cross each other, creating stress concentration points. Branched cracks are where one crack splits into multiple cracks.

8.5.2.1 Characteristics

- Multiple cracks exist within the same medium (Figure 8.8).
- The stress fields of one crack influence the stress and displacement fields of others.
- The effective stress intensity factors (SIFs) change due to interactions.

8.5.2.2 Using Fourier series, determine the stress intensity factors at the crack tips of the above crack problem

Step 1: We assume that the displacement and electric potential fields can be expressed in terms of Fourier series expansions. For a periodic

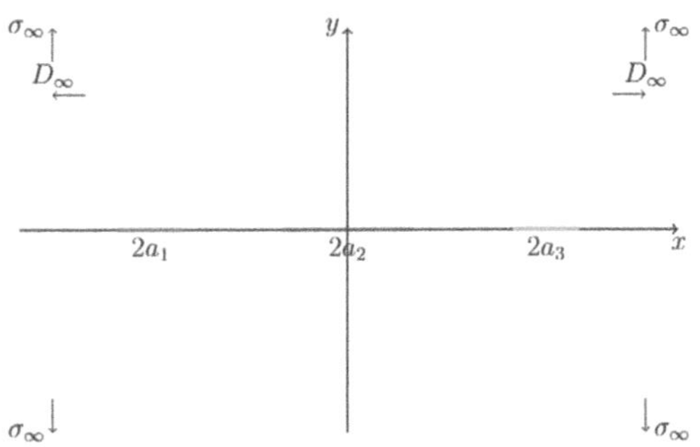

Figure 8.8 Multiple interacting cracks of length $2a_1$, $2a_2$ *and* $2a_3$ in piezoelectric material stresses (σ_∞) and electric displacement D_∞.

distribution of cracks, we expand the stress function and electric displacement as:

$$\sigma_{yy}(x,0) = \sum A_n \cos\left(\frac{n\pi x}{L}\right). \tag{8.31}$$

$$D_y(x,0) = \sum B_n \cos\left(\frac{n\pi x}{L}\right). \tag{8.32}$$

Step 2: Boundary Conditions at the Crack Faces

The crack surfaces satisfy traction-free conditions:

$$\sigma_{yy}(x,0) = 0, -a_1 < x < a_1, -a_2 < x < a_2, -a_3 < x < a_3.$$

The electric displacement condition at the crack faces is:

$$D_y(x,0) = 0, -a_1 < x < a_1, -a_2 < x < a_2, -a_3 < x < a_3.$$

Step 3: Fourier series Coefficients

By solving the Fourier expansion under these conditions, we obtain:

$$A_n = \left(\frac{2\sigma_\infty}{n\pi}\right)\left[\sin\left(\frac{n\pi a_1}{L}\right) + \sin\left(\frac{n\pi a_2}{L}\right) + \sin\left(\frac{n\pi a_3}{L}\right)\right].$$

$$B_n = \left(\frac{2D_\infty}{n\pi}\right)\left[\sin\left(\frac{n\pi a_1}{L}\right) + \sin\left(\frac{n\pi a_2}{L}\right) + \sin\left(\frac{n\pi a_3}{L}\right)\right].$$

Step 4: Stress Intensity Factors (SIFs)

The mode-I stress intensity factor at each crack tip is given by:

$$K_I^i = \sigma_\infty\sqrt{(\pi a_i)}\left[1 + \sum\left(\frac{2}{n\,\pi}\right)\sin\left(\frac{n\pi a_i}{L}\right)\right].$$

Similarly, the electric displacement intensity factor is:

$$K_D^i = D_\infty\sqrt{(\pi a_i)}\left[1 + \sum\left(\frac{2}{n\,\pi}\right)\sin\left(\frac{n\pi a_i}{L}\right)\right].$$

Figure 8.9 illustrates the impact of multiple cracks on the stress distribution in a piezoelectric material. The top-left subplot shows the spatial distribution of multiple cracks along the x-axis. Each crack is marked in red, with black dots indicating its centers. Fourier Approximation with $N = 10$. The top-right subplot presents the stress field approximation using a Fourier series expansion with $N = 10$. The oscillations indicate the interaction effects

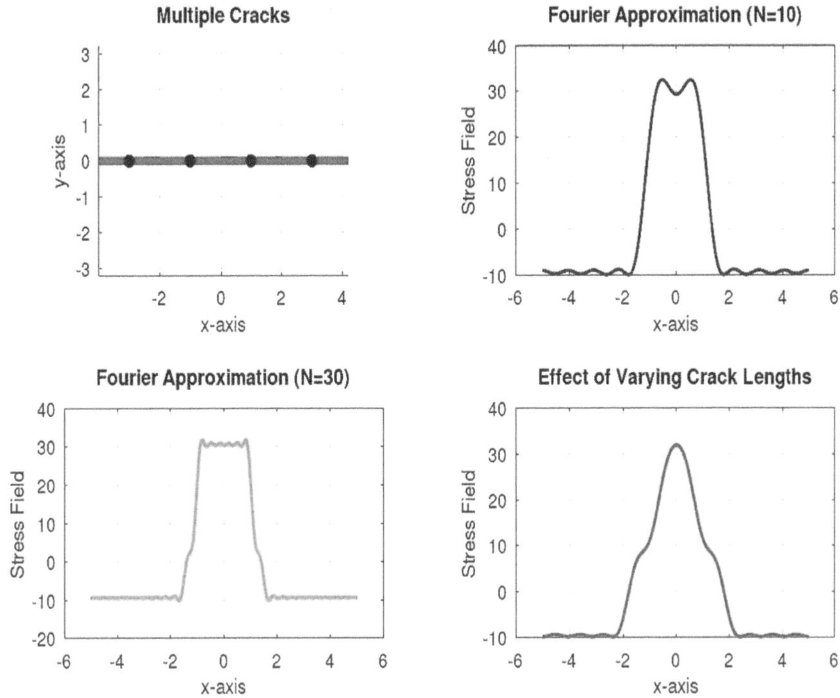

Figure 8.9 Visualization of multiple cracks and their effects on stress distribution. (Top-left) Multiple cracks along the *x*-axis with marked centers. (Top-right) Stress distribution using Fourier series approximation with *N*=10. (Bottom-left) Improved accuracy with $N = 30$. (Bottom-right) Effect of varying crack lengths on stress distribution.

of multiple cracks. The bottom-left subplot refines the approximation with $N = 30$, providing a smoother and more accurate representation of the stress distribution. The bottom-right subplot examines the influence of different crack lengths. Longer cracks tend to intensify local stress concentration, modifying the overall distribution significantly. The results highlight the importance of considering multiple cracks and varying crack lengths when analyzing stress distribution in piezoelectric materials. Increasing the number of Fourier terms improves accuracy, while different crack configurations significantly alter the stress behavior.

8.6 CONCLUSIONS

Understanding crack interactions in piezoelectric materials is crucial for predicting failure mechanisms and enhancing material reliability, as these materials experience stress concentrations influenced by both mechanical and electrical fields. Mathematical models incorporating electromechanical coupling provide valuable insights into crack behavior, with analytical methods such as the complex variable approach offering direct formulations for multiple cracks. Numerical simulations, including Finite Element and BEM, further aid in visualizing crack propagation under various loading conditions. However, to improve computational efficiency and experimental validation, further research is needed to refine models, explore Multiphysics effects, and develop fracture-resistant piezoelectric materials for advanced engineering applications.

REFERENCES

1. Govorukha V., Kamlah M., Loboda V., Lapusta Y. Interface cracks in piezoelectric materials, Smart Mater Struct 2016; 25(2): 023001. doi:10.1088/0964-1726/25/2/023001.
2. Sharma K., Bui T. Q., Zhang C., Bhargava R. R. Analysis of a subinterface crack in piezoelectric bimaterials with the extended finite element method. Eng Fract Mech May 2013; 104: 114–139.
3. Pak Y. E. Crack extension force in a piezoelectric material. ASME J Appl Mech 1990; 57: 647–653.
4. Sosa H. A., Pak Y. E. Three-dimensional eigenfunction analysis of a crack in a piezoelectric material. Int J Solids Struct 1990; 26: 1–15.
5. Pak Y. E. Linear electric-elastic fracture mechanics of piezoelectric materials. Int J Fract 1992; 54: 79–100.
6. Sosa H. A. On the fracture mechanics of piezoelectric solids. Int J Solids Struct 1992; 29: 2613–2622.
7. Kuna M. Finite element analyses of cracks in piezoelectric structures: a survey. Arch Appl Mech 2006; 76: 725–745.

Chapter 9

Novel similarity solutions of (1+1)-General Boussinesq and (1+1)-Gilson–Pickering equations

Raj Kumar, Nidhi Asthana, Dig Vijay Tanwar, and Saurabh Kapoor

9.1 INTRODUCTION

9.1.1 Aims and scope

A lot of different types of physical systems have nonlinear phenomena. These include plasma physics, fluid dynamics, fiber optics, particle motion, fluid flow, and wave behavior. They also happen in astrophysics, electrochemistry, electromagnetic theory, cosmology, acoustics, fluid dynamics, fiber-optic communications, nonlinear photonic crystals, Bose–Einstein condensates, nuclear physics, and biological systems [1, 2]. By plotting their profiles directly, analytical solutions can help explain how nonlinear partial differential equations (PDEs) change over time. So strategies for extracting analytical solutions are effective in achieving our goal. The Boussinesq equation is important because it can be used to simulate complicated fluid behaviors, make simulations of fluid flow and heat transfer easier, and help us understand convection and turbulence in many natural and man-made systems. Studying the Boussinesq equation theoretically and computationally enhances our understanding of fluid dynamics in various real-world applications. The Gilson–Pickering equations can be used to show and understand different kinds of random events, especially in the areas of biology, engineering, fluid dynamics, and reaction kinetics.

9.1.2 Origin of the problem

9.1.2.1 General Boussinesq equation (GBE)

J. Boussinesq was scrutinizing the bidirectional propagation of small-amplitude and long-wavelength capillary–gravity waves on the surface of shallow water. They [3] attained the following form of the Boussinesq equation.

$$\eta_{tt} - gh_0\eta_{xx} = gh_0\left(\frac{3\eta^2}{2h_0} + \frac{h_0^2}{3}\eta_{xx}\right)_{xx} \tag{9.1}$$

DOI: 10.1201/9781003470885-9

In this equation, where $\eta(x,t)$ represents the transversal displacement of the fluid wave on its surface. Subscript shows partial derivatives. Keulegan and Patterson [4] established the equation in 1940. In dimensionless units, Eq. (9.1) appears as

$$u_{tt} - u_{xx} - (u^2)_{xx} - u_{xxxx} = 0 \tag{9.2}$$

The function $u\,(x,t)$ represents a real-valued function, and the subscript indicates the partial derivatives in the whole article. Ill-posed is a major issue, and Eq. (9.2) is often referred to as the Bad Boussinesq equation. To remove the ill-posedness, one can change the sign of the fourth derivative term and change the u by −u.

$$u_{tt} - u_{xx} + (u^2)_{xx} + u_{xxxx} = 0, \tag{9.3}$$

Due to the terms u_{tt} and u_{xxxx}, the equation is linearly well-posed; it is known as a good Boussinesq equation [5].

Eq. (9.3) governs small nonlinear oscillations in an elastic beam, and it is also known as the nonlinear string equation [6].

$$u_{tt} + au_{xx} + b(u^2)_{xx} + cu_{xxxx} = 0 \tag{9.4}$$

where a, b, c are constants, $u(x,t)$ are real valued function.

9.1.2.2 Gilson–Pickering equation (GPE)

The nonlinear Gilson–Pickering equation (GPE) describes the propagation of waves in plasma physics [7]. In 1997, Clarkson et al. [8] studied symmetry reductions of the following form of

$$v_t - \epsilon v_{xxt} + 2kv_x - vv_{xxx} - \sigma vv_x - \beta v_x v_{xx} = 0 \tag{9.5}$$

where $\epsilon, k, \sigma, \beta$ are nonzero constants. Since it was first introduced by Gilson and Pickering in 1995, it is termed GPE [7]. Due to its large applications, it has attracted the scientific community on a very large scale. GPE can be classified into three types:

1. For $\epsilon = 1, \sigma = -1, k = 0.5$, and $\beta = 3$, it is known as the Fornberg–Whitham equation [9, 10].
2. For $\epsilon = 0, \sigma = 1, k = 0$, and $\beta = 3$, it is known as the Rosenau–Hyman equation [11].
3. If we take $\epsilon = 1, \sigma = -3, k = 0$, and $\beta = 2$, it becomes the Fuchssteiner–Fokas–Camassa–Holm equation [12, 13].

9.1.3 Literature review

(1+1)-General Boussinesq Equation (GBE): In 1872, Boussinesq derived the following form of equation, which is familiar as (1+1)-GBE.

$$u_{tt} + au_{xx} + b(u^2)_{xx} + cu_{xxxx} = 0, \tag{9.6}$$

where $u(x,t)$ represents the transversal displacement of the fluid wave on its surface. The variable x signifies the horizontal position along the direction of wave propagation.

In case of convection and buoyancy-driven flows, the GBE Eq. (9.6) model shows how a fluid with changing densities behaves. It also has other applications in a variety of fields, including astronomy, oceanography, geophysics, meteorology, and environmental engineering. By making the Navier–Stokes equations easier to understand with GBE Eq. (9.6), the approximation helps us understand how complex fluids behave, as long as changes in density are not too big and don't affect buoyancy.

An extensive literature survey addressing the GBE Eq. (9.6) includes:

For $a = 0, b = 1/2$, and $c = \pm 1$, Clarkson [14] exploited the direct method to derive the similarity reductions, and hence the exact solutions of the GBE Eq. (9.6). Clarkson [15] obtained the exact solutions of GBE Eq. (9.6) in two different ways: classical and non-classical similarity reductions. The process also generates extensive sets of exact solutions for the second and fourth Painlevé equations.

Wazwaz [16] obtained single soliton solutions employing the tanh-cot method and N soliton solutions by using Hirota's technique. Darvishi et al. [17] applied the semi-inverse variational principle to the GBE Eq. (9.6) and attained its soliton solutions. Clarkson and Dowie [18] derived the rational solutions by the inverse scattering method and showed the bell-shaped solitons. GBE Eq. (9.6) for $a = 0$, $b = c = 1$, is reduced to bilinear form [19], and a new Wronskian formulation as well as rational, positron, and complexion solutions are derived. Charlier et al. [20] also used the inverse scattering technique to extract the solutions of GBE Eq. (9.6) in terms of the Riemann–Hilbert problem. These solutions apply to generic solitonless Schwartz class solutions that maintain a bounded space average at infinity.

Some numerical results related to GBE Eq. (9.6) are imperative to discuss, notably Cruz et al. [21] used finite difference to GBE Eq. (9.6) and derived the solution for two dimensions of a wave on a porous medium to model the horizontally two-dimensional transformation of waves over porous beds on uneven bottoms. Hsu et al. [22] have discussed the second-order, fully nonlinear Boussinesq equation and used several parameters, such as the number, the relative height, and the spacing of bars, affecting Bragg's reflection. Tzimopoulos et al. [23] have applied the Boltzmann transformation and numerical method on a one-dimensional second-order unsteady nonlinear fuzzy PDE.

(1+1)-Gilson–Pickering Equation (GPE): The GPE is governed by

$$v_t + 2kv_x - v_{xxt} + 3vv_x = 2v_x v_{xx} + vv_{xxx} \tag{9.7}$$

where γ, α, β, and k are nonzero parameters. The GPE Eq. (9.7) talks about how non-Newtonian fluids and granular solids flow, especially when there is shear stress. In order to represent how these materials react to applied stress, it draws on concepts from rheology, the study of the movement of matter. In fields like food processing, pharmaceuticals, and materials engineering, where knowing how slurries, pastes, and suspensions flow is important for handling and processing, the equation is especially useful. Ak et al. [24] developed a hybrid scheme for GPE Eq. (9.7) by using the collocation method and B-splines. They [24] looked at how GPE changed the forced GPE from a weakly chaotic feature to motion that was almost periodic. Ali and Mehanna [25] find the traveling wave solutions of the GPE Eq. (9.7) by using the finite difference method. A study by Rezazadeh et al. [26] used Jacobi elliptic functions and an exponential rational function approach to find periodic waves, bell-shaped solitary waves, and shock waves. Ali et al. [27], by using the sine-Gordon expansion method to GPE Eq. (9.7), obtained the exact solitary wave solutions, such as shock waves, topological, non-topological, compound topological, and non-topological soliton wave solutions. This chapter is the first to use similarity transformations method (STM) and Lie symmetry to find different types of analytical solutions to both equations, which are GBE Eq. (9.6) and GPE Eq. (9.7).

9.1.4 Motivation

The nonlinear evolution equation has been a major source of inspiration in recent decades due to its significant applicability in the applied sciences [1–27]. The STM is a very good way to get new analytical solutions that are different from those found in [14–27] using one-parameter Lie symmetry analysis. Implementation of this method refers to some articles [28–38].

9.2 INFINITESIMALS USING SIMILARITY TRANSFORMATIONS METHOD

Sophus Lie created Lie symmetry in 1981 as an ad hoc way to integrate any system of PDEs. After applying similarity transformations, the system remains invariant. Invariance calculates infinitesimals, which are functions of the independent and dependent variables. Making use of the Lagrange equation yields similarity variables and similarity functions. Substituting the values of dependent variables in the PDEs, the STM reduces the number of independent variables in the PDEs with or without boundary constraints.

This section depicts the fundamental procedures for getting infinitesimal transformations to (1+1)-dimensional PDEs via one-parameter classical Lie symmetry. Initially, we impose an invariant condition on the PDEs and use the test problem to eliminate one derivative. This process gives an overdetermined system of PDEs, which, when solved, yields infinitesimal transformations. Furthermore, Lagrange's characteristic equation utilizes infinitesimal transformations, and the reduction of similarity generates a system of equivalent ODEs. For wide applications of Lie symmetry, one can read some articles [21, 31, 32, 39–41], textbooks [42, 43], and references therein.

Consider one parameter (ϵ) infinitesimal transformations for any $(1+1)$ PDE:

$$\begin{aligned} x^* &= x + \in \xi(x,t,u) + O(\in^2) \\ t^* &= t + \in \tau(x,t,u) + O(\in^2), \\ u^* &= u + \in \phi(x,t,u) + O(\in^2) \\ n^*_{(x^*)} &= \theta_x + \in [n_x] + O(\in^2) \end{aligned} \tag{9.8}$$

where ϵ is a parameter, $\xi(x,t,u), \tau(x,t,u)$, and $\phi(x,t,u)$ are infinitesimals for x,t, and u, respectively. $[\eta_x]$ is the extension [41, 42] of the first order.

For illustration, consider the $(1+1)$ GBE Eq. (9.6) and

$$H \equiv u_{tt} + au_{xx} + b(u^2)_{xx} + cu_{xxxx} = 0, \tag{9.9}$$

The invariant condition of Eq. (9.6) yields

$$\frac{\partial H}{\partial(u_{tt})}[\eta_{tt}] + \frac{\partial H}{\partial(u_{xx})}[\eta_{xx}] + \frac{\partial H}{\partial(u)}[\eta] + \frac{\partial H}{\partial(u_x)}[\eta_x] + \frac{\partial H}{\partial(u_{xxxx})}[\eta_{xxxx}] = 0, \tag{9.10}$$

where $[\eta_{tt}]$ is the extension of second order *etc.*

Getting a derivative (say $\theta_{tt} = -a\theta_{xx} - 2bu\theta_{xx} - 2b(\theta_x)^2 - c\theta_{xxxx}$, θ is the true solution) from Eq. (9.9) and using in Eq. (9.10), one can have an overdetermined system of PDEs in ξ, τ, and η which on solving gives infinitesimals of GBE Eq. (9.6)

$$\xi(x,t,u) = \frac{c_1}{2}x + c_3,\ \tau(x,t,u) = c_1 t + c_2 \text{ and } \eta(x,t,u) = -\ c_1\left(\frac{a}{2b} + u\right), \tag{9.11}$$

where c_1, c_2 and c_3 are arbitrary constants in Eq. (9.11).

Similarly, the following infinitesimals of GPE Eq. (9.7) can be derived as

$$\xi(x,t,v) = \frac{2b_1 k}{\alpha\gamma + 1}t + b_3,\ \tau(x,t,v) = b_1 t + c_2, \text{ and } \eta(x,t,v) = b_1\left(\frac{2\gamma k}{\alpha\gamma + 1} - v\right), \tag{9.12}$$

where b_1, b_2 and b_3 are arbitrary constants in Eq. (9.12).

9.3 SIMILARITY SOLUTIONS

9.3.1 For GBE

Making use of Eq. (9.11) in Lagrange's characteristic equations, one can have

$$\frac{dx}{c_1\frac{x}{2}+c_2}=\frac{dt}{c_1t+c_3}=-\frac{du}{c_1\left(u+\frac{a}{2b}\right)} \tag{9.13}$$

To get further reduction of Eq. (9.6), the following cases are raised:
Case 1: Assuming $c_1 \neq 0$, $\frac{c_2}{c_1}=A$, and $\frac{c_3}{c_1}=B$ provides

$$\frac{2dx}{x+2A}=\frac{dt}{t+B}=-\frac{du}{\left(u+\frac{a}{2b}\right)} \tag{9.14}$$

After solving Eq. (9.14), we get similarity variable (X) and similarity function $F(X)$ represented by

$$X=\frac{(x+2A)^2}{t+B}\text{and } u=-\frac{a}{2b}+\frac{F(X)}{t+B},\ t\neq -B \tag{9.15}$$

Utilizing Eq. (9.15) in GBE (9.6), the reduction of it gives.

$$2F+4XF'+X^2+4bFF'+8bXFF''+2bX(F')^2+c\left(12F''+48XF'''+16X^2F''''\right)=0 \tag{9.16}$$

Its solution is

$$F(X)=-\frac{X}{2b}+C_1 \tag{9.17}$$

Making use of Eq. (9.17) in Eq. (9.15), the first analytical solution of GBE Eq. (9.6) is represented by

$$u_1(x,t)=-\frac{a}{2b}-\frac{(x+2A)^2}{2b(t+B)^2}+\frac{C_1}{t+B}, \tag{9.18}$$

where C_1 is an arbitrary constant of integration.
Case 2: Treating $c_1=0, c_2\neq 0$ and $\frac{c_3}{c_2}=A_1$, in Eq. (9.13), then we have

$$\frac{dx}{1}=\frac{dt}{A_1}=\frac{du}{0} \tag{9.19}$$

Solving it, one can get $A_1 x - t = Y$, and $u=H(Y)$. Then GBE Eq. (9.6) reduces to

$$cH'''' + H'' + aH'' + 2bHH'' + 2b(H')^2 = 0 \tag{9.20}$$

Solving Eq. (9.20), and hence one can obtain the second solution of GBE Eq. (9.6) as

$$u_2(x,t) = -\frac{a}{2b} + \frac{\frac{1}{b}\left(4cc_2^2 - a - 1\right) - bcc_2^2 \tanh(c_1 + c_2 x)^2}{t + B}. \tag{9.21}$$

where C_2 is an arbitrary constant of integration.

9.3.2 For GPE

Making use of infinitesimals Eq. (9.12) in Lagrange's characteristic equation, it recasts as

$$\frac{dx}{b_1 \frac{2kt}{\alpha\gamma + 1} + b_2} = \frac{dt}{b_1 t + b_3} = \frac{dv}{b_1 \frac{2k\gamma}{\alpha\gamma + 1} - v}. \tag{9.22}$$

To get further similarity reduction of Eq. (9.7), the following cases can be raised:

Case 1: Assuming $b_1 \neq 0$, $\frac{b_2}{b_1} = B_1$, and $\frac{b_3}{b_1} = B_2$ provides

$$\frac{dx}{\frac{2kt}{\alpha\gamma + 1} + B_1} = \frac{dt}{t + B_2} = \frac{dv}{\frac{2k\gamma}{\alpha\gamma + 1} - v} \tag{9.23}$$

After solving Eq. (9.23), we get the similarity function $V(X_1)$, then

$$v = \frac{2k\gamma}{\alpha\gamma + 1} + \frac{V(X_1)}{t + B_2} \text{ where } X_1 = x - \frac{2kt}{\alpha\gamma + 1} + \left(\frac{2kB_2}{\alpha\gamma + 1} - B_1\right) log\left(t + B_2\right) \tag{9.24}$$

Utilizing Eq. (9.24) in GPE (9.7), the reduction of it gives.

$$V - \left(\frac{2kB_2}{\alpha\gamma + 1} - B_1\right)V' - \gamma V'' + \gamma\left(\frac{2kB_2}{\alpha\gamma + 1} - B_1\right)V''' + VV''' + \alpha VV' + \beta V'V'' = 0 \tag{9.25}$$

First particular solution of Eq. (9.25) for $B_1 = \frac{2kB_2}{\alpha\gamma+1}$ is

$$v_1 = \frac{2k\gamma}{\alpha\gamma+1} + \frac{1}{t+B_2}\left[-x + \frac{2kt}{\alpha\gamma+1} - \left(\frac{2kB_2}{\alpha\gamma+1} - B_1\right)log\left(t+B_2\right)\right] \tag{9.26}$$

Case 2: Treating $b_1 = 0, b_2 \neq 0$ and $\frac{b_2}{b_3} = C$, in Eq. (9.13), then we have

$$\frac{dx}{C} = \frac{dt}{1} = \frac{dv}{0} \tag{9.27}$$

Solving it, one can get $At - x = Y_1$, and $v = H(Y_1)$. Then GPE Eq. (9.7) reduces to

$$(-A+2k)H' + \epsilon AH''' - HH''' - \alpha HH' - \beta H'H''' = 0 \tag{9.28}$$

Solution is trivial.
Case 3: Treating $b_1 = 0, b_2 \neq 0$ and $\frac{b_3}{b_2} = B$, in Eq. (9.22), then we have

$$\frac{dx}{1} = \frac{dt}{B} = \frac{dv}{0} \tag{9.29}$$

Solving it, one can get $Bx - t = Z_1$, and v=$G(Z_1)$. Then GPE Eq. (9.7) reduces to

$$(-1+2kB)G' + \epsilon B^2G''' - B^3GG''' - \alpha BGG' - \beta B^4G'G''' = 0 \tag{9.30}$$

Solution is trivial.

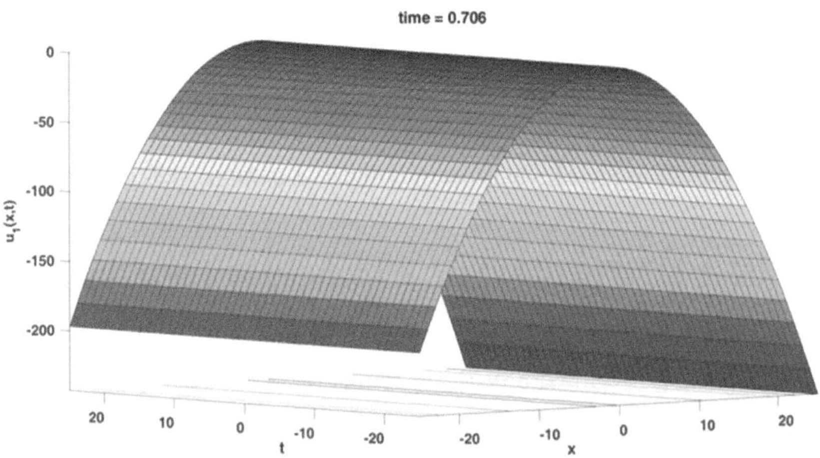

Figure 9.1 Parabolic solution for Eq. (9.18).

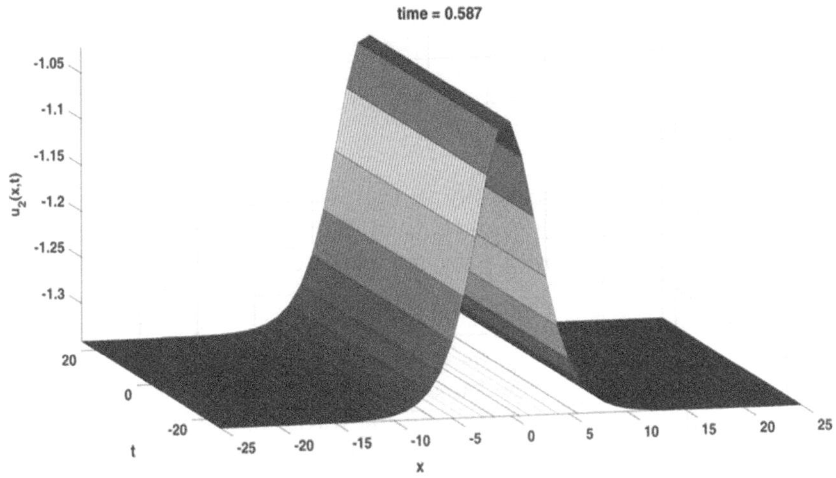

Figure 9.2 Dark soliton solution for Eq. (9.21).

9.4 ANALYSIS AND DISCUSSION

The analytic solution of GBE Eq. (9.6) is represented by Eqs. (9.18) and (9.21), while the analytic solution of GPE Eq. (9.7) is given by Eq. (9.26). The profile of the solutions Eqs. (9.18) and (9.21) are parabolic (Figure 9.1) and dark solitons (Figure 9.2), respectively. Profiles are simulated by MATLAB taking appropriate values of a = 0.1270, b = 0.9134, c = 0.6834, c_2 = 0.2785, and B = 0.9649.

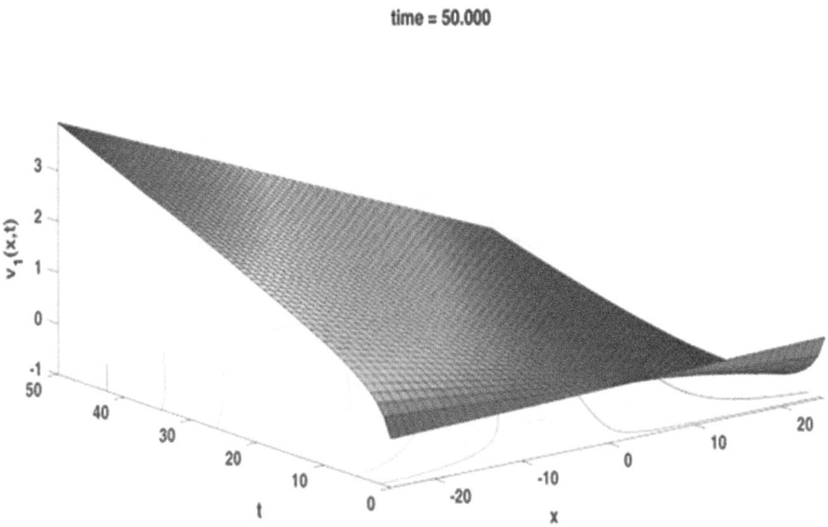

Figure 9.3 Stationary profile for Eq. (9.26).

For Eq. (9.26), stationary profile (Figure 9.3) appropriated value of $K = 1.25$, $B_1 = 0.9575$, $B_2 = 0.9649$, $\alpha = 0.1270$, and $\gamma = 0.6324$.

9.5 CONCLUSIONS

The STM is employed successfully in both the (1+1) equations. Analytical results are represented by Eqs. (9.18), (9.21), and (9.26) obtained herein and are novel. The profiles show a parabolic nature, dark soliton solutions, and stationary profiles.

The dark soliton solution of the GBE Eq. (9.6) has several significant uses across various fields, such as fluid dynamics, nonlinear optics, quantum mechanics, plasma physics, and mathematical physics. In situations where wave propagation is asymmetrical and the wavefront evolves over time, we often employ parabolic solutions. The stationary profile of the GPE Eq. (9.7) has served as a tool for engineers, oceanographers, and fluid dynamicists to understand and visualize the steady-state behavior of waves in shallow water. It provides crucial insights into wave height, energy distribution, and interaction patterns, which are essential for application in tsunami modeling, coastal engineering, flood risk assessment, and the study of wave-induced forces on structures. Moreover, the solutions derived in this work can be useful to study the NPDEs in environmental science, nonlinear dynamics, steady flow analysis, hydrostatics, numerical simulations, and climate modeling.

REFERENCES

1. M. Tajiri and H. Maesono. Resonant interactions of drift vortex solitons in aconvective motion of a plasma. Physical Review E, 55(3): 3351–3357, 1997. https://doi.org/10.1103/PhysRevE.55.3351
2. H. Xu and S. J. Liao. Laminar flow and heat transfer in the boundary-layer of non-Newtonian fluids over a stretching flat sheet. Computers and Mathematics with Applications, 57(9): 1425–1431, 2009. https://doi.org/10.1016/j.camwa.2009.01.029
3. J. Boussinesq. Théorie des ondes et des remous qui se propagent le long d'un canal rectangulaire horizontal, en communiquant au liquide contenu dans ce canal des vitesses sensiblement pareilles de la surface au fond. Journal de Mathématiques Pures et Appliquées 2e série, 17(2): 55–108, 1872.
4. G. H. Keulegan and G. W. Patterson. A criterion for instability of flow in steep channels. Eos. Transactions American Geophysical Union, 21(2): 594–596, 1940. https://doi.org/10.1029/TR021i002p00594
5. A. A. Himonas and D. Mantzavinos. The "good" Boussinesq equation on the half-line. Journal of Differential Equations, 258(9): 3107–3160, 2015. https://doi.org/10.1016/j.jde.2015.01.005
6. G. E. Fal'kovich, M. D. Spector and S. K. Turitsyn. Destruction of stationary solutions and collapse in the nonlinear string equation. Physics Letters, 99(6): 271–274, 1983. https://doi.org/10.1016/0375-9601(83)90882-4
7. C. Gilson and A. Pickring. Third-order partial differential equations. Journal of Physics A: Mathematical and General, 28: 2871–2888, 1995. https://doi.org/10.1007/978-3-540-88111-7_8

8. P. A. Clarkson and T. J. Priestley. Symmetries of a class of nonlinear fourth order partial differential equations. Journal of Nonlinear Mathematical Physics, 6(1): 66–98, 1999. https://doi.org/10.2991/jnmp.1999.6.1.6
9. G. B. Whitham. Variational methods and applications to water waves. Proceedings of the Royal Society of London. Series A. Mathematical and Physical Sciences, 299(1456): 6–25, 1967. https://doi.org/10.1098/rspa.1967.0119
10. B. Fornberg and G. B. Whitham. A numerical and theoretical study of certain nonlinear wave phenomena. Philosophical Transactions of the Royal Society A, 289(1361): 373–404, 1978. https://doi.org/10.1098/rsta.1978.0064
11. P. Rosenau and J. M. Hyman. Compactons: Solitons with finite wavelength. Physical Review Letters, 70(5): 564–567, 1993. https://doi.org/10.1103/PhysRevLett.70.564
12. R. Camassa and D. D. Holm. An integrable shallow water equation with peaked solitons. Physical Review Letters, 71: 1661–1664, 1993. https://doi.org/10.1103/PhysRevLett.71.1661
13. B. Fuchssteiner and A. S. Fokas. Symplectic structures, their Bäcklund transformations and hereditary symmetries. Physica D: Nonlinear Phenomena, 4(1): 47–66, 1981/82.
14. P. A. Clarkson and M. D. Kruskal. New similarity reductions of the Boussinesq equation. Journal of Mathematical Physics, 30(10): 2201–2213, 1989. https://doi.org/10.1063/1.528613
15. P. A. Clarkson. New exact solutions of the Boussinesq equation. European Journal of Applied Mathematics, 1(3): 279–300, 1990. https://doi.org/10.1017/S095679250000022X
16. A. M. Wazwaz. Multiple-soliton solutions for the Boussinesq equation. Applied Mathematics and Computation, 192(2): 479–486, 2007. https://doi.org/10.1016/j.amc.2007.03.023
17. M. T. Darvishi, M. Najafi and A. M. Wazwaz. Soliton solutions for Boussinesq-like equations with spatio-temporal dispersion. Ocean Engineering, 130: 228–240, 2017. https://doi.org/10.1016/j.oceaneng.2016.11.052
18. P. A. Clarkson and E. Dowie. Rational solutions of the Boussinesq equation and applications to rogue waves. Transactions of Mathematics and Its Applications, 1(1): 1–26, 2017. https://doi.org/10.1093/imatrm/tnx003
19. Y. L. Ma and B. Q. Li. Analytic rogue wave solutions for a generalized fourth-order Boussinesq equation in fluid mechanics. Mathematical Methods in the Applied Sciences, 42(1): 39–48, 2019. https://doi.org/10.1002/mma.5320
20. C. Charlier, J. Lenells and D. S. Wang. The "Good" Boussinesq equation: Long-time asymptotics. Analysis and PDE, 16(6): 1351–1388, 2023. https://doi.org/10.2140/apde.2023.16.1351
21. E. C. Cruz, M. Isobe and A. Watanabe. Boussinesq equations for wave transformation on porous beds. Coastal Engineering, 30(1–2): 125–156, 1997. https://doi.org/10.1016/S0378-3839(96)00039-7
22. T. W. Hsu, S. C. Hsiao, S. H. Ou, S. K. Wang, B. D. Yang and S. E. Chou. An application of Boussinesq equations to Bragg reflection of irregular waves. Ocean Engineering, 34(5–6): 870–883, 2007. https://doi.org/10.1016/j.oceaneng.2006.02.005
23. C. Tzimopoulos, K. Papadopoulos, B. Papadopoulos, N. Samarinas and C. Evangelides. Fuzzy solution of nonlinear Boussinesq equation. Journal of Hydroinformatics, 24(6): 1127–1147, 2022. https://doi.org/10.2166/hydro.2022.026
24. T. Ak, A. Saha and S. Dhawan. Performance of a hybrid computational scheme on traveling waves and its dynamic transition for Gilson–Pickering equation. International Journal of Modern Physics C, 30(4): 1–17, 2019. https://doi.org/10.1142/S0129183119500281

25. K. K. Ali and M. Mehanna. Traveling wave solutions and numerical solutions of Gilson–Pickering equation. Results in Physics, 28: 104596, 2021. https://doi.org/10.1016/j.rinp.2021.104596
26. H. Rezazadeh, A. Jhangeer, E. Tala-Tebue, M. S. Hashemi, S. Sharif, H. Ahmad and S.-W. Yao. New wave surfaces and bifurcation of nonlinear periodic waves for Gilson–Pickering equation. Results in Physics, 24: 104192, 2021. https://doi.org/10.1016/j.rinp.2021.104192
27. K. K. Ali, R. Yilmazer, H. M. Baskonus and H. Bulut. New wave behaviors and stability analysis of the Gilson–Pickering equation in plasma physics. Indian Journal of Physics, 95(5): 1003–1008, 2021. https://doi.org/10.1007/s12648-020-01773-9
28. D. V. Tanwar and R. Kumar. A novel class of soliton solutions and conservation laws of the generalised BS equation by Lie symmetry method. Pramana Journal of Physics, 98(113): 1–19, 2024. https://doi.org/10.1007/s12043-024-02796-1
29. D. V. Tanwar and R. Kumar. Kinks and soliton solutions to the coupled Burgers equation by Lie symmetry approach. Physica Scripta, 99: 075223, 2024. https://iopscience.iop.org/article/10.1088/1402-4896/ad51b6
30. R. Kumar and R. S. Verma. Dynamics of some new solutions to the coupled DSW equations traveling horizontally on the seabed. Journal of Ocean Engineering and Science, 9(2): 154–163, 2024. https://doi.org/10.1016/j.joes.2022.04.015
31. R. Kumar, K. S. Pandey, S. K. Yadav and A. Kumar. Some more variety of analytical solutions to (2+1)-Bogoyavlensky-Konopelchenko equation. Physica Scripta, 99: 045240 (1–11), 2024. https://doi.org/10.1088/1402-4896/ad32fd
32. R. Kumar and R. S. Verma. Dynamics of invariant solutions of mKdV-ZK arising in a homogeneous magnetised plasma. Nonlinear Dynamics, 108(4): 4081–4092, 2022. https://doi.org/10.1007/s11071-022-07389-4
33. S. Shukla, M. Kumar and R. Kumar. A study on analytical solutions and conservation laws of the cgKP occurring in straits. Physica Scripta, 98: 125238 (1–15), 2023. https://doi.org/10.1088/1402-4896/ad0bbe
34. R. Kumar, K. S. Pandey, A. Kumar and A. Kumar. Novel traveling wave solutions of Jaulent-Miodek equations and coupled Konno-Oono systems and their dynamics. CHAOS Theory and Applications, 5(4): 281–285, 2023. https://doi.org/10.51537/chaos.1322939
35. R. Kumar. On similarity solution of (3+1)-dimensional Jimbo–Miwa equations arising in fluids. Indian Journal of Natural Sciences, 14(78): 57745–57751, 2023.
36. R. Kumar, R. S. Verma and A. K. Tiwari. On similarity solutions to (2+1)-dispersive long-wave equations. Journal of Ocean Engineering and Science, 8(2): 111–123, 2023. https://doi.org/10.1016/j.joes.2021.12.005
37. R. Kumar, K. S. Pandey and A. Kumar. Dynamical behavior of the solutions of coupled Boussinesq–Burgers equations occurring at the seaside beaches. Brazilian Journal of Physics, 52: 201 (1–16), 2022. https://doi.org/10.1007/s13538-022-01195-4
38. M. Kumar, A. Kumar and R. Kumar. Similarity solutions of the Konopelchenko-Dubrovsky system using Lie group theory. Computers and Mathematics with Applications, 71(10): 2051–2059, 2016. https://doi.org/10.1016/j.camwa.2016.03.023
39. M. Kumar, D. V. Tanwar and R. Kumar. On Lie symmetries and soliton solutions of (2+1)-dimensional Bogoyavlenskii equations. Nonlinear Dynamics, 94(4): 2547–2561, 2018. https://doi.org/10.1007/s11071-018-4509-2
40. R. Kumar and A. Kumar. Some invariant solutions of coupled Konno-Oono equations arising in electromagnetic and quantum fields. Physica Scripta, 97(7): 1402–4896, 2022. https://doi.org/10.1088/1402-4896/ac71e1

41. R. Kumar and A. Kumar. Dynamical behavior of similarity solutions of CKOEs with conservation law. Applied Mathematics and Computation, 422: 126976, 2022. https://doi.org/10.1016/j.amc.2022.126976
42. G. W. Bluman and J. D. Cole. Similarity methods for differential equations. Springer, New York, 1st edition, 1974. https://doi.org/10.1007/978-1-4612-6394-4
43. P. J. Olver. Applications of Lie groups to differential equations. Springer, New York, 2nd edition, 1993.

Chapter 10

Heat and mass transfer of double diffusive convection in a vertical porous flat plate

Durgaprasad Nayak and Saurabh Kapoor

10.1 INTRODUCTION

The simultaneous action of thermal and concentration buoyancy forces significantly impacts transport phenomena across diverse engineering and environmental contexts. Industrial applications abound, including chemical distillation processes, heat exchanger design, solar energy harvesting, and thermal protection systems. In these scenarios, mass and heat transfer are intrinsically coupled, induced by gradients in temperature as well as concentration. The coupled transport of heat and mass within porous media is a phenomenon of significant engineering and geophysical relevance, finding applications in processes including geothermal reservoir dynamics, porous solid desiccation, nuclear reactor cooling, and enhanced oil recovery nuclear reactor cooling.

Recently, the study of free convection flow in magnetohydrodynamic (MHD) fluids has garnered considerable attention among researchers owing to its numerous applications in various fields, including astrophysics, aerodynamics, meteorology, geophysics, and MHD power generation and pumps. Additionally, convective flow within porous media finds practical relevance in chemical engineering processes, particularly in purification and filtration. In petroleum technology, it is employed to model the transport of oil, water, and natural gas through subterranean channels and reservoirs. Agricultural engineering also utilizes this knowledge to investigate underground water resources (refer to Hayat and Abbas [1], Rahman and Sattar [2], and Vafai and Tien [3]). Investigations of combined thermal convection over semi-infinite vertical plates, both with and without porous medium influence, are extensively documented [e.g., 4, 5, 6]. In porous medium, Chamkha [7] presented a mathematical model to analyze boundary layer flow over an inclined plate, incorporating the effects of a magnetic field. The study of convective flow, heat, and mass transfer within porous media is of significant contemporary interest due to its relevance to numerous applications, including nuclear reactor cooling, enhanced oil recovery, geothermal reservoir systems, metallurgical processing, and underground energy transport. In a porous medium, Raptis [8] investigated the impact of the permeability parameter and Grashof

DOI: 10.1201/9781003470885-10

number on the velocity field in a steady hydromagnetic natural convective flow of a conducting fluid confined between two parallel plates. Similarly, Lloyd et al. [9] examined the impact of an electromagnetic field on free convection in a porous inclined surface. Additionally, Ahmmed and Sarker [10] investigated the MHD natural convection flow past a vertical plate with temperature-dependent viscosity.

Hossain and Ahmed [11] investigated mixed convection under homogeneous heat source conditions with a strong magnetic field. Following this, Hossain et al. [12] analyzed natural convection flow over a porous flat plate subjected to a power-law surface temperature and a variable magnetic field. Wilks [13] introduced the parameter ξ, defined as $\xi = (\sigma H/\rho\infty)\ g\beta\ (T_0 - T_\infty)$, to analyze natural convective flow around a semi-infinite vertical plate with a high-strength magnetic field. They reformulated the problem using coordinate expansions based on a non-dimensional characteristic length, which helps determine the balance between buoyancy and magnetic forces along the plate.

Ahmmed and Sarker [10] conducted an early study on natural convection flow of a non-compressible fluid in a vertical flat plate. Furthermore, Kapoor [14] derived an analytical closed-form solution to investigate unsteady hydromagnetic free convection flow with simultaneous mass and heat transfer in an incompressible, viscous Boussinesq fluid. The analysis conducted by the researchers included the effects of radiative heat transfer and a first-order chemical reaction occurring between the diffusing species and the fluid. A comprehensive analysis of the literature regarding natural convection in porous media is presented in the book authored by Nield and Bejan [15]. Sparrow and Gregg [16] conducted research on the flow of a viscous, non-compressible fluid in a vertical surface, considering situations involving forced, natural, and combined (mixed) convection. Wilks and Hunt [17, 18] developed a set of ongoing mathematical relationships that link different types of similarity solutions for mixed convection flow. Tingwei et al. [19] investigated how buoyancy forces affect mixed convection along a vertical flat plate where the heat flux is constant. Hossain and Ahmed [11] explored free convection in a porous vertical plate subject to a magnetic field, considering a surface temperature that varies according to a power law. Then, Ahmmed and Sarker [10] have also studied the MHD free convection flow of non-compressible fluid in a vertical flat plate. Kapoor et al. [20] analytically solved the MHD free convection flow of non-compressible fluid over a vertical flat plate. Furthermore, the author [14] extended their study by adopting chemical reaction parameters in thermosolutal natural convection flow in a vertical flat plate. In 2021, Kapoor S. numerically investigated the thermosolutal MHDs free connective flow through a vertical plane in non-Darcy permeable medium.

Keeping in view of the above, it has been observed that the cross diffusion during double diffusive convection has not received much attention. Initially, Dufour effects were disregarded due to their perceived small magnitude compared to Fourier's and Fick's laws. However, exceptions exist where the heat transfer induced by concentration gradients (Dufour effect) becomes significant. Consequently, this chapter aims to investigate the study of Dufour effects on thermosolutal convection flow along a vertical flat plate.

10.2 MATHEMATICAL MODEL

This study investigates a steady-state, two-dimensional laminar natural convective boundary layer flow. We consider a non-compressible fluid with viscosity dependence flowing over a semi-infinite vertical impermeable flat plate. The fluid is electrically conducting, and the flow occurs through a porous medium with uniformly distributed transverse magnetic field of strength B_0. Within the Boussinesq approximation, the Dufour effect, which describes how a concentration gradient can induce a temperature gradient, manifests mathematically as a term in the energy equation. The physical model is represented in Figure 10.1. The analysis is restricted to the X–Y plane, where the boundary layer flow is presumed to develop solely in the Y-direction.

10.2.1 Equation of continuity

$$\frac{\partial u}{\partial x}+\frac{\partial v}{\partial y}=0 \tag{10.1}$$

10.2.2 Equation of momentum

$$u.\frac{\partial u}{\partial x}+v.\frac{\partial v}{\partial y}=\nu.\frac{\partial^2 u}{\partial x^2}+g\beta\left(T-T_\infty\right)-\frac{\sigma u B_0^2}{\rho_\infty}-\frac{\mu}{k_p}.u \tag{10.2}$$

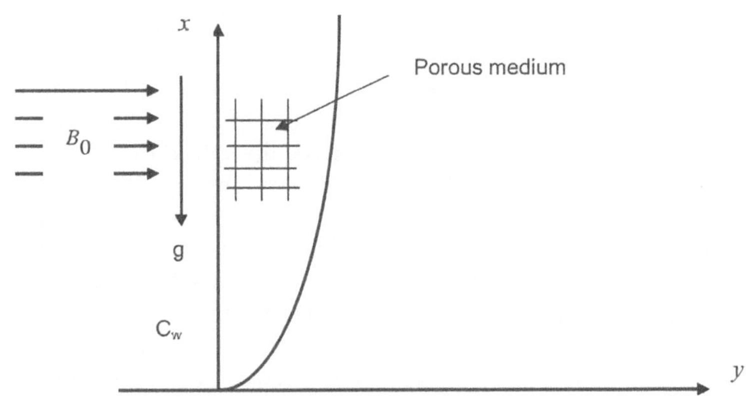

Figure 10.1 The flow configuration and coordinate system.

10.2.3 Equation of energy

$$u\frac{\partial T}{\partial x}+v\frac{\partial T}{\partial y}=\frac{k}{\rho_\infty c_p.}\frac{\partial^2 T}{\partial y^2}+\frac{D.K_T}{c_s.c_p}.\frac{\partial^2 s}{\partial y^2} \tag{10.3}$$

10.2.4 Equation of solute

$$u\frac{\partial S}{\partial x}+v\frac{\partial S}{\partial y}=\epsilon\frac{\partial^2 S}{\partial y^2} \tag{10.4}$$

10.2.5 Boundary conditions

$$y=0\text{: } u=v=0,\ T=T_0,\text{ and } S=S_{0:}y\to\infty\text{: } u\to 0,\ T\to T_\infty \tag{10.5}$$

10.2.6 Similarity transformation parameter

$$\begin{gathered}Y=\frac{yV_0}{v}\ U=\frac{u}{U_0},\quad \theta=\frac{T-T_\infty}{T_0-T_\infty}\ \phi=\frac{S-S_\infty}{S_0-S_\infty},\ Gr_T=\frac{g\beta_T\left(T_0-T_\infty\right)v}{U_0^2V_0^2},\\ Gr_S=\frac{g\beta_s\left(S_0-S_\infty\right)v}{U_0^2V_0^2}\ P_r=\frac{\mu c_p}{k}\ M=\frac{\sigma v B_0^2}{\rho_\infty V_0^2}\ Du=\frac{DK_T}{c_s c_p v}\left(\frac{S_0-S_\infty}{T_0-T_\infty}\right)\end{gathered} \tag{10.6}$$

10.3 THE PHYSICAL MODEL OF THE PROBLEM

Here, u represents the velocity components along the x-axis. Similarly, v is the velocity in the y-direction respectively, where the coordinate x runs along the vertical flat plate, while y extends perpendicular from it. The coefficient of thermal expansion is denoted by β, the symbol T denotes the fluid temperature within the boundary layer, while g represents the gravitational acceleration and k represents the thermal conductivity of the fluid.

The symbol T_∞ denotes the temperature of the ambient fluid and C_p denotes specific heat capacity where the pressure remains same. Furthermore, fluid ρ_∞ denotes the density of the ambient fluid, and v represents the kinematic fluid viscosity. Based on the continuity equation, the velocity normal to the plate is assumed to be of the form $v = -V_0$.

10.4 THE TRANSFORMED GOVERNING EQUATION

$$U''(Y)+U'(Y)-\left(M+\frac{1}{Da}\right)U(Y)+Gr_T\theta(y)+Gr_s\phi(y)=0 \tag{10.7}$$

$$\theta''(Y) + Pr\theta'(Y) + Pr.Du.\phi''(Y) = 0 \tag{10.8}$$

$$\phi''(Y) + Sc\Phi'(Y) = 0 \tag{10.9}$$

$$\begin{aligned} &at\ Y = 0:\ U = 0,\ \theta = \phi = 1 \text{ and } at\ Y \to \infty:\ U(Y) \to 0, \\ &\theta(Y) \to 0\ \phi(Y) \to 0 \end{aligned} \tag{10.10}$$

10.5 ANALYTICAL SOLUTION

The system of ordinary differential equations defined by Equations (10.7)–(10.9), along with the boundary condition (10.10), can be solved and represented by Equations (10.11)–(10.13), respectively.

$$\begin{aligned} U = \left(\frac{Pr.Du.Sc.Gr_T}{Sc + Pr}\right).&\left[\frac{e^{-\left(\frac{1+\sqrt{1+4M\left(1+\frac{1}{Da}\right)}}{2}\right)Y} - e^{-PrY}}{Pr^2 - Pr - \left(M + \frac{1}{Da}\right)} + \frac{e^{-ScY} - e^{-\left(\frac{1+\sqrt{1+4M\left(1+\frac{1}{Da}\right)}}{2}\right)Y}}{Sc^2 - Sc - \left(M + \frac{1}{Da}\right)}\right] \\ &+ Gr_T.\frac{e^{-\left(\frac{1+\sqrt{1+4M\left(1+\frac{1}{Da}\right)}}{2}\right)Y} - e^{-Pr.Y}}{Pr^2 - Pr - \left(M + \frac{1}{Da}\right)} + Gr_s.\frac{e^{-\left(\frac{1+\sqrt{1+4M\left(1+\frac{1}{Da}\right)}}{2}\right)Y} - e^{-Sc.Y}}{Sc^2 - Sc - \left(M + \frac{1}{Da}\right)} \end{aligned} \tag{10.11}$$

$$\theta(y) = e^{-Pr.Y} + \left(\frac{Pr.Du.Sc}{Sc + Pr}\right).\left(e^{-Pr.Y} - e^{-ScY}\right) \tag{10.12}$$

$$\phi(Y) = e^{-ScY} \tag{10.13}$$

10.6 RESULTS AND DISCUSSION

Figures 10.2–10.7 display the effect of velocity on different parameters. Velocity distribution field for the various fluids, including hydrogen (Sc = 0.22), helium (Sc = 0.30), ammonia (Sc = 0.78), and water vapor (Sc = 0.60), is illustrated in Figure 10.2. As the Schmidt number increases, there is an increase in momentum diffusivity, i.e., viscosity of the fluid increases, leading to less velocity of the fluid. Figure 10.3 shows the effect of Prandtl number

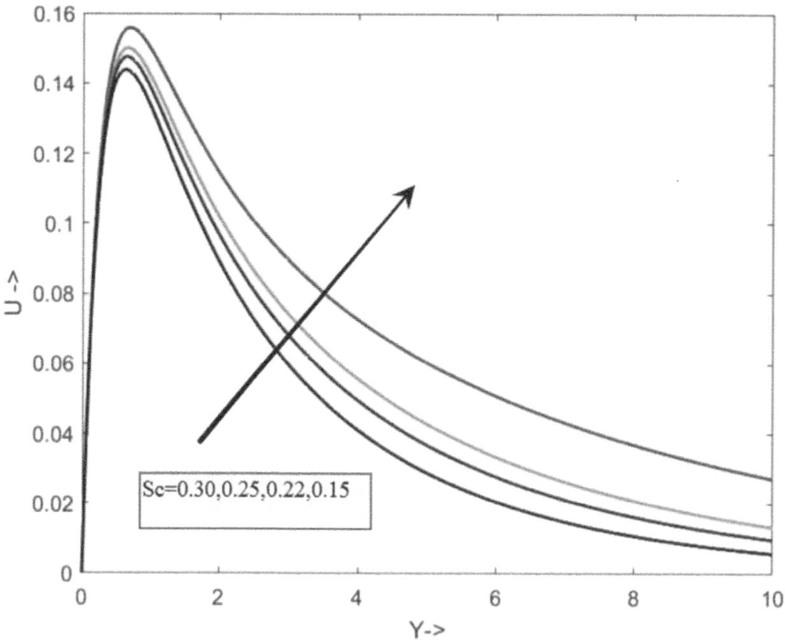

Figure 10.2 Velocity profiles for different values of Sc with Gr_s = 1.5, Pr = 0.71, Da = 0.1.

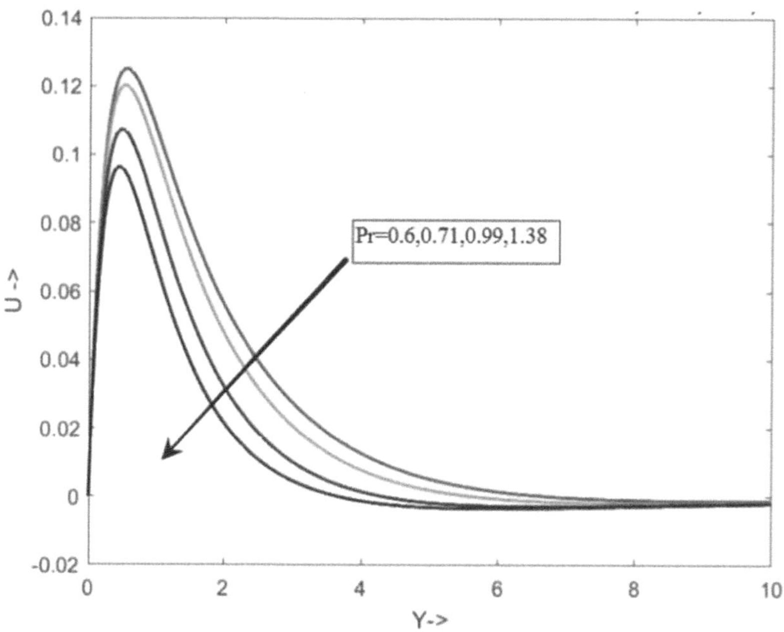

Figure 10.3 Velocity profiles for different values of Pr with Du = 1, Sc = 0.22, Da = 0.1.

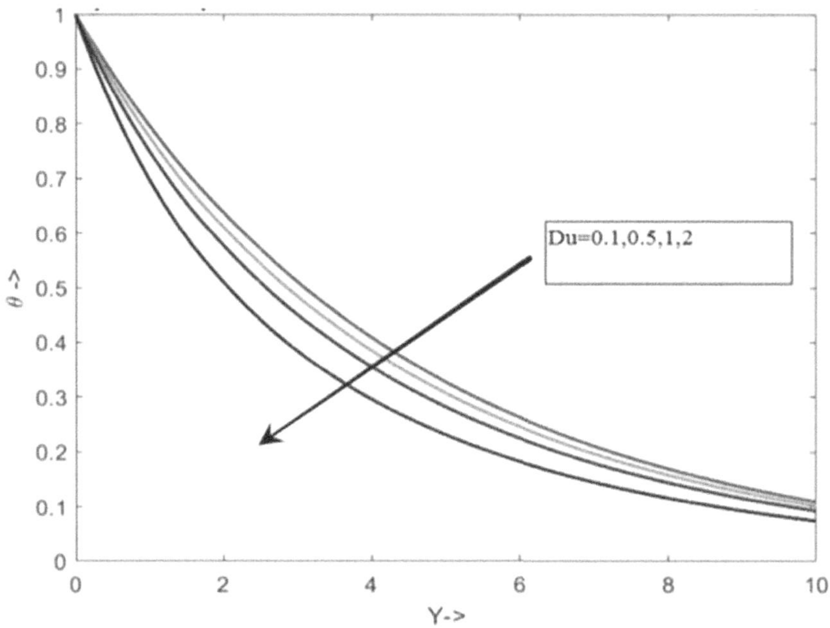

Figure 10.4 Velocity profiles for different values of Da with $Gr_T = 1$, Pr = 0.71, Sc = 0.22.

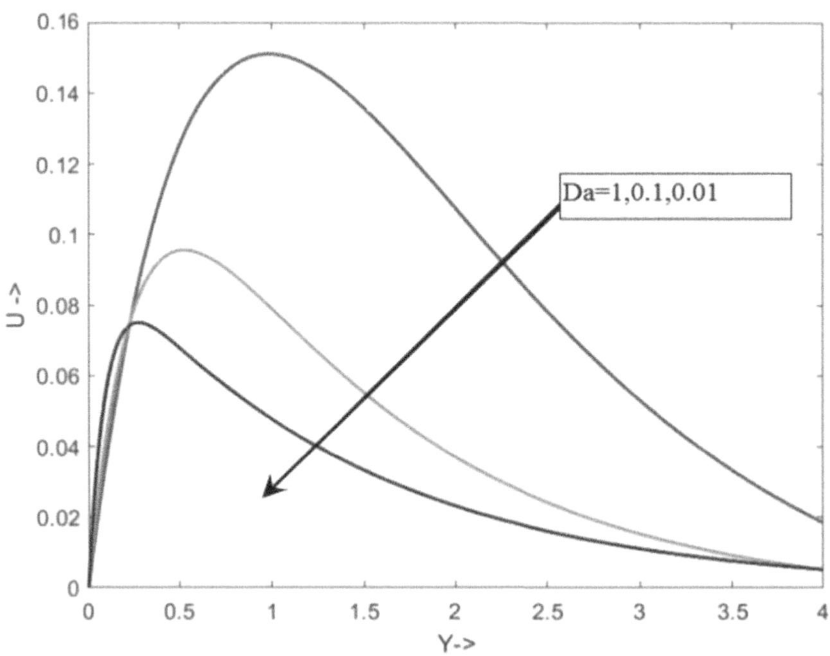

Figure 10.5 Velocity profiles for different values of Du with Sc = 0.22, Pr = 0.71, Da = 0.1.

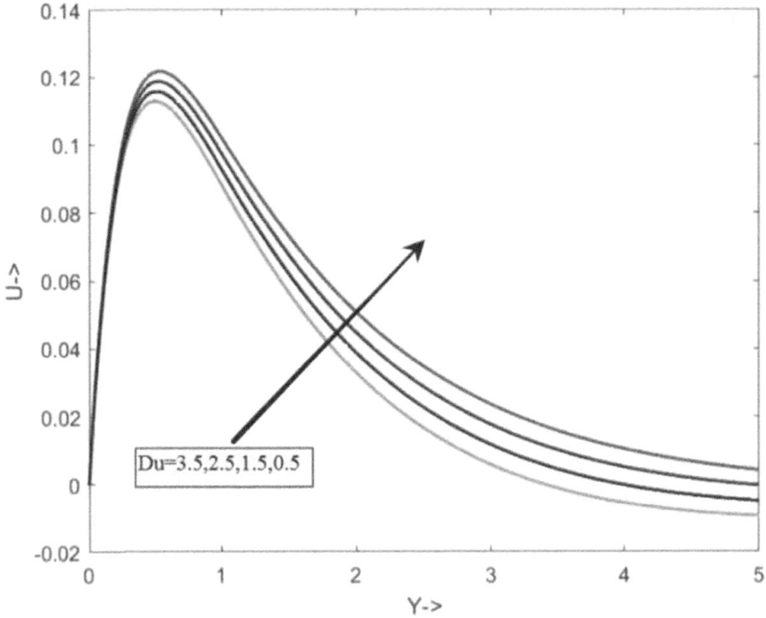

Figure 10.6 Velocity profiles for different values of Gr_T with Sc = 0.22, Pr = 0.71, M = 1.

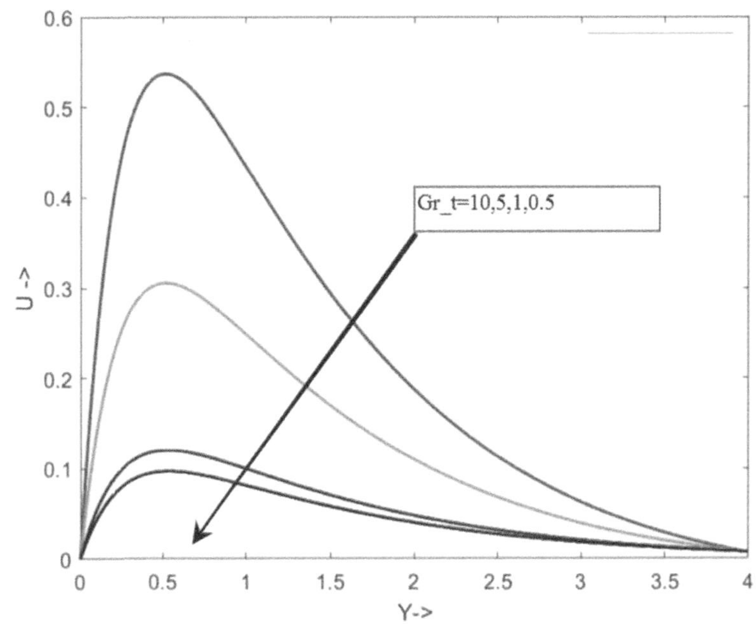

Figure 10.7 Velocity profiles for different values of Gr_S with Sc = 0.22, Pr = 0.71, Da = 0.1, M = 1.

on velocity with Y. It can be noticed that the velocity of fluid increases with a small Prandtl number. This happens due to a small Prandtl number having high thermal conductivity.

We observe from Figure 10.4 that an increase in Darcy number leads to an increase in dimensionless velocity *U*, with respect to an increase in *Y*. This observation arises from the correlation between larger Darcy numbers and higher permeability porous media. Elevated permeability signifies diminished resistance from the porous fibers to fluid flow, consequently leading to an acceleration in transport phenomena.

The influence of the Dufour parameter on the velocity profiles is demonstrated in Figure 10.5. An increment on the Dufour number indicates a comprehensive fall in velocity field, which declines considerably as the Dufour number rises.

Figure 10.6 illustrates the influence of the thermal Grashof number on the velocity field. It is evident that increasing the thermal Grashof number results in a higher velocity profile, attributed to the strengthening of buoyancy forces.

In Figure 10.7, it is observed that fluid velocity increases when the solutal Grashof number increases, which agrees with the natural phenomena because the buoyancy force assists the flow.

We observe from Figure 10.8 the impact of Prandtl number on temperature with *Y*. The figure reveals an inverse relationship between the temperature field and increasing values of the Prandtl number. Specifically, a higher

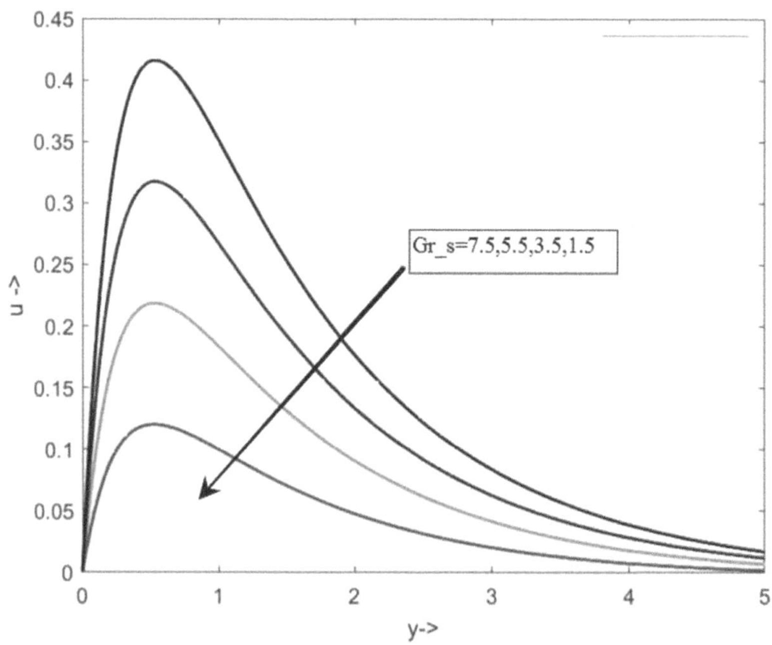

Figure 10.8 Temperature profiles for different values of Pr with Sc = 0.22, Du = 1.

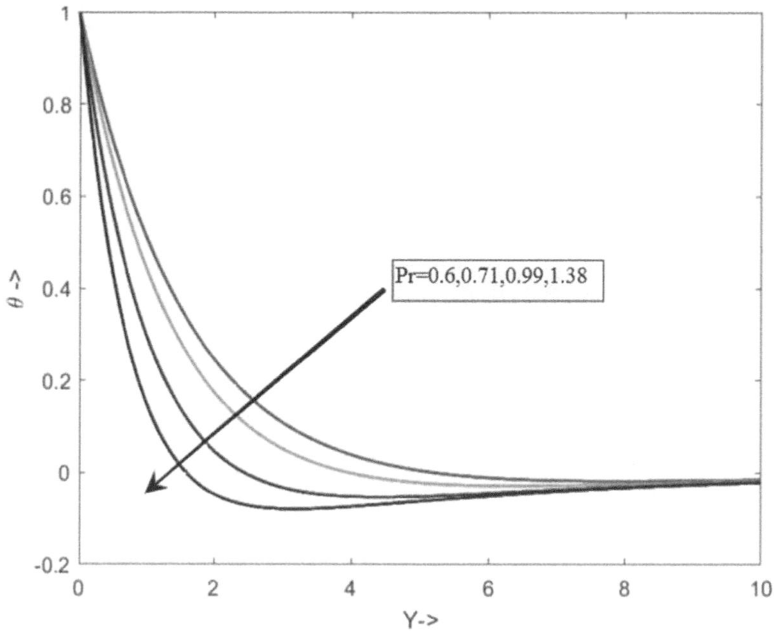

Figure 10.9 Temperature profiles for different values of Du with Sc = 0.22, Pr = 0.71.

Prandtl number corresponds to a decrease in temperature. This happens due to a small Prandtl number having high thermal conductivity.

Figure 10.9 illustrates the impact of the Dufour parameter on the temperature distribution. As depicted, the temperature profiles are sensitive to variations in the Dufour number. Specifically, an increase in Dufour results in a significant elevation of the temperature.

Figure 10.10 graphically illustrates the change of concentration on Schmidt number. It indicates that an increase in the Schmidt number results in a reduction of the concentration field. Since lower Schmidt number have higher mass diffusivity which hikes the concentration field.

Table 10.1 Computational values of U for different parameter

Pr	*Sc*	*Du*	*Da*	Gr_t	Gr_s	*M*	*U*
	0.15						0.15723
0.71	0.22						0.14838
	0.25		0.1				0.14289
	0.3			1			0.13827
0.6							0.12756
0.71							0.11922
0.99		1	0.1				0.11346
1.38							0.09627

(Continued)

Table 10.1 (Continued) Computational values of U for different parameter

Pr	*Sc*	*Du*	*Da*	Gr_t	Gr_s	*M*	*U*
			0.1				0.69824
			0.01				0.01845
			0.001	1	1.5		0.00823
			0.0001			1	0.00067
		0.5		1			0.12147
		1.5					0.11823
0.71	0.22	2.5					0.11567
		3.5					0.11329
				0.5			0.08436
		1	0.1	1			0.09235
				5			0.2536
				10			0.14623
					0.5		0.0523
					1.5		0.08735
				1	3.5		0.18937
					5.5		0.26798

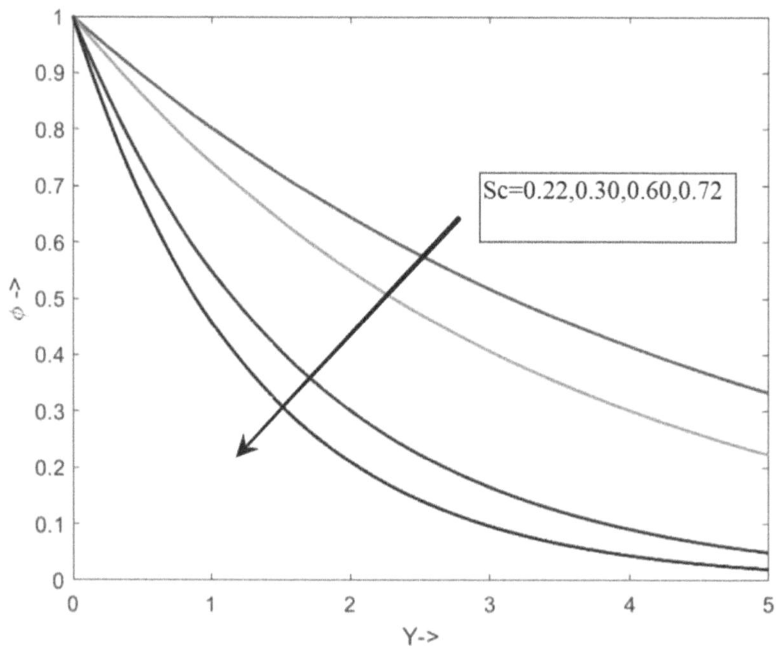

Figure 10.10 Concentration profiles for different values of Sc.

Table 10.2 Computational values of θ for different parameters

Sc	*Pr*	*Du*	θ
	0.6		0.6090
0.22	0.71		0.5740
	0.99	1	0.5100
	1.38		0.1476
	0.71	1.5	0.3877
		2	0.4137
		2.5	0.4396

Table 10.3 Computational values of ϕ for different values of Sc

Pr	*Sc*	Φ
	0.22	0.8025
	0.30	0.7408
0.71	0.60	0.5488
	0.78	0.4584

10.7 CONCLUSION

This study explores the behavior of MHD thermosolutal natural convective flow occurring through a vertical flat plate within a porous medium. We derived local non-similarity equations to model this flow, assuming constant viscosity and thermal conductivity. Numerical calculations are performed specifically for assisting flow conditions, focusing on fluids with low Prandtl numbers, which are characteristic of liquid metals, as well as fluids with small Schmidt numbers.

Velocity, temperature, and concentration profiles are visualized graphically for the case of constant viscosity. The key observations of this study are summarized as follows:

- The velocity profile decreases with increase in Prandtl number, Schmidt number, and Dufour number, while the velocity increases with increase Darcy number, solutal Grashof number, and thermal Grashof number.
- The temperature of the fluid increases with increase in Dufour number, and temperature decreases with increase in Prandtl number.
- The concentration decreases when there is an increase in Schmidt number.

REFERENCES

1. Hayat, T., and Z. Abbas. "Heat transfer analysis on the MHD flow of a second-grade fluid in a channel with porous medium." *Chaos, Solitons & Fractals* 38.2 (2008): 556–567.

2. Rahman, M. M., and M. A. Sattar. "Magnetohydrodynamic convective flow of a micropolar fluid past a continuously moving vertical porous plate in the presence of heat generation/absorption." *Journal of Heat Transfer* 128.2 (2006): 142–152.
3. Vafai, K., and C. L. Tien. "Boundary and inertia effects on flow and heat transfer in porous media." *International Journal of Heat and Mass Transfer* 24.2 (1981): 195–203.
4. Schlichting, H., and K. Gersten. *Boundary-layer theory*, Springer, New York (2016).
5. Sparrow, E. M., and R. D. Cess. "The effect of a magnetic field on free convection heat transfer." *International Journal of Heat and Mass Transfer* 3.4 (1961): 267–274.
6. Riley, N. "Magnetohydrodynamic free convection." *Journal of Fluid Mechanics* 18.4 (1964): 577–586.
7. Chamkha, A. J., and A. M. Aly. "MHD free convection flow of a nanofluid past a vertical plate in the presence of heat generation or absorption effects." *Chemical Engineering Communications* 198.3 (2010): 425–441.
8. Raptis, A., and N. Kafousias. "Heat transfer in flow through a porous medium bounded by an infinite vertical plate under the action of a magnetic field." *International Journal of Energy Research* 6.3 (1982): 241–245.
9. Lloyd, J. R., and E. M. Sparrow. "Combined forced and free convection flow on vertical surfaces." *International Journal of Heat and Mass Transfer* 13.2 (1970): 434–438.
10. Ahmmed, S. F., and M. S. Sarker. "MHD natural convection flow of viscous incompressible fluid from a vertical flat plate." *Journal of Physical Sciences* 13 (2009): 77–85. www.vidyasagar.ac.in/journal.
11. Hossain, M. A., and M. Ahmed. *MHD forced and free convection boundary layer flow near the leading edge.* No. IC–88/140. International Centre for Theoretical Physics, Trieste (Italy) (1988).
12. Hossain, M. A., and I. Pop. "Magnetohydrodynamic boundary layer flow and heat transfer on a continuous moving wavy surface." *Archives of Mechanics* 48 (1996): 813–823.
13. Wilks, G. "Magnetohydrodynamic free convection about a semi-infinite vertical plate in a strong cross field." *Zeitschrift für angewandte Mathematik und Physik ZAMP* 27 (1976): 621–631.
14. Kapoor, S. "MHD double diffusive convective flow past a low-heat-resistance sheet with Soret – Dufour effects under chemically reaction." *Journal of Basic and Applied Engineering Research* 6.2 (2019): 122–126.
15. Nield, D. A., and A. Bejan. *Convection in porous media.* Vol. 3. Springer, New York (2006).
16. Sparrow, E. M., and J. L. Gregg. "Buoyancy effects in forced-convection flow and heat transfer." *Journal of Applied Mechanics* 26.1 (1959): 133–134.
17. Hunt, R., and G. Wilks. "Continuous transformation computation of boundary layer equations between similarity regimes." *Journal of Computational Physics* 40.2 (1981): 478–490.
18. Hossain, M. A., and H. S. Takhar. "Radiation effect on mixed convection along a vertical plate with uniform surface temperature." *Heat and Mass Transfer* 31.4 (1996): 243–248.
19. Tingwei, G., R. Bachrun, and M. Daguenet. "Influence of natural-convection on forced-convection above a vertical plane surface subject to a radiation flux." *International Journal of Heat and Mass Transfer* 25.7 (1982): 1061–1065.
20. Kapoor, S., P. Alam, R. Gupta, L. M. Tiwari, and S. Aggarwal. "Analytical study of MHD natural convective flow of incompressible fluid flow from a vertical flat plate in porous medium." 2011 Fourth International Conference on Modeling, Simulation and Applied Optimization, Kuala Lumpur, Malaysia, 2011, pp. 1–6.

Chapter 11

Finite amplitude analysis and thermoconvective layer of Maxwell fluid in porous medium in LTNE (local thermal non-equilibrium) model with cross-diffusion

A. K. Sahoo and V. Dabral

11.1 INTRODUCTION

In a porous medium with a horizontal layer of Maxwell viscoelastic fluid, double diffusive convection is examined when temperature and concentration gradients (Dufour and Soret effects) are present. The Darcy model is taken into consideration for the porous medium. The emergence of instabilities is investigated using a linear stability analysis based on the normal mode technique for the Maxwell viscoelastic fluid layer confined between two free-free barriers. To examine the effects of the Lewis number, Dufour parameter, Soret parameter, and solutal Rayleigh number on stationary convection, graphs have been made, and based on these computations, we can now analyze the critical conditions that trigger the transition from a quiescent state to convective motion, either through a steady pattern or through time-dependent oscillations. This work also does a finite amplitude analysis for flow stability.

The viscoelastic Maxwell fluid model has been extensively researched, yet these fluids have been described in a variety of ways with constitutive equations that vary substantially in complexity [1–3]. There has been some success with the Maxwell fluid in characterizing polymeric liquids, which makes it easier to analyze and more significant for experiments. The four-parameter Maxwell model's relaxation and retardation functions were established by Friedrich [4]. Song and Jiang [5] analyzed the experimental data of viscoelastic gum using fractional calculus and got good findings.

Using the generalized Oldroyd-B model, Qi and Xu [6, 7] examined Stokes' first problem and a few unstable unidirectional flows for a viscoelastic fluid. The generalized viscoelastic fluid's magnetohydrodynamic (MHD) flows were examined by Zheng et al. [8 9]. Regarding heated generalized second-grade fluids with fractional derivatives, Shen et al. [10, 11] investigated the rate at which vortex velocity decreases, and temperature diffusion occurs in certain fluids, as well as a Rayleigh–Stokes problem involving these fluids. Ezzat has recently proposed several novel models for energy constitutive equations [12].

DOI: 10.1201/9781003470885-11

In Cortell's [13] analysis, the impact of viscous dissipation and radiation on the thermal boundary layer along a nonlinearly stretching sheet was investigated. Ezzat et al. [14] also explored the combined heat and mass transfer in the context of unsteady MHD flow of a perfectly conducting micropolar fluid, considering thermal relaxation effects.

Tiwari et al. [15] worked to understand the influence of Prandtl number for MHD natural convective fluid flow past a low heat resistance sheet, which is physically modeled as a boundary layer flow in viscous media, and solved it using finite difference methods. Using a uniform transverse magnetic field in porous media, Kapoor et al. [16] investigated the heat transfer and two-dimensional laminar natural convection flow of an electrically conducting, viscous, incompressible fluid across a vertical impermeable flat plate. Double-diffusive mixed convection in a vertical pipe under a local thermal non-equilibrium (LTNE) state has been investigated by Bera et al. [17]. In this chapter, Rawat et al. [18] investigated the two-dimensional flow, heat, and mass transfer of chemically reacting micropolar fluid over a nonlinear stretching sheet with variable heat flux in a non-Darcy porous medium.

In this chapter, Sahoo et al. [19] investigated the linear stability mechanism of chemically reacting cross-diffusive free thermosolutal convection of viscoelastic Maxwell fluid in a horizontal layer in a porous medium with heat source effect. Here Kapoor et al. [20] investigated the stability analysis in the case of the linear as well as nonlinear thermo-solutal chemically reactive couple stress fluid under uniform magnetic field convection.

In this chapter, a horizontal layer of Maxwell viscoelastic fluid inside a porous medium has been studied to see how the Dufour and Soret effects affect the onset of instability.

11.2 MATHEMATICAL FORMULATIONS OF THE PROBLEM

Between the planes $z = 0$ and $z = d$ in a porous medium with porosity ε and medium permeability k_1, taking into account an infinite horizontal layer

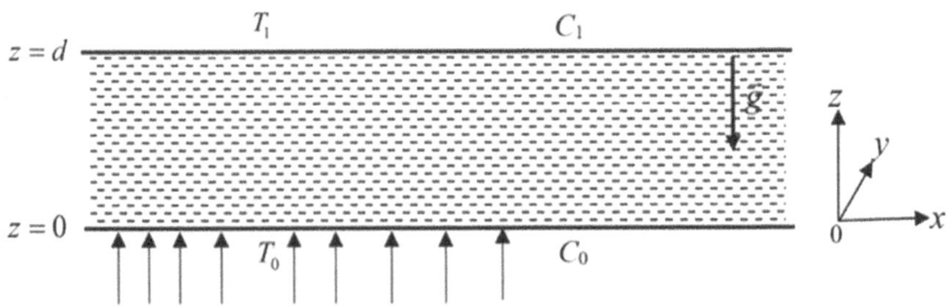

Figure 11.1 Physical configuration.

of Maxwell viscoelastic fluid, with a thickness "d," and subject to gravity $g(0,0,-g)$. as mentioned in Figure 11.1. In order to maintain a consistent temperature and concentration distribution at the fluid layer's boundaries, this layer of fluid is heated and soluted. At $z = 0$, the temperature (T) and concentration (C) are assumed to be T_0 and C_0, and the temperature and concentration difference across the borders to be T_1 and C_1.

The Darcy velocity vector, temperature, viscosity, hydrostatic pressure, density, concentration of solute, thermal diffusivity, solute diffusivity of the fluid, coefficient of thermal expansion, and coefficient of expansion for an analogous solvent are represented by the variables $q(u, v, w)$, T,μ, p, ρ, C, κ, κ' α, and α' accordingly.

11.2.1 Assumptions

The following assumptions form the basis of the mathematical equations that describe the physical model.

In the Boussinesq hypothesis, all thermophysical parameters are constant with the exception of density. When solving for momentum, we use Darcy's model with the time derivative. It is presumed that the porous medium is homogeneous and isotropic. In a fluid layer, no chemical reaction may occur. Matrix fluid and solid are at a temperature point of thermal equilibrium. When compared to other forms of heat transfer, radiation transfer across wall surfaces is almost nonexistent.

11.2.2 Governing equations

A viscoelastic fluid in a porous medium can be described by the following equations:

$$\nabla . q = 0 \tag{11.1}$$

$$\left(1+\lambda\frac{\partial}{\partial t}\right)\left(\frac{\rho_0}{\varepsilon}\frac{\partial q}{\partial t}+\nabla p-\rho g\right)+\frac{\mu}{K}q=0 \tag{11.2}$$

$$\varepsilon\left(\rho_0 c\right)_f\frac{\partial T_f}{\partial t}+\left(\rho_0 c\right)_f\left(q.\nabla\right)T_f=\varepsilon k_f\nabla^2 T_f+h\left(T_s-T_f\right) \tag{11.3}$$

$$\left(1-\varepsilon\right)\left(\rho_0 c\right)_s\frac{\partial T_s}{\partial t}=\left(1-\varepsilon\right)k_s\nabla^2 T_s-h\left(T_s-T_f\right) \tag{11.4}$$

$$\varepsilon\frac{\partial C}{\partial t}+q.\nabla C=k'\nabla^2\, C+D_{CT}\nabla^2 T, \tag{11.5}$$

$$\rho=\rho_0\left[1-\beta_T\left(T_f-T_1\right)+\beta_S\left(C-C_1\right)\right] \tag{11.6}$$

where $q(u,v,w)$ denotes the Darcy velocity, g represents the acceleration due to gravity, the pressure by p, the viscosity is μ, the relaxation time which is denoted by λ, T stands for temperature, and C for concentration. ρ indicates density, and K and ε for permeability and porosity of the medium, respectively. D_{CT} is Soret coefficients; $\sigma = (\rho c_p)_m / (\rho c_p)_f$ is the thermal capacity, wherever $(\rho c_p)_m = (1-\varepsilon)(\rho c)_s + \varepsilon(\rho c)_f$ is the saturated medium's volumetric heat capacity and $(\rho c_p)_f$ is the volumetric heat capacity of the fluid overall, with the subscripts f, m, and s signifying the characteristics of the specified fluid, porous matrix, and solid, respectively. The Soret coefficient is denoted as D_{CT}. Solutal diffusivity of the medium is represented by k. β_S and β_T solute and thermal expansion coefficients in the medium. At the fluid's periphery, we presume that both temperature and concentration remain constant.

Consequently, the boundary conditions are as follows:

$$\begin{aligned} &w = 0,\ T = T_o,\ C = C_o \ \ at\ z = 0 \text{ and } w = 0, \\ &\quad T = T_1,\ C = C_1\ at\ z = d \end{aligned} \tag{11.7}$$

11.2.3 Steady state and its solutions

As a representation of the steady state,

$$\begin{aligned} &u = v = w = 0,\ p = p(z),\ T = T_s(z),\ C = C_s\ (z), q_s = (0,0,0), \\ &\quad p = p_s(z),\ T = T_s(z),\ C = C_s\ (z),\ \rho = \rho_s(z),\ h = 0 \end{aligned} \tag{11.8}$$

The concentration and temperature at the basic states fulfill the equations.

$$\frac{d^2 T_{fb}}{dz^2} = 0,\ \frac{d^2 T_{sb}}{dz^2} = 0,\ \frac{d^2 C_b}{dz^2} = 0 \tag{11.9}$$

With the following boundary conditions

$$\begin{aligned} &T_{fb} = T_{sb} = T_l;\ C_b = C_l\ \ at\ z = 0 \text{ and} \\ &\quad T_{fb} = T_{sb} = T_u;\ C_b = C_u\ \ at\ z = d \end{aligned} \tag{11.10}$$

so that the solutions for the conduction state are given by

$$T_{fb} = T_{sb} = T_l - \Delta T\left(\frac{Z}{d}\right), \tag{11.11}$$

$$C_b = C_l - \Delta C\left(\frac{z}{d}\right) \tag{11.12}$$

11.2.4 Perturbation solution

We applied infinitesimal disturbances of the following kinds to the initial state in order to investigate the system's stability:

$$q = 0 + q', \ T_f = T_{fb} + T_f', \ T_s = T_{sb} + T_s', \ C = C_b + C', \\ p = p_b + p', \ \rho = \rho_b + \rho' \tag{11.13}$$

where the perturbed quantities are denoted by the primes. Ignoring the higher-order terms of the perturbed quantities and substituting (11.13) into (11.1)–(11.6), the formulae governing the perturbations are obtained as follow:

$$\nabla . q' = 0 \tag{11.14}$$

$$\left(1 + \lambda_1 \frac{\partial}{\partial t}\right)\left(\frac{\rho_0}{\varepsilon}\frac{\partial q'}{\partial t} + \nabla p' + \rho_0\left(\beta_T T' - \beta_C C'\right)g\right)\Bigg) + \frac{\mu}{K} q' = 0 \tag{11.15}$$

$$\varepsilon \left(\rho_0 c\right)_f \frac{\partial T_f'}{\partial t} + \left(\rho_0 c\right)_f \left(q'.\nabla\right) T_f' + \left(\rho_0 c\right)_f \ w'\left(\frac{d\, T_{fb}}{dz}\right) = \varepsilon k_f \nabla^2 T_f' + h\left(T_S' - T_f'\right) \tag{11.16}$$

$$(1-\varepsilon)\left(\rho_0 c\right)_s \frac{\partial T_S'}{\partial t} = (1-\varepsilon) k_s \nabla^2 T_S' - h\left(T_S' - T_f'\right) \tag{11.17}$$

$$\varepsilon \frac{\partial C'}{\partial t} + \left(q'.\nabla\right) C' - w' \frac{\Delta C}{d} = k' \nabla^2 \ C' + D_{CT} \nabla^2 T', \tag{11.18}$$

Curl is applied twice to Eq. (11.15), removing p' from it. The resulting equation is then rendered. Other equations are dimensionless using the following transformations as:

$$\left(x'', y'', z''\right) = \frac{1}{d}(x', y', z'), \ \left(u'', v'', w''\right) = \frac{\left(\rho_0 c\right)_f d}{k}(u', v', w'), \\ T_f' = (\Delta T) T_f'', \ T_S' = (\Delta T) T_S'' \ t'' = \frac{d^2 \left(\rho_0 c\right)_f}{k_f} t', \ C'' = \frac{C'}{\Delta c'}, \ p'' = \frac{k_1 d^2}{\mu k} p' \tag{11.19}$$

$$\nabla . q = 0 \tag{11.20}$$

$$\left(1 + \lambda \frac{\partial}{\partial t}\right)\left(\frac{d^2}{\gamma v} \nabla^2 w + Ra_T \frac{\partial T_f}{\partial t} - Ra_C \frac{\partial C}{\partial t}\right) - \nabla^2 w = 0 \tag{11.21}$$

$$\left(\frac{\partial}{\partial \mathrm{t}} - \nabla^2\right) T_f - \nabla^2 \mathrm{C} - w = \mathrm{H}\left(\mathrm{T_S} - \mathrm{T_f}\right) \tag{11.22}$$

$$\left(\beta_T \frac{\partial}{\partial \mathrm{t}} - \nabla^2\right) T_S = \gamma \mathrm{H}\left(\mathrm{T_f} - \mathrm{T_S}\right) \tag{11.23}$$

$$\left(\eta \frac{\partial}{\partial \mathrm{t}} - \frac{1}{\mathrm{Le}} \nabla^2\right) \mathrm{C} - S_r \nabla^2 \mathrm{C} - w = 0 \tag{11.24}$$

Here non-dimensional parameters are as follows:

$Ra = \frac{\alpha\ dk_1\ \Delta T g \rho_0}{\mu K}$ is the thermal Rayleigh number, $Rs = \frac{\alpha'\ dk_1\ \Delta C g \rho_0}{\mu k'}$ is the solutal Rayleigh number, Va is the Vadasz number, $Le = \frac{k}{k'}$ represents the Lewis number, $F = \left(\frac{k}{\sigma d^2}\right)\lambda$ denotes the stress relaxation parameter. $H = \frac{hd^2}{\varepsilon k_f}$ represents the interphase heat transfer coefficient, and $Sr = \frac{D_{CT}}{k}\frac{\Delta T}{\Delta C}$ is the Soret parameter.

The boundary conditions that are nondimensional, are mentioned below:

$$w = T = C = 0 \ at \ z = 0,\ z = 1 \tag{11.25}$$

11.3 NORMAL MODES AND STABILITY ANALYSIS

Suppose that the perturbed quantities are of the following kind and analyze the disturbances into the normal modes.

$$\begin{pmatrix} w \\ \mathrm{T_f} \\ \mathrm{T_S} \\ \mathrm{C} \end{pmatrix} = e^{\omega t} \begin{pmatrix} \psi \sin(ax) \\ \Theta \cos(ax) \\ \Phi \cos(ax) \\ \Sigma \cos(ax) \end{pmatrix} \sin(\pi z) \tag{11.26}$$

The growth rate of disturbances is denoted by n, whereas the wave numbers along x and y directions are represented by k_x and k_y, respectively. By using (11.26), (11.20)–(11.24) become into

$$\begin{bmatrix} \delta^2\left(\frac{\omega}{Va} + 1\right) & -aRa_T & 0 & aRa_S \\ -a & \omega + \delta^2 + H & -H & 0 \\ 0 & -\gamma H & \alpha\omega + \delta^2 + \gamma H & 0 \\ -a & 0 & S_r \delta^2 & \frac{\delta^2}{Le} + \omega \end{bmatrix} \begin{bmatrix} \psi \\ \Theta \\ \Phi \\ \Sigma \end{bmatrix} = \begin{bmatrix} 0 \\ 0 \\ 0 \\ 0 \end{bmatrix} \tag{11.27}$$

$$\delta^2 = \pi^2 + a^2$$

The following matrix equation is obtained by replacing solution (11.13) in Eq. (11.11), each equation from $z = 0$ to $z = 1$ is integrated by parts, as the nontrivial solution of the above matrix necessitates that

$$Ra_T = \frac{\delta^2\left(\frac{\omega}{Va}+1\right)\left[\left(\alpha\omega+\delta^2+\gamma H\right)\left(-\alpha\omega+\delta^2+\gamma H\right)-H^2\gamma\right]}{a^2\left(\alpha\omega+\delta^2+\gamma H\right)} - \frac{a^2\left[\left(\alpha\omega+\delta^2+\gamma H\right)^2-H^2\gamma\right]+\left[a\gamma HS_r\delta^2\right]}{a^2\left(\alpha\omega+\delta^2+\gamma H\right)\left(\omega+\frac{\delta^2}{Le}\right)}Ra_S \tag{11.28}$$

Lewis number *Le*, solutal Rayleigh number *Rs*, and Soret parameter *Sr* are all independent of the stress relaxation parameter, as is evident from (11.28) as well. Additionally, the dimensionless wave number *a* determines the stationary Rayleigh number *Ra*. For stationary convection, the Maxwell viscoelastic fluid acts similarly to a typical Newtonian fluid.

11.4 FINITE AMPLITUDE STEADY CONVECTION

This section focuses on just two terms and analyzes the nonlinear analysis using a simplified Fourier series representation. Although the stability condition of the motionless solution may be determined using the linear stability analysis, and the corresponding eigenfunctions that characterize the convective flow qualitatively, it is not able to furnish details regarding the convection amplitude values or the heat transfer rate. We conduct nonlinear analysis to get this extra knowledge, which helps us comprehend the physical mechanism with the least amount of mathematical analysis and advances our grasp of the complete nonlinear situation. The subject of thermal convection in porous media with finite amplitude and local thermal equilibrium (LTE) conditions is well covered in the literature for systems with one or two components. The methodology developed by these authors has been applied in this work to examine the effects of double diffusive convection on a porous layer caused by LTNE.

The finite amplitude steady-state convection is described by a minimal double Fourier series, which is given by

$$\psi = A\sin(ax)\sin(\pi z) \tag{11.29}$$

$$T_f = B_1\cos(ax)\sin(\pi z) + B_2\sin(2\pi z) \tag{11.30}$$

$$T_S = B_3 \cos(ax)\sin(\pi z) + B_4 \sin(2\pi z) \tag{11.31}$$

$$C = B_5 \cos(ax)\sin(\pi z) + B_6 \sin(2\pi z) \tag{11.32}$$

where the system dynamics can be used to determine the steady-state amplitudes A, B_i, which are constants. The following nonlinear system equations are obtained by equating the coefficients of like terms and substituting Eqs. (11.29)–(11.32) through the coupled nonlinear system of partial differential Eqs. (11.20)–(11.24) into the stable section.

$$\left(\pi^2 + a^2\right)A + aRa_T B_1 - aRa_S B_5 = 0 \tag{11.33}$$

$$aA + \left[\left(\pi^2 + a^2\right) + H\right]B_1 - HB_3 + \pi a A B_2 = 0 \tag{11.34}$$

$$2\left[4\pi^2 + H\right]B_2 - 2HB_4 - \pi a A B_1 = 0 \tag{11.35}$$

$$\gamma H B_1 - \left[\left(\pi^2 + a^2\right) + \gamma H\right]B_3 = 0 \tag{11.36}$$

$$\gamma H B_2 - \left[4\pi^2 + \gamma H\right]B_4 = 0 \tag{11.37}$$

$$aA + \frac{1}{Le}\left(\pi^2 + a^2\right)B_5 + \pi a A B_6 = 0 \tag{11.38}$$

$$\frac{8\pi^2}{Le}B_6 - \pi a A B_5 = 0 \tag{11.39}$$

The utilization of steady-state solutions is advantageous as they forecast the possibility of a finite amplitude solution for Rayleigh number subcritical values and the lowest Ra_T values for which a stable solution can be found. These values fall below the limits of instability to either an overstable infinitesimal perturbation or a marginal state. If A is the only amplitude left, the outcome is

$$A\left\{\left(\pi^2 + a^2\right) - a^2 Ra_T\left[\frac{\left(\pi^2 + a^2\right)\left\{\pi^2 + a^2 + H(1+\gamma)\right\}}{\left(\pi^2 + a^2 + \gamma H\right)} + \frac{a^2\left(4\pi^2 + \gamma H\right)}{4\pi^2 + H(1+\gamma)}\left(\frac{A^2}{8}\right)\right]^{-1}\right.$$
$$\left. + a^2 Ra_S\left[\frac{\left(\pi^2 + a^2\right)}{Le} + a^2 Le\left(\frac{A^2}{8}\right)\right]^{-1}\right\} = 0 \tag{11.40}$$

while it is unstable when Ra_T is sufficiently large, we know that pure conduction, which is represented by the solution $A=0$, is a viable solution. The residual answers are provided by

$$\frac{A^2}{8}=\frac{1}{2x_1}\left[-x_2+\sqrt{x_2^2-4x_1x_3}\right] \tag{11.41}$$

where

$$x_1=\frac{a^4\delta^2 Le\left(4\pi^2+\gamma H\right)}{4\pi^2+H(1+\gamma)} \tag{11.42}$$

$$\begin{aligned} x_2=&\frac{a^2\delta^4 Le\left[\delta^2+H(1+\gamma)\right]}{\delta^2+\gamma H}+\frac{a^2\delta^4 Le\left(4\pi^2+\gamma H\right)}{Le\left[4\pi^2+H(1+\gamma)\right]}\\ &+\frac{a^4 Ra_S\left(4\pi^2+\gamma H\right)}{\left[4\pi^2+H(1+\gamma)\right]}-a^4 LeRa_T \end{aligned} \tag{11.43}$$

$$x_3=\left(\frac{\delta^6}{Le}+a^2\delta^2 Ra_S\right)\left(\frac{\delta^2+H(1+\gamma)}{\delta^2+\gamma H}\right)-\frac{a^2\delta^2}{Le}Ra_T \tag{11.44}$$

The above formula for the finite amplitude Rayleigh number, Ra_T^F, which denotes the start of finite amplitude steady motions, can be obtained by eliminating the radical from the expression.

The following is a representation of the finite amplitude Rayleigh number:

$$Ra_T^F=\frac{1}{2y_1}\left[-y_2+\sqrt{y_2^2-4y_1y_3}\right] \tag{11.45}$$

where

$$y_1=a^4Le^2\left(\delta^2+\gamma H\right)\left[4\pi^2+H(1+\gamma)\right] \tag{11.46}$$

$$\begin{aligned} y_2=-2\,a^2\Big[&a^2LeRa_S\left(\delta^2+\gamma H\right)\left(4\pi^2+\gamma H\right)+\delta^4\Big\{H^2\left[Le(1+\gamma)+Le\right]\\ &\left[Le(1+\gamma)-\gamma\right]+4\pi^2\delta^2\left(Le^2-1\right)+H\left(\delta^2+4\pi^2\right)\left[Le^2(1+\gamma)-\gamma\right]\Big\}\Big] \end{aligned} \tag{11.47}$$

$$\begin{aligned} y_3=&\frac{1}{Le^2\left(\delta^2+\gamma H\right)\left[4\pi^2+H(1+\gamma)\right]}\Big[a^2LeRa_S\left(4\pi^2+\gamma H\right)\times\left(\delta^2+\gamma H\right)\\ &-\delta^4\Big\{H^2\left[Le(1+\gamma)+Le\right]\left[Le(1+\gamma)-\gamma\right]+4\pi^2\delta^2\left(Le^2-1\right)\\ &-H\left(\delta^2+4\pi^2\right)\left[Le^2(1+\gamma)-\gamma\right]\Big\}\Big]^2 \end{aligned} \tag{11.48}$$

11.5 HEAT AND MASS TRANSPORTS

We can determine the heat and mass transmission once we know the amplitude. Heat and mass transfer quantification are crucial to the study of convection via porous materials. This is because the effect of convection on heat transfer makes its commencement easier to detect as the Rayleigh number increases. Heat and mass are transported solely by conduction in the fundamental condition. If the rates of mass transfer per unit area and heat transfer per unit area for the fluid phase are denoted by $\bar{J}$ and $\bar{H}$, respectively, then

$$\bar{H} = -k_f \left\langle \frac{\partial T_{f\ total}}{\partial z} \right\rangle_{z=0} \tag{11.49}$$

$$\bar{J} = -D \left\langle \frac{\partial C_{total}}{\partial z} \right\rangle_{z=0} \tag{11.50}$$

where the angular bracket denotes a horizontal average and

$$T_{f\ total} = T_l - \Delta T\left(\frac{Z}{d}\right) + T_f\ (x,z) \tag{11.51}$$

$$C_{total} = C_l - \Delta C\left(\frac{Z}{d}\right) + C(x,z) \tag{11.52}$$

Eqs. (11.29)–(11.32) can be substituted into Eqs. (11.51) and (11.52), and the resulting equations can then be used in Eqs. (11.49) and (11.50). This yields

$$\bar{H} = \frac{k_f \Delta T}{d}\left(1 - 2\pi B_2\right) \tag{11.53}$$

$$\bar{J} = \frac{D\Delta C}{d}\left(1 - 2\pi B_6\right) \tag{11.54}$$

The Sherwood number and the fluid phase Nusselt number are calculated by

$$Nu = \frac{\bar{H}}{\frac{k_f \Delta T}{d}} = \left(1 - 2\pi B_2\right) \tag{11.55}$$

$$Sh = \frac{\bar{J}}{\frac{D\Delta C}{d}} = \left(1 - 2\pi B_6\right) \tag{11.56}$$

After writing B_2 and B_6 in terms of A using Eqs. (11.33–11.39) and then inserting them into Eqs. (11.55) and (11.56), we get:

$$Nu = 1 + 2a^2 \frac{A^2}{8}\left[(\delta^2 + \gamma H)(4\pi^2 + \gamma H) / \left\{\delta^2\left[\delta^2 + H(\gamma+1)\right]\right.\right.$$
$$\left.\left.\left[4\pi^2 + H(\gamma+1)\right] + a^2(4\pi^2 + \gamma H)\times(\delta^2 + \gamma H)\left(\frac{A^2}{8}\right)\right\}\right] \tag{11.57}$$

$$Sh = 1 + 2a^2 \frac{A^2}{8}\left[\frac{Le^2}{\delta^2 + a^2 Le^2\left(\frac{A^2}{8}\right)}\right] \tag{11.58}$$

In Eqs. (11.57) and (11.58), the convective contribution to heat and mass transport is represented by the second component on the right side. It is clear that the convective heat and mass transmission line separates from the conductive heat transfer line at the critical value of the thermal Rayleigh number, as $Nu = Sh = 1$ for all $Ra_T \le Ra_{T,c}^{St}$. It is noteworthy that our approach with finite amplitude is applicable to thermal Rayleigh numbers in the vicinity of the convection threshold. Thus, an upper bound value of 3 limits is imposed on both the Nusselt number and the Sherwood number in the current investigation. Only by adding more terms to the Fourier series representation, which permits the wave number to vary in response to changes in the thermal Rayleigh number, can better results be achieved. For the limiting scenarios H →0 and H →∞, we now want to derive the quantification of heat and mass transports. Eq. (11.57) produces when H →0.

$$Nu = 1 + 2a^2 \frac{A^2}{8}\left[\frac{1}{a^2 + \delta^2\left(\frac{A^2}{8}\right)}\right]$$

While Eq. (11.58) stays the same, Eq. (11.37) provides $\frac{A^2}{8}$ where $x_1 = a^4\delta^2 Le$, $x_2 = \frac{a^2}{Le}\left[\delta^4\left(1 + Le^2\right) + a^2 Le\gamma\left(Ra_S - Ra_T Le\right)\right]$

$x_3 = \frac{a^2}{Le}\left[\delta^4 + a^2\left(LeRa_S - Ra_T\right)\right]$

Furthermore, when H →∞, Eqs. (11.57) reads

$$Nu = 1 + 2a^2 \frac{A^2}{8}\left[\frac{1}{a^2 + \delta^2\left(1 + \frac{1}{\gamma}\right)^2\left(\frac{A^2}{8}\right)}\right]$$

and Eq. (11.58) remains invariant, with $\frac{A^2}{8}$ given by Eq. (11.41) where

$$x_1 = \left(\frac{\gamma}{\gamma+1}\right)\alpha^4\delta^2 Le$$

$$x_2 = \frac{\alpha^2}{Le\gamma(\gamma+1)}\left[\delta^4\left\{\gamma^2 + Le^2(\gamma+1)^2\right\} + \alpha^2 Ley\left\{\gamma Ra_S - Ra_T Le(\gamma+1)\right\}\right]$$

$$x_3 = \frac{\delta^2}{\gamma Le}\left[\delta^4(\gamma+1) + a^2\left\{LeRa_S(\gamma+1) - \gamma Ra_T\right\}\right]$$

11.6 RESULTS AND DISCUSSION

By taking into account a thermal nonequilibrium, in a horizontal fluid-saturated sparsely packed porous layer, linear and nonlinear double diffusive convection are analyzed analytically. Using the linear theory, the onset criteria for both oscillatory and marginal convection are obtained. The minimum form of the Fourier series is used to construct analytically the expression for the finite amplitude Rayleigh number. Investigations are conducted into how solute diffusion and temperature non-equilibrium affect the system's stability. It is discovered that the majority of the time, finite amplitude motions cause the instability to begin before stationary or oscillatory convection.

The influence of the Vadasz number in Figure 11.2 is clearly defined. As the value of the Vadasz number increases, the critical Rayleigh number decreases. This result indicates a destabilizing effect on the pure conduction state. Physically, the increase in Vadasz number signifies an increase in horizontal permeability, allowing the fluid to move more easily in the horizontal plane. This reduction in the resistance to fluid motion enhances advective transport, thereby lowering the threshold (Rayleigh number) required for the onset of convection.

The onset of convection is encouraged by increasing the conductivity ratio , as demonstrated by the reduction in the critical Rayleigh number shown in Figure 11.3. An increase in the fluid's thermal conductivity directly leads to an increase in γ resulting in a destabilizing effect.

Increasing the thermal conductivity has a stabilizing effect on the critical wave number, which effectively delays the onset of convection, particularly for large values of layer thickness. This behavior is evident in Figure 11.4. Conversely, for smaller values of layer thickness, the solid thermal conductivity, only weakly affects the convection.

The variable viscosity has a significant, destabilising effect on the system, which is evident in Figure 11.5. As the parameter, (which quantifies the viscosity variation) increases, the critical Rayleigh number decreases for every value of layer thickness. This outcome is physically reasonable, as fluids whose

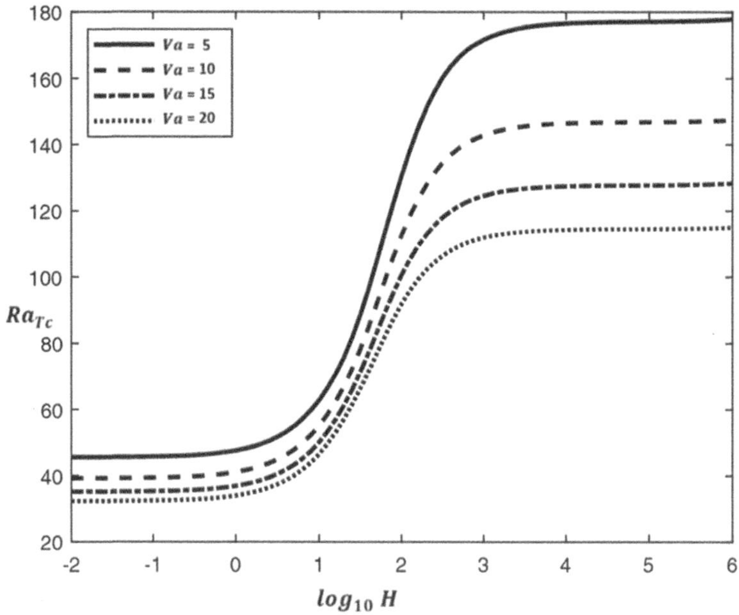

Figure 11.2 The variation of the critical Rayleigh number with *H* for different values of Vadasz number, *Va*.

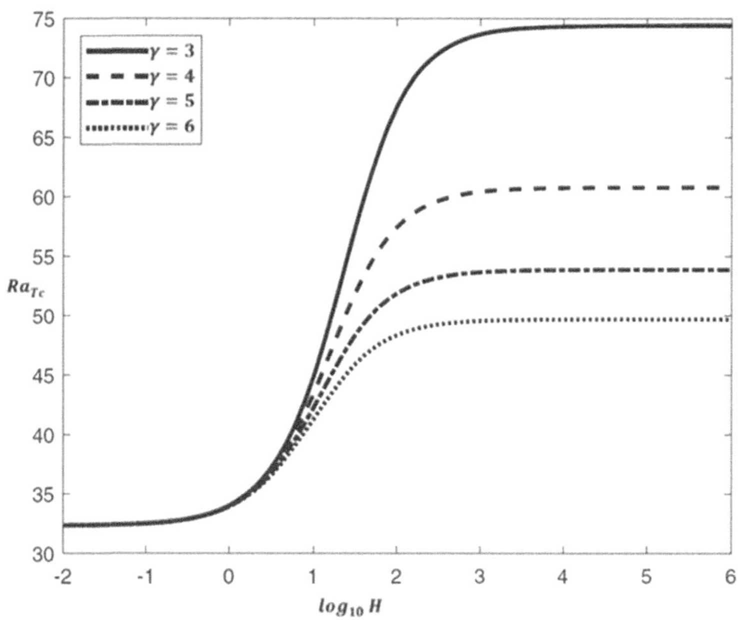

Figure 11.3 The variation of the critical Rayleigh number with *H* for different values of conductivity ratio, γ.

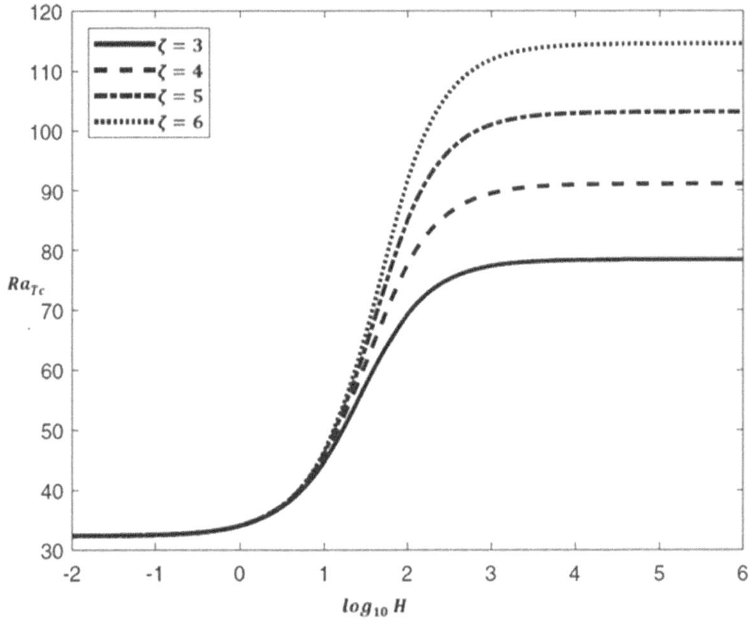

Figure 11.4 The variation of the critical Rayleigh number with H for different values of thermal conductivity, (symbol of conductivity to be mentioned).

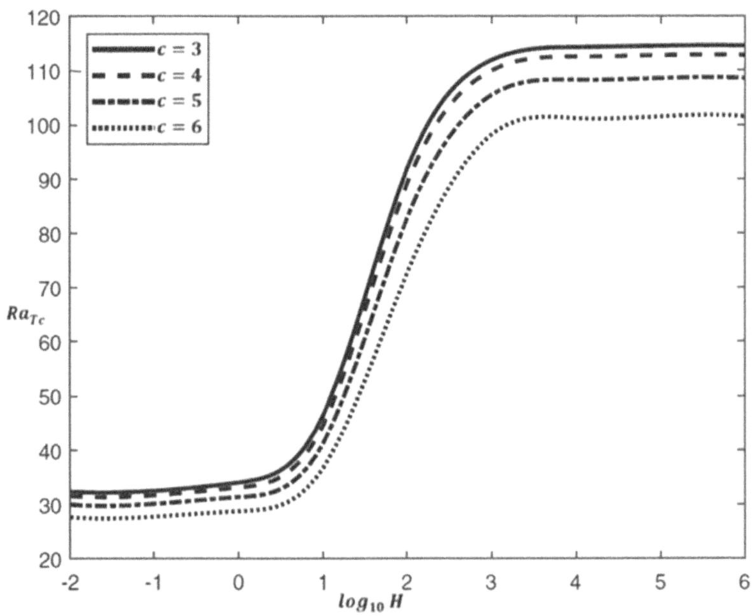

Figure 11.5 The variation of the critical Rayleigh number with H for different values of viscosity variation, c.

viscosity is lower near the hot boundary are more likely to become unstable sooner, as the reduced viscosity diminishes the resistance to fluid motion.

When the interface heat transfer coefficient is large, convection becomes the oscillatory mode first, while for small to moderate values of finite amplitude motions occur before oscillatory motions. This suggests that in the case of a densely packed porous media. The critical Rayleigh number is found to decrease with decreasing for stationary, oscillatory, and finite amplitude convection. Reduction in lessens the viscous effects at the borders, allowing the fluid to flow more freely and lowering the critical Rayleigh number. The magnitude of is connected to these effects. Specifically, the graphical representation demonstrates that the fastest growing disturbances, characterized by the critical wavenumber, are most pronounced at moderate values of exhibiting a peak before diminishing for all three convection (stationary, oscillatory, and finite amplitude) modes considered. Nonetheless, the crucial wavenumber is unaffected by both small and high. It is crucial to determine the mass and heat transfer across the layer while studying double diffusive convection. Here, the effect of increasing the Rayleigh number on mass and heat transmission allows for a quicker detection of the commencement of convection.

11.7 CONCLUSION

Analytical studies are conducted on linear and nonlinear double diffusive convection in a horizontal fluid-saturated sparsely packed porous layer where the fluid and solid phases are not in LTE. A thorough investigation has been conducted into how the boundary and LTNE work together to influence the beginning of convection with double diffusion in a porous layer. The linear theory states that the interphase heat transfer coefficient, Rayleigh number, diffusivity ratio, Lewis number, solute's Darcy number, and Vadasz number all affect the stationary and oscillatory convection thresholds.

The nonlinear theory indicates that motions with finite amplitude will occur. We discovered that the convective instability manifests as an oscillating and finite amplitude mode rather than a stationary one as heat processes and solute diffusion are in competition. Interestingly, despite the substantial influence of intermediate interphase heat transfer coefficients on the characteristics of stationary, oscillatory, and finite amplitude convection, the system tends to conform to LTE behavior in the limiting cases of very large and very small values of this coefficient.

Double diffusive convection's beginning will be prevented by the presence of a stabilizing solute gradient. Reduction in *Da* lowers the critical Rayleigh number by lessening the significance of viscous effects at the borders, which is correlated with the size of *Da*. Porosity-modified conductivity ratio and Vadasz number, increasing the system's instability. The values of *Nu* and *Sh* grow with each of the following parameters: Ra_S, *H*, *Le*, and *Da*.

Nomenclature

a: Wave number
C: Solute concentration
d: Thickness
Da: Darcy number
D_{CT}: Soret coefficients
F: Stress relaxation parameter
g: Gravitational force
H: Interphase heat transfer coefficient
$\bar{H}$: The rate of heat transport per unit area for the fluid
$\bar{J}$: The rate of mass transport per unit area for the fluid
K: Permeability
k_x: Wave numbers along x-axis
k_y: Wave numbers along y-axis
Le: Lewis number
n: Growth rate of disturbances
Nu: Nusselt number
p: Hydrostatic pressure
q: Darcy velocity vector
Ra: Rayleigh thermal number
Rs: Solutal Rayleigh number
Ra_T^F: Finite amplitude Rayleigh number
S_r: Soret parameter
Sh: Sherwood number
T: Temperature
Va: Vadasz number

Greek symbols

α: Thermal expansion coefficient
α': Solvent expansion coefficient
β_T: Coefficient of thermal expansion
β_S: Coefficient of solute expansion
ε: Porosity
k: Solutal diffusivity
κ: Thermal diffusivity
κ': Solute diffusivity
λ: Relaxation time
μ: Viscosity
ρ: Density
σ: Thermal capacity

Subscript

b: Basic state
c: Critical state
f: Fluids
m: Porous matrix
s: Solid
0: Reference value

Superscript

$'$: Perturbed quantity

REFERENCES

1. Fetecau, C., Athar, M. and Fetecau, C. Unsteady flow of a generalized Maxwell fluid with fractional derivative due to a constantly accelerating plate. *Computers and Mathematics with Applications*, **57**, 596–603 (2009).
2. Hayat, T. and Sajid, M. Homotopy analysis of MHD boundary layer flow of an upper-convected Maxwell fluid. *International Journal of Engineering Science*, **45**, 393–401 (2007).
3. Tan, W. C. and Masuoka, T. Stability analysis of a Maxwell fluid in a porous medium heated from below. *Physics Letters A*, **360**, 454–460 (2007).
4. Podlubny, I. *Fractional differential equations*, Academic Press, New York (1999)
5. Guo, B. L., Pu, X. K. and Huang, F. H. *Fractional partial differential equations and their numerical solutions*, Science Press, Beijing (2011)
6. Bagley, R. L. and Torvik, P. J. A theoretical basis for the application of fractional calculus to viscoelasticity. *Journal of Rheology*, **27**, 201–210 (1983).
7. Friedrich, C. Relaxation and retardation functions of the Maxwell model with fractional derivatives. *Rheologica Acta*, **30**, 151–158 (1991).
8. Song, D. Y. and Jiang, T. Q. Study on the constitutive equation with fractional derivative for the viscoelastic fluids modified Jeffreys model and its application. *Rheologica Acta*, **27**, 512–517 (1998).
9. Qi, H. T. and Xu, M. Y. Stokes' first problem for a viscoelastic fluid with the generalized Oldroyd-B model. *Acta Mechanica Sinica*, **23**, 463–469 (2007).
10. Qi, H. T. and Xu, M. Y. Some unsteady unidirectional flows of a generalized Oldroyd-B fluid with fractional derivative. *Applied Mathematical Modeling*, **33**, 4184–4191 (2009).
11. Zheng, L. C., Liu, Y. Q. and Zhang, X. X. A new model for plastic-viscoelastic magnetohydrodynamic (MHD) flow with radiation thermal transfer. *International Journal of Nonlinear Sciences and Numerical Simulation*, **14**, 435–441 (2013).
12. Zheng, L. C., Liu, Y. Q. and Zhang, X. X. Exact solutions for MHD flow of generalized Oldroyd-B fluid due to an infinite accelerating plate. *Mathematical and Computer Modelling*, **54**, 780–788 (2011).
13. Cortell, R. Effects of viscous dissipation and radiation on the thermal boundary layer over a nonlinearly stretching sheet. *Physics Letters A*, **371**, 631–636 (2008).
14. Ezzat, M., El-Bary, A. A. and Ezzat, S. Combined heat and mass transfer for unsteady MHD flow of perfect conducting micropolar fluid with thermal relaxation. *Energy Conversion and Management*, **52**, 934–945 (2011).
15. Tiwari, L. M., Kapoor, S., Agarwal, S. and Dabral, V. Effect of Prandtl number in free convective MHD flow past a low-heat-resistance sheet in porous media. *National Workshop-Cum-Conference on Recent Trends in Mathematics and Computing 2011*, RTMC, 6, 26–30 (May 2012).
16. Kapoor, S., Alam, P., Gupta, R., Tiwari, L. M. and Aggarwal, S., "Analytical study of MHD natural convective flow of incompressible fluid flow from a Vertical Flat Plate in Porous Medium", *IEEE Conference Proceeding* (2011) 1–6.
17. Bera, P., Kapoor, S. and Khandelwal, M. K. Double-diffusive mixed convection in a vertical pipe: A thermal non-equilibrium approach. *International Journal of Heat and Mass Transfer*, **55**, 7079–7092 (2012).
18. Rawat, S., Kapoor, S. and Bhargava, R. MHD flow heat and mass transfer of micropolar fluid over a nonlinear stretching sheet with variable micro inertia density, heat flux and chemical reaction in a non-Darcy porous medium. *Journal of Applied Fluid Mechanics*, 9(1), 321–331 (2016).

19. Sahoo, A. K. and Kapoor, S. Stability analysis of chemically reacting cross diffusive thermosolutal convective flow of Maxwell fluid passing through porous medium with heat source effect. *Mathematics in Science, Engineering, and Aerospace*, **15**(2) (2024).
20. Kapoor, S., Sahoo, A. K. and Dabral, V. Linear and weakly nonlinear stability of thermo-solutal magnetoconvective chemically reacting couple stress fluid in porous medium. *International Journal of Mathematical, Engineering and Management Sciences*, **9**(6), 1483–1509 (2024).

Chapter 12

Linear stability of thermosolutal free convection of couple stress fluid in porous zone

Siprarani Sahoo and Saurabh Kapoor

12.1 INTRODUCTION

The linear stability analysis of double-diffusive phenomena, combined with coupled heat and mass transfer driven by natural convection in fluid-saturated porous media, holds significant importance due to its relevance in various engineering, geophysical, and natural systems. These include applications such as geothermal energy extraction, thermal energy storage and recovery systems, and petroleum reservoirs. Instability arises due to disruptions in the balance among external forces, inertial effects, and viscous stresses acting on the fluid. Many important developments in stability theory are given by Chandrasekhar [1], Nield and Bejan [2], and Drazin [3].

Double-diffusive convection occurs when both temperature and concentration differences exist in a fluid, and they diffuse at different rates. This is a complex fluid movement that usually does not happen in single-component fluids. Extensive theoretical and experimental research has been conducted on the phenomenon of convection in fluid layers that contain two or more agencies. Platten and Legros [4], Huppert and Turner [5], and Turner [6–8] all provided excellent assessments of these research. The obvious distinction between single-component and multi-component systems has sparked research into two- and multi-component convection. Even when the fundamental condition is hydrostatically stable—that is, when density declines with height—convection sets in, unlike single-component systems. When it comes to producing pure pharmaceuticals, storing liquid gas, solidifying molten alloys, and geothermal heating lakes and magmas, double-diffusive convection plays a significant role.

Rana et al. [9] discussed the thermal convection in couple stress fluid in hydromagnetics saturating porous medium and found that couple stress parameter has stabilizing effect on the system. Subsequently, Rana et al. [10] conducted detailed investigation on the stability characteristics of double-diffusive convection in a couple stress nanofluid, exploring the influence of various physical parameters on the onset of stability. Chand et al. [11] investigated the thermal instability in a layer of couple stress nanofluid-saturated porous medium and also studied the thermal instability in horizontal layer of couple stress nanofluid in a porous medium for more realistic boundary

DOI: 10.1201/9781003470885-12

conditions. Recently in 2023, analytical and numerical simulation of non-linear couple stress nanofluid with fractal-fractional order is discussed by Murtaza et al. [12]. The investigation shows that couple stress slows down the velocity of fluid. Bifurcations, chaos, and heat transmission in couple stress nanofluids are discussed in [13].

Gaikwad and Kouser [14] investigated double-diffusive convection in couple stress fluid with internal heat source in a saturated porous layer using linear and weakly non-linear stability analysis. Malashetty et al. [15] analytically investigate the double-diffusive convection in couple stress liquid under Soret effect using both linear and non-linear stability analysis. The study highlights the effect of governing parameters such as couple stress parameter, solute Rayleigh number, Soret parameter, diffusive ratio, and Prandtl number on stationary, oscillatory, and finite amplitude convection and results show that stability is enhanced by positive Soret number whereas negative one enhanced the instability and the couple stress parameter has quite large stabilizing effect. In [16], Malashetty et al. discussed Soret effect on double-diffusive convection of couple stress fluid saturated in a Darcy porous medium. Malashetty et al. [17] also provided a thermal non-equilibrium model to understand the onset of convection in couple stress fluid saturated in a porous layer.

Simulation of the onset of time-dependent convection in horizontal porous layer saturated by couple stress nanofluid through linear and non-linear stability analysis is well observed by Umavathi and Beg [18]. The study shows that the couple stress parameter has stabilizing effect on both stationary and oscillatory convections. Linear and non-linear stability convection of a reacting viscous incompressible fluid under couple stress effect in a plain layer is studied by Harfash and Maften.

Linear stability analysis of cross diffusive chemical reacting fluid in a Darcy porous medium is well discussed by Sahoo et al. [19]. Study shows that during stationary convection, Lewis number, Soret and Dufour parameters all exhibit stabilizing and destabilizing properties. Rudraiah and Malashetty [20] (Figure 12.1) studied the coupled molecular diffusion effect on double-diffusive convection using linear and non-linear stability analysis in a horizontal porous medium. Malashetty and Basavaraja [21] discussed the thermal/gravity effect on the arise Rayleigh–Benard convection in couple stress fluid. Using both linear and non-linear stability analysis, Siddheshwar and Pranesh [22] investigated the Rayleigh–Benard situation in Boussinesq–Stokes suspensions. Sharma et al. [23] explained the particle suspension effect on couple stress fluid in presence of magnetic field.

Research into non-Newtonian fluids is highly valued because of its increasing significance in contemporary technology and industry. Since Newtonian fluids fail to adequately depict the feature of fluid movement with suspended particles, the theory of polar fluids has garnered a lot more interest. Numerous industrial processes, including polymer fluid extrusion, liquid crystal solidification, and bath cooling of metal plate, depend heavily on the unique properties of polar fluids. Couple stress fluid stands out among non-Newtonian fluids

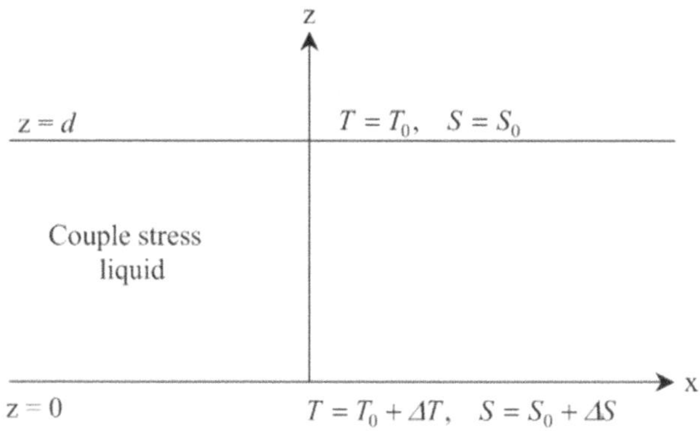

Figure 12.1 Physical configuration.

with characteristics like polar effects. Fluids with mechanically important microstructure are modeled by the theory of polar fluids and related theories. Stokes [24] provided the constitutive equations for couple stress fluids. The simplest theory for microfluids is the one put out by Stokes [24]. It accounts for polar phenomena such as couple stress, body couples, and non-symmetric tensors. In fluids containing very big molecules, couple stresses manifest with a discernible amplitude. When the stress is non-linearly proportional to the gradient of velocity between 13 and 15, only a handful of scholars have investigated Rayleigh–Benard convection in fluids. In this chapter, we use Stokes' pair stress model to talk about the thermo-solutional impact of double-diffusion convection. In the present work, a comprehensive linear stability analysis is carried out for the flow of a couple stress fluid in horizontal porous layer. This investigation aims to understand the onset of convective instabilities influenced by various parameters such as couple stress parameter, solute Rayleigh number, diffusivity ratio, and Prandtl number.

12.2 MATHEMATICAL FORMULATION

12.2.1 Equation of basic state

Without a body couple, the momentum and continuity equations controlling the motion of an incompressible couple stress fluid are provided by [24].

The Equation of Continuity:

$$\nabla . q = 0 \tag{12.1}$$

The Momentum Equation:

$$\rho_0\left[\frac{\partial q}{\partial t}+(q.\nabla)q\right]=-\nabla p+\mu\nabla^2 q-\mu_1\nabla^4 q+\rho g \tag{12.2}$$

The influence of one region on its environment in a polar fluid is best characterized by a combination of two forces, rather than by a single force alone. An incompressible fluid's response to couple stresses is seen by the third term on the right side of Eq. (12.2).

In this case, the couple stress property is controlled by the material constant μ_1, which has momentum dimension (MLT^{-1}). Stokes' theory characterizes the rheological behavior of an incompressible viscous couple stress fluid using two material constants: μ and μ_1. Since the dimensions of μ and μ_1 are $ML^{-1}T^{-1}$, the ratio $\left(\frac{\mu_1}{\mu}\right)^{\frac{1}{2}}$ has the dimension of length.

Couple stress has a significant influence for higher non-dimensional number values $a = \frac{l}{d}$ where d is the flow geometry's average dimension and l is the material constant, $l = \left(\frac{\mu_1}{\mu}\right)^{\frac{1}{2}}$. If l is determined by the liquid's molecular dimensions, it will differ significantly among liquids. A polymer chain's length, for instance, can exceed that of a water molecule's diameter by a factor of millions. Therefore, it is reasonable to anticipate that couple stresses will be noticeable in liquids composed of large or complex molecules.

The empirical equations that describe the relationship between the heat flux J_T and the matter flux J_C in a binary mixture and the thermal and solute gradients are the following:

$$J_T = -k\nabla T - \rho Ts\left(\frac{\partial \lambda}{\partial S}\right)D'\nabla S \tag{12.3}$$

$$J_C = -\rho D\left[S_T\nabla T + \nabla S\right] \tag{12.4}$$

where

D = diffusion constant
k = thermal conductivity
ρ = density
C = concentration
T = temperature
λ = chemical potential of solute

Since we're working with liquids in this study, we'll substitute Eq. (12.3) with

$$J_T = -k\nabla T \tag{12.5}$$

From a mathematical perspective, we may express the temperature and concentration equation as follows, using Eqs. (12.1) and (12.2) without sacrificing generality:

$$\frac{\partial T}{\partial t} + (q.\nabla)T = k\nabla^2 T \tag{12.6}$$

$$\frac{\partial s}{\partial t}+(q.\nabla)S = D\nabla^2 S \tag{12.7}$$

The equation of state is the following:

$$\rho = \rho_0\left[1-\beta_T\left(T-T_0\right)+\beta_S\left(S-S_0\right)\right] \tag{12.8}$$

Rajagopal et al. [25] recently demonstrated that the Oberbeck–Boussinesq approximation for the Navier–Stokes problem is valid. It is necessary to inquire about the present situation's approximation state. The purpose of particles in suspension is to make the fluid carrying the particles more viscous; this approach appears to be consistent with the Oberbeck–Boussinesq approximation. Based on this reasoning, we accept the estimation for couple stress fluid.

Two parallel infinite stress-free barriers contain a horizontal couple stress liquid layer with a thickness of d. The *x*-axis follows the bottom edge, whereas the *z*-axis goes straight up. A salt solution is added to the heated couple stress liquid from below, ΔT represents the temperature difference and ΔS represents the concentration difference between the boundary and the surrounding wall. Figuratively speaking, the setup is depicted in the diagram.

12.3 BASIC STATE

In the basic state, fluid is quiescent, which is described as the following:

$$q_b = (0,0,0),\ \ p = p_b(z),\ \ T = T_b(z),\ \ S = S_b(z),\ \ \rho = \rho_b(z) \tag{12.9}$$

Using (12.9), Eqs. (12.1), (12.2) and (12.6)–(12.8) yields,

$$\frac{dp_b}{dz} = \rho_b g,\ \frac{d^2T_b}{dz^2} = 0,\ \frac{d^2S_b}{dz^2} = 0,\ \rho_b = \rho_0\left[1-\beta_T\left(T_b-T_0\right)+\beta_s\left(S_b-S_0\right)\right] \tag{12.10}$$

12.4 PERTURBED STATE

We superpose perturbations on the basic state as follows:

$$q = q_b + q',\ \ T = T_b(z)+T',\ \ S = S_b(z)+S',\ \ p = P_b(z)+p',\ \ \rho = \rho_b(z)+\rho' \tag{12.11}$$

The primes denote disturbances. Utilizing the fundamental state Eq. (12.10) and substituting (12.11) into Eqs. (12.1)–(12.2) and (12.6)–(12.8), we derive,

$$\nabla.q' = 0 \tag{12.12}$$

$$\rho\left[\frac{\partial q'}{\partial t}+(q'.\nabla)q'\right]=-\nabla p'+\mu\nabla^2 q'-\mu_1\nabla^4 q'+\rho' g \tag{12.13}$$

$$\frac{\partial T'}{\partial t}+(q'.\nabla)T'+w'\frac{\partial T_b}{\partial z}=k\nabla^2 T' \tag{12.14}$$

$$\frac{\partial S'}{\partial t}+(q'.\nabla)S'+w'\frac{\partial S_b}{\partial z}=D\nabla^2 S' \tag{12.15}$$

$$\rho'=-\rho_0\left[\beta T'-\beta_s S'\right] \tag{12.16}$$

The stream function Ψ is defined by focusing solely on two-dimensional disturbances

$$(u',\ w')=\left(\frac{\partial\psi}{\partial z},-\frac{\partial\psi}{\partial x}\right) \tag{12.17}$$

which fulfill the continuity. The following non-dimensional parameters are used to non-dimensionalize Eq. (12.12), the resultant equation (Eqs. (12.14) and (12.15)), and the stream function Ψ after removing pressure from Eq. (12.13).

$$(x^*,\ z^*)=\left(\frac{x}{d},\frac{z}{d}\right),\ t^*=\frac{t}{\frac{d^2}{k}},\ \psi^*=\frac{\psi}{k},\ T^*=\frac{T'}{\Delta T},\ S^*=\frac{S'}{\Delta S} \tag{12.18}$$

We obtain,

$$\frac{1}{P_r}\frac{\partial}{\partial t}(\nabla^2\psi)=-R\frac{\partial T}{\partial x}+R_s\frac{\partial S}{\partial x}+\nabla^4\psi-C\nabla^6\psi+\frac{1}{P_r}\frac{\partial(\psi,\Delta^2\psi)}{\partial(x,z)} \tag{12.19}$$

$$\frac{\partial T}{\partial t}=-\frac{\partial\psi}{\partial x}+\nabla^2 T+\frac{\partial(\psi,T)}{\partial(x,z)} \tag{12.20}$$

$$\frac{\partial S}{\partial t}=-\frac{\partial\psi}{\partial x}+\tau\nabla^2 S+\frac{\partial(\psi,S)}{\partial(x,z)} \tag{12.21}$$

where

$P_r=\frac{\nu}{k}$, Prandtl number
$R=\frac{\beta_T g\Delta T d^3}{\nu k}$, thermal Rayleigh number
$C=\frac{\mu_1}{\mu d^2}$, couple stress parameter
$\tau=\frac{D}{k}$, diffusivity ratio

To make things easier to understand, the asterisks have been removed. For boundary conditions that are stress free, isothermal, and vanishing couple stress, namely, we solve Eqs. (12.19)–(12.21).

$$\psi = \frac{\partial^2 \psi}{\partial z^2} = \frac{\partial^4 \psi}{\partial z^4} = T = S = 0, \quad at \quad z = 0,1 \tag{12.22}$$

12.5 LINEAR STABILITY

We provide the useful linear stability analysis here. In this investigation, we will assume that the solutions to Eqs. (12.19)–(12.21) are periodic waves of the form and will disregard the Jacobians.

$$\begin{pmatrix} \Psi \\ T \\ S \end{pmatrix} = e^{\sigma t} \begin{pmatrix} \Psi \sin(\pi \alpha x) \\ \Theta \cos(\pi \alpha x) \\ \phi \cos(\pi \alpha x) \end{pmatrix} \sin(\pi z) \tag{12.23}$$

where α is the horizontal wave number and σ is the growth rate, which is often a complex number $(\sigma = \sigma_r + i\sigma_i)$.

Substituting Eq. (12.23) to Eqs. (12.19)–(12.21) in their linearized form yields

$$\left(\frac{\sigma}{Pr} + \eta K^2 \right) K^2 \Psi = -R\pi\alpha\Theta + R_s \pi\alpha \tag{12.24}$$

$$(\sigma K^2)\Theta = -\pi\alpha \tag{12.25}$$

$$(\sigma + \tau K^2)\Phi = -\pi\alpha \tag{12.26}$$

After simplification,

$$\begin{pmatrix} \left(\frac{\sigma}{Pr} + \eta K^2 \right) K^2 & R\alpha\pi & -R_s\pi\alpha \\ \alpha\pi & (\sigma + K^2) & 0 \\ \alpha\pi & 0 & (\sigma + \tau K^2) \end{pmatrix} \begin{pmatrix} \Psi \\ \theta \\ \varnothing \end{pmatrix} = \begin{pmatrix} 0 \\ 0 \\ 0 \end{pmatrix} \tag{12.27}$$

Zero as the determinant of the coefficient matrix is required for non-trivial solutions. Consequently, we obtain by making the determinant of the coefficient matrix equal to zero

$$\Rightarrow\left(\frac{\sigma}{Pr}+\eta K^2\right)\left(\sigma+K^2\right)\left(\sigma+\tau K^2\right)K^2-R\pi^2\alpha^2\left(\sigma+\tau K^2\right)+R_s\pi^2\alpha^2\left(\sigma+K^2\right)=0$$

$$\Rightarrow R=\left[\pi^2\alpha^2\left(\sigma+\tau K^2\right)\right]^{-1}\left[\left(\sigma+K^2\right)\left(\sigma+\tau K^2\right)\left(\frac{\sigma}{Pr}+\eta K^2\right)K^2+R_s\pi^2\alpha^2\left(\sigma+K^2\right)\right] \tag{12.28}$$

where

$\eta=1+CK^2$, $K^2=\pi^2\left(\alpha^2+1\right)$ and η is representative of the viscosity of the fluid. Here, $\eta=1$ pertains to Newtonian fluids. Observing the η's expression, it has been seen that it is consistent with Einstein's findings, hence it clarifies that the suspended particles increase the viscosity. Compared to the case of Newtonian fluids, the enforcement of the Oberback–Boussinesq approximation in suspensions is superior.

If σ is real, then marginal stability is achieved. When $\sigma=0$, in this case Eq. (12.28) provides the stationary Rayleigh number, R^{st} at the stability threshold in the form,

$$R^{st}=\left[\pi^2\alpha^2\left(\tau K^2\right)\right]^{-1}\left[\left(K^2.\tau K^2.\eta K^2.K^2\right)+R_s\pi^2\alpha^2K^2\right]=\frac{\eta\tau K^8+R_s\pi^2\alpha^2K^2}{\pi^2\alpha^2\tau K^2}$$

$$R^{st}=\frac{\eta\tau K^6}{\pi^2\alpha^2\tau}+\frac{R_s}{\tau} \tag{12.29}$$

The minimum value of Rayleigh number R^{st} is obtained at the critical wavenumber, $\alpha=\alpha_c$.

where α_c satisfies the equation,

$$3C\pi^2\left(\alpha^2\right)^2+2\left(1+C\pi^2\right)\alpha^2-\left(1+C\pi^2\right)=0 \tag{12.30}$$

We observe the dependency of critical wave number α_c on the couple stress parameter C.

Further, considering a single-component system, $R_s=0$, Eq. (12.29) provides

$$R^{st}=\frac{\eta K^6}{\pi^2\alpha^2} \tag{12.31}$$

With the critical value given by,

$$R_C^{st}=\frac{\pi^4\left(1+\alpha_C{}^2\right)\left[1+C\pi^2\left(1+\alpha_C{}^2\right)\right]}{\{^2\alpha_C} \tag{12.32}$$

Eq. (12.30) should be used to get the critical wave number α_c. Siddheshwar and Pranesh [22] provided these identical numbers for a system with only one component.

12.6 OSCILLATORY STATE

The existence of other limitations like rotation, salinity gradient, and magnetic field is known to be necessary for the occurrence of oscillatory movements. For the oscillatory mode $\sigma_r = 0$ and $\sigma_i \neq 0$. By substituting $\sigma = i\omega$ (ω is real) into Eq. (12.28) and rearranging the terms, we obtain the oscillatory Rayleigh number R^{osc} at the stability threshold that is expressed as

$$R^{osc} = \left[(1+\tau)\left(\eta^2 Pr + (1+\tau)\eta + \frac{\tau}{Pr}\right)K^6 + \pi^2\alpha^2 R_s(\eta Pr + \tau)\right] \times \left[\pi^2\alpha^2(1+\eta Pr)\right]^{-1} \tag{12.33}$$

and non-dimensional frequency in the form

$$\omega^2 = -\eta K^6(1+\tau)(\tau) + \frac{\tau K^6}{Pr}(\tau) - \eta\tau K^6 + \pi^2\alpha^2(\tau-1)R_s \times K^2\left[\eta + \frac{(1+\tau)}{Pr} - \frac{\tau}{Pr}\right]^{-1} \tag{12.34}$$

12.7 RESULTS AND DISCUSSIONS

Linear stability theory is applied to analyze double-diffusive convection within a saturated porous layer containing couple stress fluid. The standard normal mode analysis serves as the foundation for this linear theory. For various parameter values, including the pair stress parameter, diffusivity ratio, solute Rayleigh number, and Prandtl number, the formulas for stationary and oscillatory Rayleigh numbers are calculated and shown in Figures 12.2–12.7.

Figure 12.2 displays the neutral stability curve for various couple stress parameter values in a range between 2 and 10, with $\tau = 0.3$ and $R_s = 100$ used as fixed values. The graph illustrates that as the couple stress parameter C increases, there is a corresponding rise in the Rayleigh number. This behavior suggests that the presence of couple stress in the fluid system enhances the driving forces for convection, leading to an onset of instability. Consequently, an increase in the couple stress parameter C serves to destabilize the system more quickly, making it more susceptible to convective motion. This represents that the effect of couple stress plays a significant role in influencing the stability of the flow in the system.

Figure 12.3 shows the dynamic relationship between the critical Rayleigh number and critical wave number for various solute Rayleigh number values (10^3, 5×10^3, 10^4, and 10^5) with a fixed couple stress parameter C = 2 and

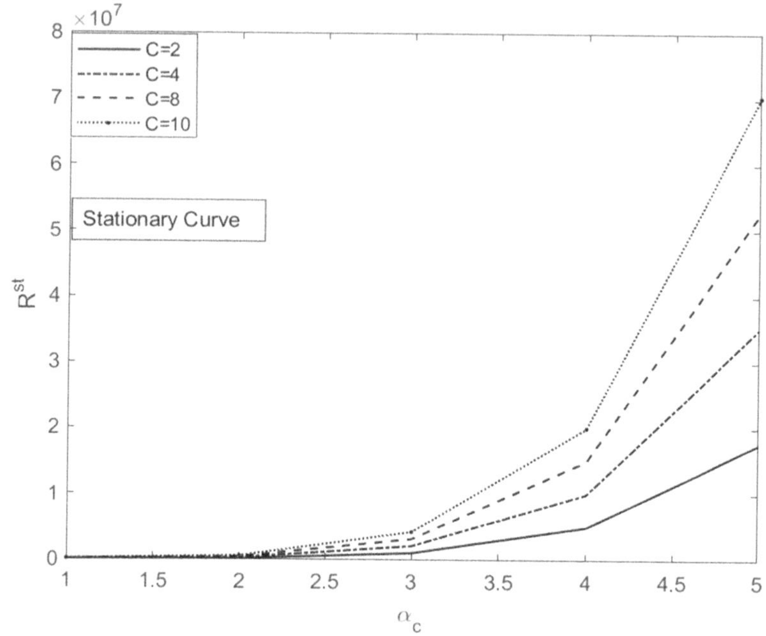

Figure 12.2 Variation of R^{st} with α_c for different values of C with R_s = 100, τ = 0.3.

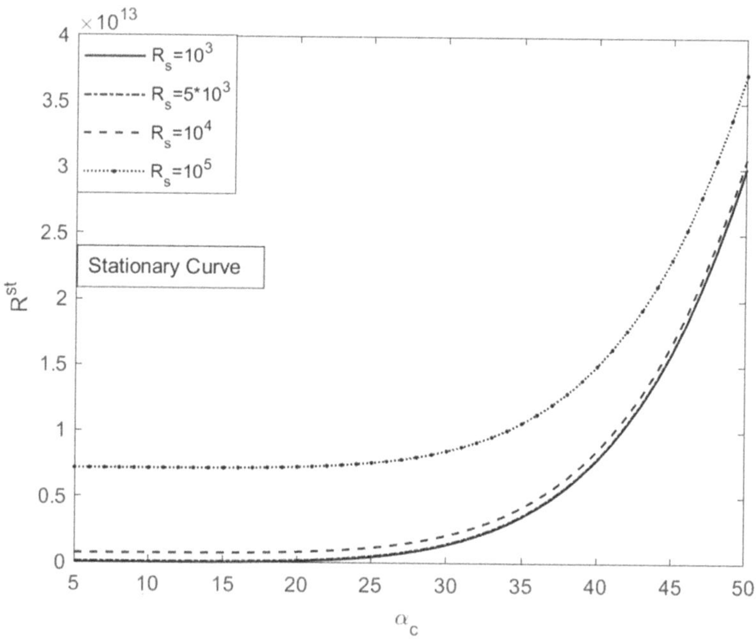

Figure 12.3 Variation of R^{st} with α_c for different values of R_s with C = 2, τ = 0.3.

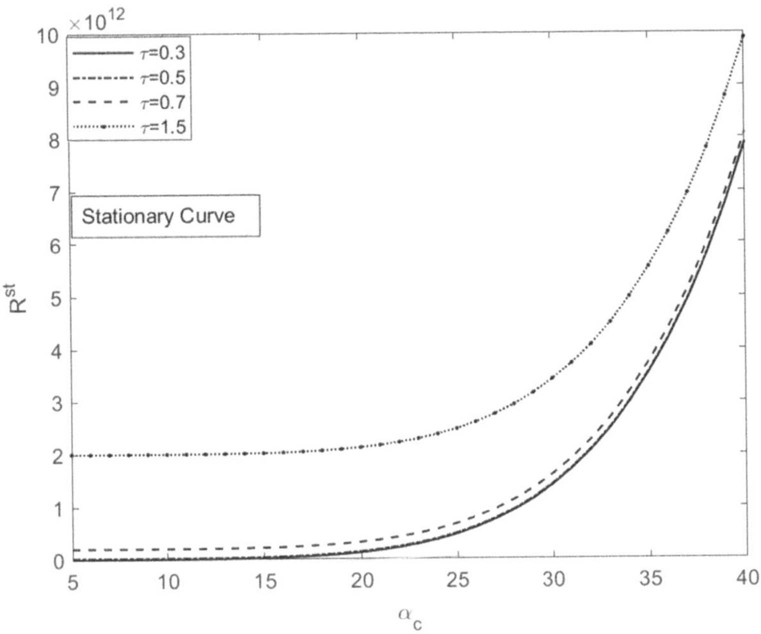

Figure 12.4 Variation of R^{st} with α_c for different values of τ with $C = 2$, R_s=100.

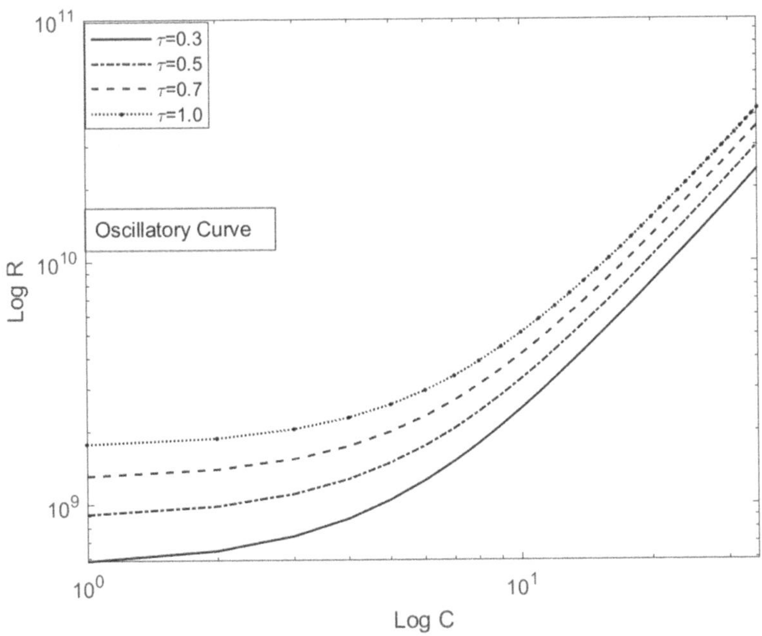

Figure 12.5 Variation of R^{osc} with C for different values of τ with $R_s = 10^3$, Pr = 5, $\alpha^2 = 0.35$.

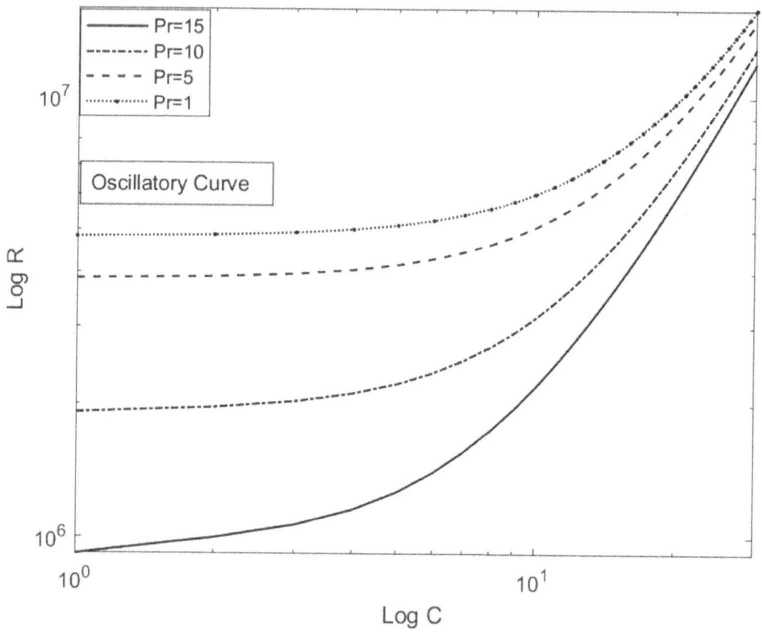

Figure 12.6 Variation of R^{osc} with C for different values of Pr with $R_s = 10^3, \alpha^2 = 0.35, \tau = 0.3$.

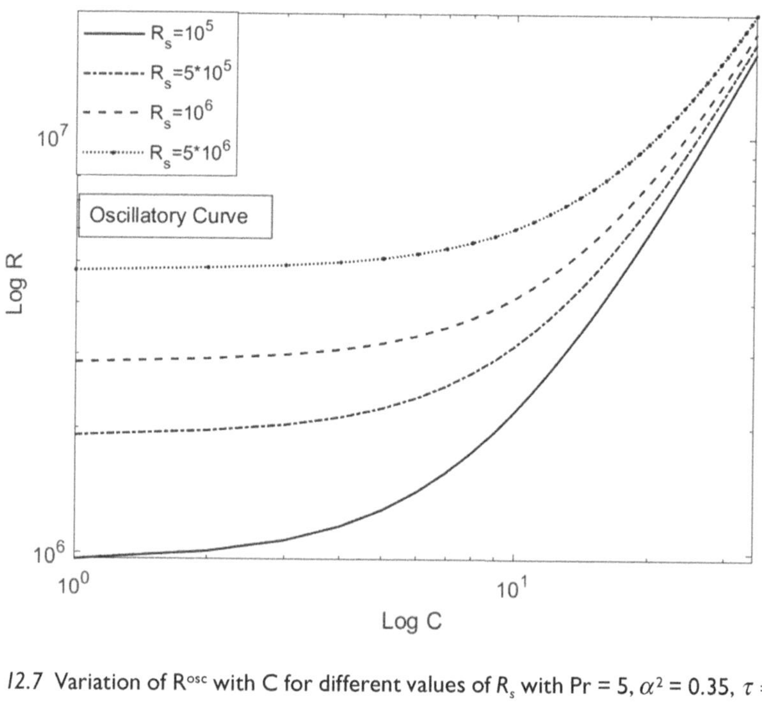

Figure 12.7 Variation of R^{osc} with C for different values of R_s with Pr = 5, $\alpha^2 = 0.35, \tau = 0.3$.

diffusivity ratio $\tau = 0.3$. The observation, increase in critical Rayleigh number with the rise of the solute Rayleigh number, as depicted in the graph, suggests a stabilizing effect. The presence of solute gradients, which contribute to mass transfer effects, seems to counteract the thermal effects and stabilize the system. This stabilizing influence means that the system is less prone to convective instability, with higher values of the solute Rayleigh number effectively rising the threshold at which instability begins to occur.

Figure 12.4 shows how the neutral stability curves are affected by the diffusivity ratio τ in a range between 0.3 and 1.5 for constant values of C = 2 and the solute Rayleigh number $R_s = 100$. It can be observed that the critical Rayleigh number increases progressively as the diffusivity ratio rises, indicating that the system's stability is influenced by the diffusivity ratio. This indicates that changes in the diffusivity ratio have a significant impact on the onset of instability, with higher diffusivity ratios requiring a higher thermal driving force to trigger convection. Furthermore, the stationary curve is growing at a quicker rate for $\tau = 1.5$ compared to τ in the range of 0.3–0.7.

In Figure 12.5, the impact of the diffusivity ratio τ in a range (03–1.0) on the oscillatory curve is shown once again for the fixed values $R_s = 10^3$, Pr = 5, and $\alpha^2 = 0.35$. The figure clearly demonstrates that the oscillatory Rayleigh number grows as the diffusivity ratio rises. As the diffusivity ratio, which represents the relative rates of heat and mass diffusion, becomes larger, the system requires a higher thermal driving force to reach the point of oscillatory instability. This leads to that higher diffusivity ratio makes the system more sensitive to perturbations, leading to a more pronounced and earlier onset of oscillatory convection.

Figure 12.6 shows the oscillatory Rayleigh number's change with couple stress parameter for various Prandtl number (Pr) values 1, 5, 10, and 15, at $R_s = 10^3$, $\tau = 0.3$, and $\alpha^2 = 0.35$. It can be observed that the oscillatory Rayleigh number exhibits an inverse relationship with the Prandtl number. This indicates that, as the Prandtl number (Pr), which represents the ratio of momentum diffusivity to thermal diffusivity, becomes larger, the system needs a lower thermal driving force to reach the point of oscillatory instability. This trend emphasizes the critical role of the Prandtl number in controlling the dynamics of heat and mass transfer in the system.

Figure 12.7 represents the fluctuation of the critical Rayleigh number with a couple stress parameter with a variation in solute Rayleigh number 10^5, 5×10^5, 10^6, and 5×10^6 with fixed $\tau = 0.3$, Pr = 5, and $\alpha^2 = 0.35$ for an oscillatory curve. In case of oscillatory stability, it is observed that the critical Rayleigh number increases as the solute Rayleigh number rises. This leads to a stabilizing influence of solute gradient on the system. As the solute Rayleigh number becomes larger, representing a stronger influence of solute concentration gradients relative to diffusive effects, a greater thermal driving force i.e., a higher Rayleigh number is required to initiate oscillatory convection.

12.8 CONCLUSION

Linear stability analysis is employed to investigate the onset of double-diffusive convection in a couple stress fluid saturated within a horizontal porous layer. The momentum equation is modeled using a modified form of Darcy's equation, which includes the inertia term and the time derivative term. A function of the governing parameters yields the formulae for the stationary and oscillatory Rayleigh numbers. Stationary and oscillatory convection are illustrated visually in relation to the couple stress parameter, solute Rayleigh number, diffusivity ratio, and Prandtl number. Upon completing the required calculations, we arrive at the following results:

- An increase in couple stress (C) parameter leads to rise in critical Rayleigh number, which in turn has a stabilizing effect on the stationary curve.
- The solute Rayleigh number exhibits an increasing trend toward critical Rayleigh numbers in both oscillatory and stationary convection, reflecting stabilizing effect on system.
- The critical Rayleigh number for stationary and oscillatory convection both rise as the magnitude of the diffusivity ratio increases.
- Prandtl number has a reverse trend toward critical Rayleigh number as with increase in Prandtl number, the oscillatory curve for critical Rayleigh number decreases.

REFERENCES

1. S. Chandrasekhar; Hydrodynamic and Hydromagnetic Stability; Dover Publications Inc.; New York, 1981.
2. D. A. Nield, A. Bejan; Convection in Porous Media; Springer; New York, 1992.
3. P. G. Drazin; Introduction to Hydrodynamic Stability; Cambridge University Press; Cambridge, 2002.
4. J. K. Platten, J. C. Legros; Convection in Liquids; Springer; Berlin, 1984.
5. H. E. Huppert and, J. S. Turner; Double diffusive convection; Journal of Fluid Mechanics; Volume 106, Pages 299–329, (1981).
6. J. S. Turner; Buoyancy Effects in Fluids; Cambridge University Press; London, 1973.
7. J. S. Turner; Double-diffusive phenomena; Annual Review of Fluid Mechanics; Volume 6, Pages 37–56, (1974).
8. J. S. Turner; Multicomponent convection; Annual Review of Fluid Mechanics; Volume 17, Pages 11–44, (1985).
9. G. C. Rana; The onset of thermal convection in couple-stress fluid in hydromagnetics saturating a porous medium; Bulletin of Polish Academy of Sciences – Technical Sciences; Volume 62, Pages 357–362, (2014).
10. G. C. Rana, R. Chand, V. Sharma; Stability analysis of double-diffusive convection in a couple stress nanofluid; Technische Mechanik; Volume 38, Pages 246–255, (2018).

11. R. Chand, G. C. Rana, D. H. Yadav; Thermal instability in a layer of couple stress nanofluid saturated porous medium; Journal of Theoretical and Applied Mechanics; Volume 46, Pages 69–84, (2017).
12. S. Murtaza, Z. Ahmad, I. E. Ali, Z. Akhtar, F. Tchier, H. Ahmad, S.-W. Yao; Analysis and numerical simulation of fractal-fractional order non-linear couple stress nano fluid with cadmium telluride nanoparticles; Journal of King Saud University – Science; Volume 35, Issue 4, Page 102618, (2023).
13. D. Purbasha, G. C. Layek; Bifurcations, chaos, and heat transmissions in couple-stress nanoliquid flows; International Journal of Bifurcations and Chaos; Volume 35, (2025).
14. S. N. Gaikwad, S. Kouser; Double diffusive convection in a couple stress fluid saturated porous layer with internal heat source; International Journal of Heat and Mass Transfer; Volume 78, Pages 1254–1264, (2014).
15. M. S. Malashetty, S. N. Gaikwad, M. Swamy; An analytical study of linear and non-linear double diffusive convection with Soret effect in couple stress liquids; International Journal of Thermal Sciences; Volume 45, Issue 9, Pages 897–907, (2006).
16. M. S. Malashetty, I. Pop, P. Kollur, W. Sidram; Soret effect on double diffusive convection in a Darcy porous medium saturated with a couple stress fluid; International Journal of Thermal Sciences; Volume 53, Pages 130–140, (2012).
17. M. S. Malashetty, I. S. Shivakumar, S. Kulkarni; The onset of convection in a couple stress fluid saturated porous layer using a thermal non-equilibrium model; Physics Letters A; Volume 373, Pages 781–790, (2009).
18. J. C. Umavathi, O. A. Beg; Simulation of the onset of convection in a porous medium layer saturated by a couple-stress nanofluid; Microfluidics and Nanofluids; Volume 25, Article Number 53, (2021).
19. A. K. Sahoo, S. Kapoor, S. Sahoo; Linear stability of cross diffusive chemically reacting Maxwell fluid in Darcy porous medium; Mathematics in Engineering, Science and Aerospace; Volume 13, Issue 2, Pages 487–498, (2024).
20. N. Rudraiah, M. S. Malashetty; The influence of coupled molecular diffusion on double diffusive convection in a porous medium; ASME Journal of Heat Transfer; Volume 108, Pages 872–878, (1986).
21. M. S. Malashetty, D. Basavaraja; Effect of thermal/gravity modulation on the onset of Rayleigh-Benard convection in couple stress fluid; Journal of Transport Phenomena; Volume 7, Pages 31–44, (2005).
22. P. G. Siddheshwar, S. Pranesh; An analytical study of linear and non-linear convection in Boussinesq–Stokes suspension; International Journal of Non-Linear Mechanics; Volume 39, Pages 165–172, (2004).
23. R. C. Sharma, M. Sharma; Effect of suspended particles on couple-stress fluid heated from below in presence of rotation and magnetic field; Indian Journal of Pure and Applied Mathematics; Volume 35, Issue 8, Pages 973–989, (2004).
24. V. K. Stokes; Couple stresses in fluids; Physics of Fluids; Volume 9, Pages 1709–1716, (1996).
25. K. R. Rajagopal, M. Ruzicka, A. R. Srinivasa; On the Oberbeck-Boussinesq approximation; Mathematical Models and Methods in Applied Sciences; Volume 6, Issue 8, Pages 1157–1167, (1996).

Chapter 13

Mixing of complex fluids in mechanically agitated vessels

Challenges and advances

Zainab T. Al-Sharify, Shahad Z. Al-Najjar, Ahmed Kadhim Hussein, Sachindra Kumar Rout, and Helen Onyeaka

13.1 INTRODUCTION

Mechanical agitation of complex fluids in stirred containers plays a vital role in various industries such as medicinal products, nuclear reprocessing, food processing, wastewater management, crystallization, fermentation, catalysis, and mineral processing. Many industrial processes involve non-Newtonian fluids exhibiting unique rheological properties, including shear-thickening characteristics. Efficient mixing of these materials is vital, especially given their use in applications such as flexible protective gear and vibration-damping technologies aimed at improving safety. However, a comprehensive understanding of their flow behavior and mixing mechanisms is still lacking, largely because of the opaque nature of these fluids and the complexity of their flow patterns.

Single-phase or multiphase non-Newtonian fluids are mixed with mechanically agitated reactors for further use in various industries. These mechanically driven agitated systems consist of diverse particle types and concentrations. However, it is still a current research area of the mixing behavior of such fluids. Viscoplastic and shear-thickening fluids (STFs), in particular, demonstrate unique mixing behavior. Cavern and pseudo-cavern formations are frequently observed in these fluids, creating stagnant regions that hinder homogeneity and reduce process efficiency. Impeller type and configuration critically shape the internal flow field.

The performance improvement contributes directly to better product consistency and efficiency. Inadequate mixing design can increase energy consumption and production costs. As such, mixing of complex fluids continues to be a key research area in chemical engineering [1, 2].

A persistent challenge is predicting fluid behavior under different mixing conditions—especially in non-Newtonian systems where viscosity changes with shear rate. The absence of robust, predictive models often leads to inefficiencies and economic losses. Smith et al. [3] noted that poor mixing system design has cost industries billions of dollars globally. Addressing this issue requires deeper insight into critical variables such as vessel geometry, impeller design, power input, and fluid rheology [4].

 DOI: 10.1201/9781003470885-13

While prior research has targeted power reduction through just-drawdown speed optimization and improved impeller/baffle combinations, fewer studies have addressed the internal flow patterns or spatial distribution of floating particles. Because most complex fluid suspensions are opaque, visualization techniques have been limited to external observations and cloud depth measurements. As a result, the mechanisms dictating particle circulation within the systems are yet to be explored. This research gap highlights the need for more experimental work on both single-phase and multiphase flow behavior.

This chapter delves into the mixing of complex fluids within mechanically agitated vessels. It outlines foundational concepts, recent developments, and unresolved challenges, offering insights into the factors that govern mixing efficiency.

13.2 BACKGROUND ON MIXING COMPLEX FLUIDS IN MECHANICALLY AGITATED VESSELS

13.2.1 Importance of mixing in industrial processes

Mixing plays a vital role, which involves combining multiple phases—liquid, solid, or gas—into a uniform mixture. In sectors such as pharmaceuticals and food processing, product quality and consistency are closely linked to the effectiveness of the mixing stage. Whether developing drug formulations or creating specific food textures, achieving a high level of homogeneity is essential to ensure the uniform distribution of ingredients.

In the petrochemical and oil refining industries, mixing enhances reaction rates and supports efficient phase separation, thereby improving overall process performance [2]. Beyond product quality, efficient mixing also contributes to energy conservation. By carefully controlling power input, it is possible to lower energy consumption and reduce mechanical strain on mixing equipment—ultimately cutting down on maintenance costs and extending equipment lifespan. As noted by Al-Sharify [1], well-optimized mixing systems not only improve process outcomes but also support economic and environmental sustainability.

13.2.2 Challenges in mixing of complex fluids

Although mixing is a critical step in many industrial processes, it becomes particularly challenging when dealing with complex fluids. Many of these fluids are non-Newtonian, meaning their viscosity varies with shear rate. Examples include shear-thickening and shear-thinning fluids, viscoelastic substances, and particle-laden suspensions. In particular, suspensions often exhibit viscosity changes that depend on both shear rate and particle concentration, complicating efforts to model their behavior accurately [2]. As a result, conventional models developed for Newtonian fluids often fall short when applied to these systems.

A frequently encountered issue during the mixing of non-Newtonian fluids is the formation of caverns or pseudo-caverns—regions where fluid movement is limited. These tend to form under low Reynolds number conditions, commonly found when mixing highly viscous materials. Cavern formation leads to inefficient mixing and the creation of stagnant zones, which reduce the overall effectiveness of the process [5]. To address this, researchers have investigated the geometry and dynamics of caverns to improve mixing control [6]. However, accurately predicting cavern behavior remains difficult, particularly in opaque fluids where direct flow visualization is limited.

13.2.3 Influence of operational parameters

The efficiency of mixing in mechanically agitated vessels is influenced by a range of operational parameters. These are the geometrical parameters of the vessel, the impeller speed, the power required, and the fluid properties. Few of these parameters—such as impeller speed or vessel size—can be customized with relative ease. However, fluid properties may not be adjusted; altering them may lead to minor changes in product quality [7–9]. This is relevant in biological applications, where excessive agitation can harm delicate organisms

The impeller section is also an important parameter in determining how mixing occurs significantly [2]. Impellers that produce axial flow can improve overall fluid movement, while radial flow is better suited for generating shear, which leads to mixing in more viscous systems.

13.2.4 Recent advancements in mixing technology

Over the last decade, a notable progress is reported in the study of complex fluid mixing. The use of numerical and experimental techniques has contributed to an understanding of how these fluids behave under various mixing conditions. Methods like particle image velocimetry (PIV) [2] have made it possible to visualize flow patterns even in opaque fluids, offering valuable insights into mixing performance.

Artificial intelligence and machine learning, integrated with computational fluid dynamics (CFD), are beginning to transform how mixing parameters areoptimized in real time. These technologies have shown promise in predicting the drawdown behavior of floating particles and improving their spatial distribution [1].

13.3 FLOW AND MIXING OF COMPLEX FLUID

The mixing of complex fluids plays a central role in various industrial applications, particularly in food processing, where it serves multiple functions. These include blending ingredients, dispersing powders or immiscible liquids, and introducing gases into liquids for purposes such as mass transfer or foam formation [10].

To ensure better mixing, the proper equipment design should be used for a particular duty as the fluid flow pattern depends on the geometry of the impeller, vessel, and baffles but also the fluid rheology. In industry, the mechanically agitated vessel is commonly encountered in mixing [11].

13.3.1 Mechanically agitated vessels

Mechanical agitation that uses rotating mixers is a widespread mixing technique used to achieve mixing above 50% of total chemical production in the world [12]. Mixing is typically accomplished in stirred vessels, which are ordinarily cylindrical.

Figure 13.1 presents a typically stirred vessel equipped with a Rushton turbine impeller. The base of the stirred vessel varies depending on the applications it is designed for, sometimes being angled on the edges and other times being rounded. For efficient mixing in a stirred vessel and to prevent vortex formation, especially when mixing low viscosity fluids, *baffles* are fitted to the vessel walls. Typically, there are four baffles, separated by equal spacing, used to reduce the swirling of the fluid in the vessel. The width of each baffle

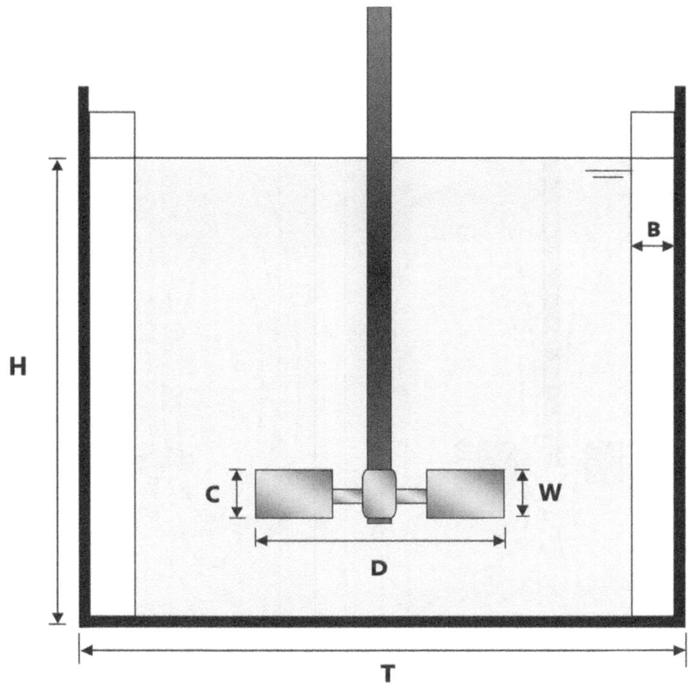

Figure 13.1 Typical configuration of the mechanically agitated vessel.

is usually 1/10 of the vessel diameter. For two-phase flow of viscous suspensions, the baffles may vary in position and clearance away from the wall or are set at a specific angle [11].

13.3.2 Types of impeller and their effect on mixing efficiency

Various types of Impellers are available which affect the quality of mixing, and there is a wide range of impeller designs used in industry. The selection of the optimum impeller design depends on the fluid viscosity and on the sensitivity of the system to the applied mechanical shear. The mixing performance of various types of impellers is one of the major interests in mixing studies. The study of the flow field generated by an impeller is essential in understanding the mixing performance inside the stirred tank. The flow pattern in the turbulent regimes of Newtonian fluids for radial and axial flow impellers is shown in Figure 13.2. The radial impeller flow pattern normally generates two circulation loops, while the axial impeller generates only one. The larger the loops, the better the material transport inside the stirred vessel, meaning significantly more efficient mixing [13]. Liu [14] compared the flow pattern generated when mixed water and shear-thinning fluids, using four types of impellers, including the Rushton turbine, pitch blade turbine, and two hydrofoil impellers (A310 and A340). She found that the A340 and pitch blade

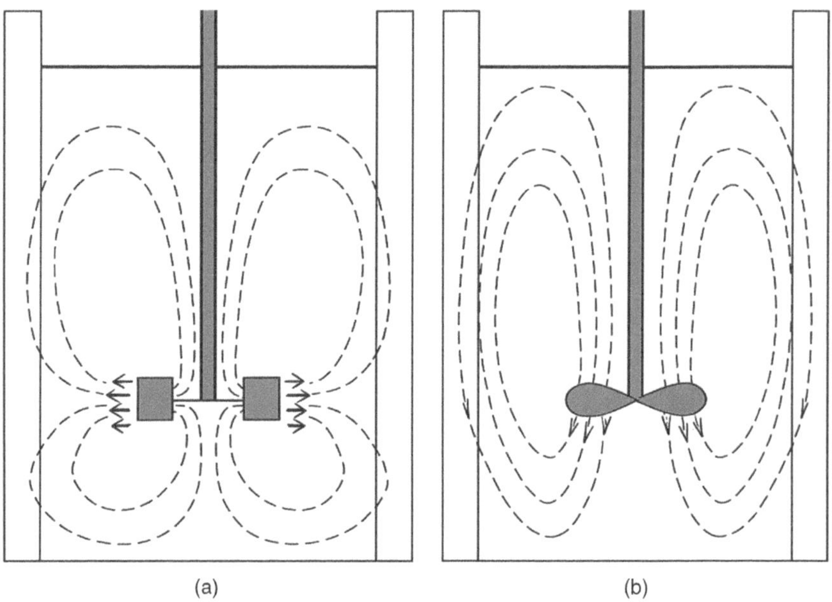

Figure 13.2 Typical flow patterns generated in stirred vessels mixed by (a) radial and (b) axial impeller [15].

turbines generated similar flow patterns when mixing the fluid in both an up and down pumping mode.

13.4 IMPACT OF IMPELLER OFF-BOTTOM CLEARANCE ON FLUID FLOW AND MIXING EFFICIENCY

Theoff-bottom clearance of the impeller plays a crucial role in determining the flow patterns within the mixing system. Typically, the standard off-bottom clearance is set at one-third of the vessel diameter (T/3). However, research has demonstrated that increasing the clearance to half the vessel diameter (T/2) results in improved flow patterns, particularly when mixing highly viscous single-phase fluids [16–20]. The mixing can be enhanced by lowering the impeller's clearance from the vessel bottom [21]. Zhang et al. [22] reported that increasing impeller clearance beyond one-third of the tank diameter (T/3) improved fluid homogeneity in high-viscosity single-phase flows. But, Yang and Li [23] reported that in two-phase systems with dense solid suspensions, reducing impeller clearance can improve the particles in suspension, minimizing settling, and promoting more uniform distribution. Building on this, Patel et al. [24–27] reported that placing the impeller lower in complex fluid systems—such as slurries—can lead to better axial mixing and reduced power consumption by improving the overall efficiency of energy use.

13.5 MASS TRANSFER STUDIES IN MIXING PROCESSES

Mass transfer between solid and liquid phases in stirred systems is vital to many industrial operations, including chemical production, pharmaceuticals, and wastewater management. Typically, mass transfer coefficients are estimated using Frössling-type correlations, which rely on the slip velocity between solid particles and the surrounding fluid. However, in agitated suspensions, this slip velocity is not uniform and is affected by factors such as vessel flow patterns and particle size, resulting in spatial variations throughout the vessel. Although several investigations have attempted to address this, accurately measuring slip velocity remains difficult [20].

13.5.1 Theoretical background

In solid–liquid systems, the mass transfer coefficient is calculated using empirical correlations; many of them are based on the classical equation. In these correlations, the Reynolds number depends on the slip velocity. In context, in mechanically agitated systems, slip velocity can fluctuate significantly due to changes in mixing intensity and flow dynamics [28].

13.5.2 Experimental approach and methodology

An extensive survey on mass transfer in solid–liquid mixing, focusing on the behavior of dissolved particles in water, was conducted by Al-Najjar [20]. The study evaluated various correlations for calculating mass transfer coefficients and compared them with traditional models. In Al-Najjar's experiments, specially prepared particles, with characteristics similar to plastic particles, were used to facilitate comparisons between dissolved and non-dissolved particle behaviors. The results from Positron emission particle tracking (PEPT) experiments were compared with theoretical models, revealing inconsistencies when using terminal settling velocity as an approximation for slip velocity. Similar techniques have been applied in more recent studies to assess mass transfer in multiphase reactors [29].

13.6 FLUID RHEOLOGY AND ITS ROLE IN PREDICTING FLOW BEHAVIOR

Rheometers are essential tools in rheology, used to measure the relationship between applied stress and the resulting deformation in fluids. These measurements are vital for understanding the flow behavior of a wide range of fluids, from simple liquids to complex mixtures. The data obtained from rheometer experiments help create models that predict fluid flow under various conditions. These models typically focus on how a fluid element deforms when exposed to stress, offering valuable insights for process optimization across industries.

Singh et al. [30] proposed an improved method for measuring the rheological properties of shear-thinning fluids, enhancing predictions for applications like coatings and food processing. Gupta and Zhao [31] proposed a new class of models to predict the flow behavior of non-Newtonian fluids, considering factors such as temperature and shear rate. For Newtonian fluids, the relationship between shear rate ($\dot{\gamma}$) and shear stress (τ) is straightforward, showing a linear correlation between shear rate and shear stress.

$$\tau = \mu\gamma \tag{13.1}$$

Non-Newtonian fluids show an increase in rheological complexity when compared to Newtonian fluids, as their viscosity is not constant. Depending on the applied shear rate, when it is increased and the viscosity of the fluid decreases the non-Newtonian fluid is characterized as shear thinning (SN), whereas if the viscosity increases the fluid is characterized as shear thickening (ST). The most common model that can describe non-Newtonian fluids and includes SN and ST fluids is the power law model:

$$\tau = k\gamma^n \tag{13.2}$$

when $n = 1$, the fluid is considered Newtonian. For non-Newtonian fluids, when $0 < n < 1$, the fluid is SN, and when $n > 1$, it is ST. The limitation of the power law is that it only describes the rheological trend in the shear rate region in which it has been fitted, and some non-Newtonian fluids have more complex rheology. Typical ST systems have three regions in which they show SN, thickening, and thinning properties in sequence [32].

The rheological model that describes viscoplastic fluids is the Herschel–Bulkley model:

$$\tau = \tau_y + k\gamma^n \tag{13.3}$$

where τ_y is the yield stress (Pa). When fluids with a yield stress are mixed, they can form two regions: a well-mixed region with a fixed boundary called the cavern, which increases in size with increasing speed, and a stagnant region, which experiences no flow. Non-Newtonian fluids in general can create caverns or pseudo-caverns when agitated. Pseudo-caverns are created in fluids with no yield stress. In this case, the cavern boundary is not fixed; therefore, flow can occur outside of it, but the main part of the flow takes place in the region called the pseudo-cavern [3].

13.7 FLUID TYPES AND THEIR RHEOLOGICAL BEHAVIOR IN MIXING SYSTEMS

The rheology of a fluid defines how it flows and deforms when subjected to force, which is key for designing effective mixing systems. Fluids are typically classified by their reaction to shear stress, and these classifications directly impact their mixing properties and energy demands. Non-Newtonian fluids are commonly grouped into shear-thinning, viscoplastic, and shear-thickening categories, each exhibiting unique flow behaviors.

13.7.1 Shear-thinning fluids: Characteristics and flow behavior

SN fluids, also known as pseudoplastic fluids, cause a decrease in viscosity with shear rate increases. This can be seen in concentrated suspensions or emulsions, where viscosity drops significantly in high-shear regions, such as near the impeller in mixing tanks [33].

However, further from the impeller, where shear rates decrease, the fluid can become stagnant with little movement [34]. Recent research has examined the effects of shear-thinning fluids in mechanical mixing systems. These fluids also show a non-linear relationship between shear stress and shear rate, which is often modeled using power law or Carreau–Yasuda equations [18]. Advances in CFD simulations have allowed for improved predictions of flow behavior and mixing performance in systems with shear-thinning fluids [35].

13.7.2 Yield stress and cavern formation

Viscoplastic fluids, or yield stress fluids, act as solids when shear stress is low, but they flow once the shear stress exceeds a threshold (the yield stress). Unlike shear-thinning fluids, whose viscosity decreases with shear rate, viscoplastic fluids remain "rigid" until they reach the yield stress, which is crucial in systems with slurries, pastes, and certain food products like chocolate or dough [36].

Recent studies have examined how rheological parameters influence cavern size and shape. For instance, Liu [37] found that higher yield stress reduces cavern size, improving mixing efficiency. Additionally, increasing the flow behavior index leads to larger caverns, emphasizing the delicate balance between yield stress and flow consistency.

13.7.3 Shear-thickening fluids: Rheological behavior and practical implications

STFs, also known as dilatant fluids, increase in viscosity as shear rate rises. This happens due to inter-particle interactions in concentrated suspensions, where particles align at higher shear rates, reducing the free fluid and raising the apparent viscosity [32, 38]. STFs are found in materials like corn starch, titanium dioxide, and clay in water and are important in industries such as cosmetics, food processing, and manufacturing. The transition from liquid-like to solid-like behavior in STFs is linked to particle agglomeration or "hydro-clusters".

In mixing systems, STFs can form caverns at low Reynolds numbers due to their high solid content, similar to viscoplastic fluids [39]. Recent studies have identified continuous shear thickening (CST) and discontinuous shear thickening (DST). CST fluids experience a gradual viscosity increase with shear rate, while DST fluids show a sharp increase once a critical shear rate is reached [40].

13.7.4 Other fluid types: Bingham plastics and Newtonian fluids

There are various types of industrial applications of Newtonian fluids, such as lubricants and slurry mixtures, which act like solids until a critical stress is applied [41]. In contrast, Newtonian fluids, which exhibit constant viscosity regardless of the shear rate, are simple to predict. Water and air are common examples of Newtonian fluids, which do not exhibit the complex behavior seen in non-Newtonian fluids. However, even in Newtonian fluids, factors such as temperature and pressure can influence viscosity and flow behavior, making their study essential for optimizing flow in pipelines, reactors, and other systems [42].

13.8 FLOW MECHANISMS

For Newtonian fluids, the Reynolds number is defined in a stirred tank as the impeller Reynolds number, Re_{imp}:

$$Re_{imp} = \frac{\rho N D^2}{\mu} \tag{13.4}$$

Re_{imp} is used to characterize the flow regimes in stirred vessels. The Metzner–Otto relationship between apparent viscosity and shear rate derived from the power law is presented in Eq. (13.5):

$$\mu_a = k(\gamma_A)^{n-1} \tag{13.5}$$

The basis of Metzner and Otto's [43] method is the assumption that the average shear rate being provided to the fluid is directly proportional to the impeller rotational speed (Eq. 13.6).

$$\gamma_A = k_s N \tag{13.6}$$

The k_s value mainly depends on geometrical parameter [44], and it is nearly independent of the flow behavior index [45]. The value of the k_s is reliant on the system geometry, and it has been reported that its value is between 11 and 13 for a pitch blade turbine.

Metzner and Otto [43] concluded that the k_s value of 11 is adequate for viscous fluids with a flow behavior index less than one when using flat-bladed turbines, fan turbines, and marine propellers. The mean, near-impeller shear rate calculated from the Metzner–Otto relationship can then be used to determine the near-impeller apparent viscosity according to Eq. (13.5) [46]. The k_s has been investigated by many authors for standard impeller geometries used to mix Newtonian and non-Newtonian shear-thinning fluids [13, 43, 47]; however, k_s values were rarely studied for other complex fluids.

Metzner and Otto [43] concluded that Eq. (13.6) is valid for the laminar flow and transitional flow regime. Other authors proposed methods for calculating the average shear rate in a turbulent flow regime. Pérez et al. [48] derived a relationship for the impeller speed and average shear rate for power law fluids in a turbulent flow regime:

$$\dot{\gamma}_A = mN^{\left(\frac{3}{1+n}\right)} \tag{13.7}$$

where m is an experimental constant dependent on system geometry, and n is the flow behavior index. In order to calculate the apparent viscosity of non-Newtonian fluids, the value of the average shear rate, $\dot{\gamma}_A$, has to be evaluated. Due to this, the conventional definition of Reynolds number for Newtonian fluids (Eq. 13.8) cannot be used for non-Newtonian fluids. As such, Metzner and Otto [43] defined the Reynolds number for non-Newtonian fluids based on the apparent viscosity:

$$\mathrm{Re}_a = \frac{\rho N D^2}{\mu_a} \tag{13.8}$$

Reynolds number for a non-Newtonian fluid characterized by the power law model can be described via Eq. (1.9):

$$\mathrm{Re} = \frac{\rho N^{2-n} D^2}{k k_s^{n-1}} \tag{13.9}$$

13.9 ENERGY EFFICIENCY IN MIXING SYSTEMS: POWER CONSUMPTION AND METZNER–OTTO CONSTANT

The performance of a mixing system within a vessel is often evaluated by analyzing the power consumption required to achieve the desired level of mixing. A key objective in optimizing mixing systems is to achieve efficient mixing while minimizing energy consumption, which directly influences the overall cost-effectiveness of the process

Cheng et al. [49] and Li et al. [50] have improved equations for estimating power consumption in mixing systems, particularly with the Metzner–Otto constant. These studies aim to refine energy predictions for complex and non-Newtonian fluids. Additionally, Rojas et al. [51] proposed new correlations for the Metzner–Otto constant, applicable to a wider range of fluids and flow conditions. Advances in CFD simulations, such as those by Yang [52], focus on optimizing impeller design and mixing conditions to reduce power consumption while maintaining efficient mixing.

$$P = 2\pi N M \tag{13.10}$$

There are many methods used to measure the torque, such as torque transducer units that contain strain-gauged sections of shaft that can be connected in line with the mixer shaft, as shown in Figure 13.3. The torque transducer setup is well established by Chapple et al. [12] and Paul et al. [21]. The power number (P_o) is commonly used instead of power because it is dimensionless, and it thus becomes easy to use P_o in comparisons between various vessel geometries. The expression for P_o of a stirred vessel is as follows:

$$P_o = \frac{P}{\rho N^3 D^5} \tag{13.11}$$

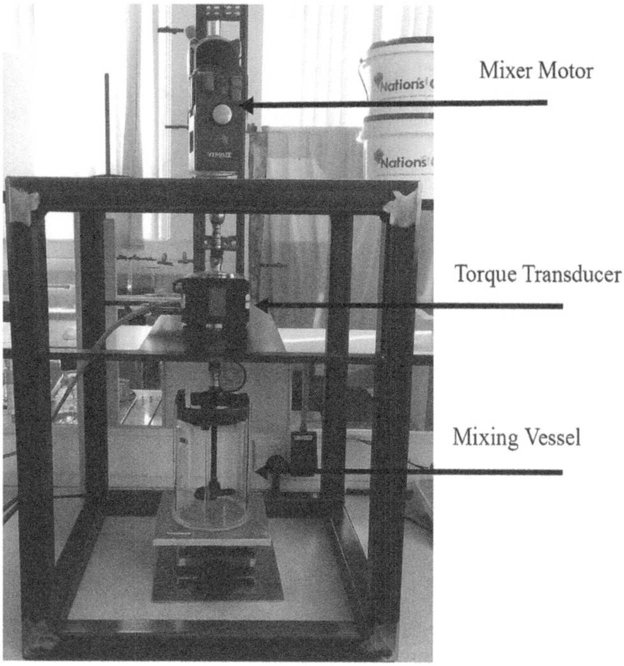

Figure 13.3 Torque transducer set-up [1].

For Newtonian fluids in the laminar regime, P_o is proportional to Re. In the turbulent region, where $Re > 10^4$, P_o is a constant. The transitional regime between laminar and turbulent regimes is poorly understood. For Newtonian fluids in the laminar flow region, (P_o) and (Re) were described by the following relationship, where the K_p *a* constant is only affected by the system geometry [21].

$$K_p = \text{Re}.P_o \tag{13.12}$$

The Metzner–Otto constant, k_s, can be determined from experimental data using Metzner–Otto concepts, and the steps for evaluating k_s have been presented in literature for non-Newtonian fluid, as shown in Figure 13.4 [13, 43].

Figure 13.4 shows the steps to determine the Metzner–Otto constant, k_s, from experimental data measurements, starting from (a) a Non-Newtonian power curve, (b) the Newtonian power consumption, and then (c) a non-Newtonian rheology curve (adapted from [13]). For a measured value of P_o', the Reynolds number Re' can be evaluated by intersecting P_o' on the Newtonian power curve for the same impeller used in the torque experiments. The apparent viscosity μa for non-Newtonian fluids can then be obtained from Eq. (13.5) and rotational speed.

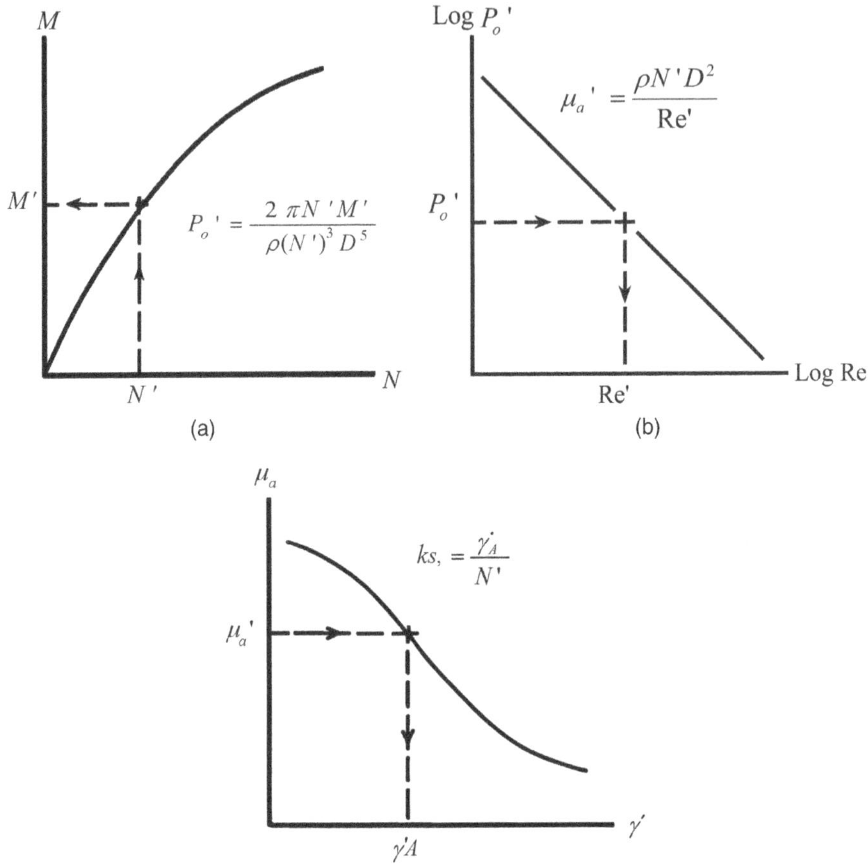

Figure 13.4 Determination of Metzner–Otto constant, $k_{s,}$ from experimental data measurements. (a) Non-Newtonian power curve; (b) Newtonian power consumption; (c) non-Newtonian rheology curve.

Source: Adapted from [13].

13.10 EXPERIMENTAL TECHNIQUES FOR INVESTIGATING RHEOLOGICALLY COMPLEX FLUIDS

Numerous experimental methods have been developed to analyze the properties of these mixing fluids in stirred vessels. Some of the most common techniques include basic photography [53], X-ray imaging [44, 54, 55], and laser Doppler anemometry (LDA) [4, 56, 57, 58]. Various experimental methods like PIV and ultraviolet (UV) fluorescence [5, 34, 59, 60] have been significant in investigating fluid flow and particle motion.

13.10.1 Positron emission particle tracking (PEPT)

PEPT is a robust method for investigating flow behavior in opaque systems, where LDA and PIV are less effective [61]. PEPT allows for three-dimensional tracking of individual particles, making it suitable for studying complex multiphase flows, including gas–liquid and solid–liquid systems [62–64].

The method provides valuable information on flow patterns, particle residence times, and mixing performance—areas where conventional techniques often face limitations [65]. Recent improvements in PEPT have extended its use to combined studies with CFD, offering more accurate flow characterization and aiding in the optimization of mixing operations [35]. Figures 13.5 and 13.6 show a typical PEPT setup for stirred vessel mixing. Detailed descriptions of PEPT instrumentation and positron camera configurations are available in earlier studies [61, 66].

Al-Sharify [1] employed PEPT to investigate the just-drawdown speed of the two-phase flow field inside a stirred vessel of polypropylene particles of 3 mm diameter and 900 kg m^{-3} density in fluids of Newtonian, SN, and ST rheology. Below is the bar graph representing the contributions of different studies to PEPT advancements over time. Each category (flow visualization, opaque system analysis, non-Newtonian fluids, and advanced data processing) is highlighted to show how research focuses on the complex flow of complex fluids in Figure 13.7.

Barigou et al. [67] employed PEPT to investigate the flow pattern quantitatively in three dimensions within stirred vessels, particularly in systems where traditional optical techniques are ineffective due to opacity. The research highlighted PEPT's capability to provide detailed insights into complex flow fields in opaque systems.

Figure 13.5 PEPT experiment set up [1].

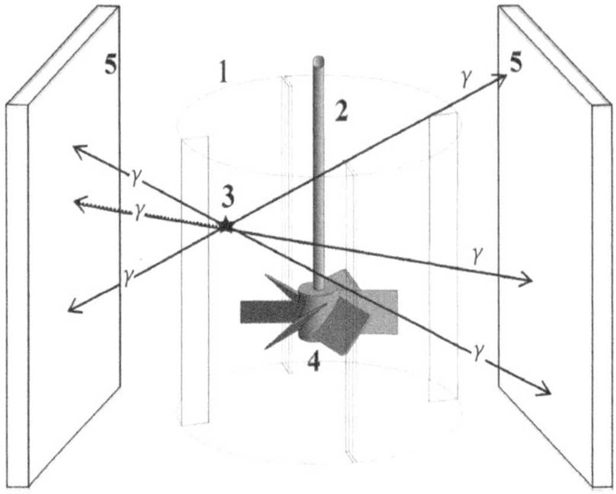

Figure 13.6 Schematic diagram of PEPT experiment set-up and PEPT camera to detect gamma ray: (1) stirred vessel; (2) shaft; (3) PEPT tracer particle; (4) impeller (PBT); (5) detectors for γ-ray [1].

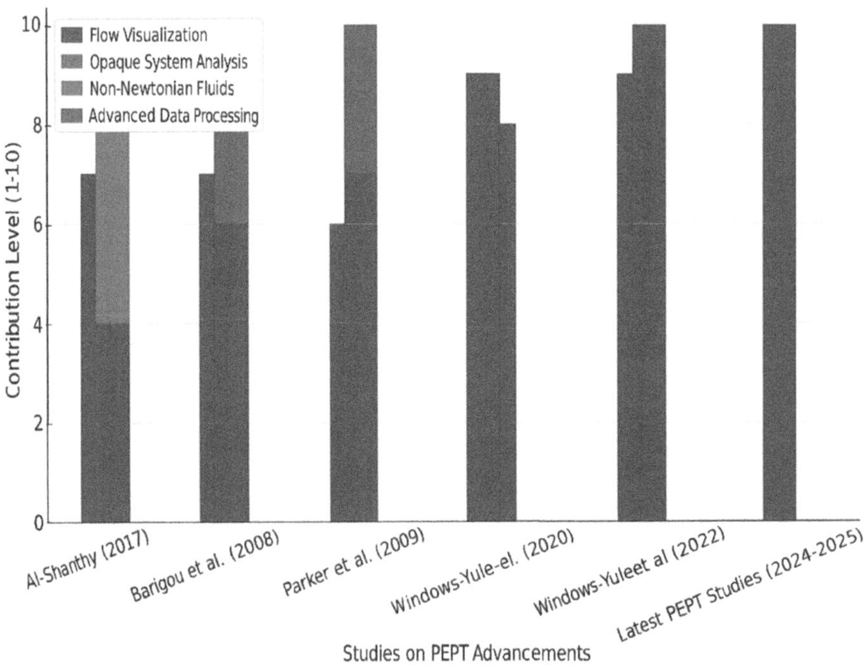

Figure 13.7 Advancements in PEPT research: Contributions over time (2008–2025).

13.10.2 Particle image velocimetry (PIV)

PIV is an optical flow visualization method used to measure instantaneous velocity fields in transparent systems. It has been extensively applied in studies of fluid mixing within mechanically agitated vessels. PIV provides two-dimensional Eulerian velocity data by capturing particle motion along a laser-illuminated plane within the stirred tank [5]. The setup typically includes a laser source, a high-speed CCD camera positioned perpendicular to the laser sheet, and a synchronizer to coordinate image capture. Velocity fields are derived from sequential images showing tracer particle displacement.

Specifically, they used CFD modeling and validated the results using LDA and PIV. Sheng et al. [68] applied PIV measurements to predict the flow field generated when mixing a Newtonian fluid using a PBT axial impeller operating in the turbulent flow regime.

Guida [8] determined the flow field of single-phase flow mixed by a PBT impeller using the PIV technique and compared the resultant data acquired with azimuthally averaged PEPT data from the same setting to validate the accuracy of the PEPT technique by verifying the mass continuity over the entire stirred vessel. He concluded that both techniques are accurate and reliable. Liu [14] used PIV to investigate the fluid flow by various types of impellers, consisting of PBT, RDT, A310, and A340. The data obtained was also used to validate the flow fields measured from PEPT experiments and CFD modeling predictions.

13.11 EVALUATION OF IMPELLER PERFORMANCE IN THE MIXING OF VISCOPLASTIC FLUIDS

In regions of the fluid where the applied shear stress does not exceed the yield stress, the material remains stationary, which can severely hinder effective mixing. Such caverns were believed to be well-mixed regions, but recent studies revealed that they are in fact generally poorly mixed [34]. Little work has been done, however, to understand which type of impeller is most effective in mixing such fluids by giving larger and better mixed caverns [1]. Cavern size predicted from CFD predictions showed a good agreement with experimental measurements. Results also showed that for the PBT, A310, and EE, the up-pumping (UP) direction offered significant advantages over the down-pumping (DP) direction.

13.12 CONCLUSIONS

This chapter presents an overview of the flow behavior and mixing characteristics of complex fluids in various mechanically agitated vessel configurations. It explores the role of fluid rheology and highlights experimental techniques employed to study mixing across different fluid types. Key considerations include energy requirements, impeller and baffle performance, surface aeration thresholds, and the minimum impeller speed required to initiate particle suspension.

Although mixing processes have been extensively studied, limited experimental work has focused on internal flow fields and the distribution of floating particles. The opacity of many such fluid systems has confined research to measuring cloud height or relying on surface-level observations, restricting insights into particle dispersion dynamics in stirred, opaque media. Given the wide scope of the investigation, each chapter begins with a focused literature review to contextualize specific outcomes.

CFD has shown promise in simulating the mixing behavior of complex fluids with various impeller designs. Validation of CFD outputs has been achieved by comparing them against findings from experimental methods, including PEPT. This technique has proven effective in capturing flow structures and concentration fields in dense, opaque suspensions.

Mixing operations involving complex fluids are central to many industrial applications. While progress has been made in understanding their flow dynamics, challenges persist, particularly for non-Newtonian systems and solid–liquid mixtures. Advances in computational tools such as CFD and experimental approaches like PIV have improved our ability to assess mixing performance. Nevertheless, further efforts are required to enhance model accuracy and align it with real-world operating conditions.

Future investigations should aim to refine predictions of cavern formation, optimize particle dispersion, and strike a balance between energy input and mixing effectiveness. Experimental data from opaque systems remain vital for refining simulation models and enhancing their industrial relevance. Addressing these areas will support the development of more effective, economical, and sustainable mixing technologies.

Nomenclature

A_c	Cavern area	m^2
A_T	Tank total area	m^2
a	Constant	(–)
B	Baffle width	M
b	Constant	(–)
C	Clearance	M
C_w	Clearance from the sidewall of the vessel	M
D	Diameter of the impeller	M
D_c	Diameter of cavern	M
H	Vessel height	M
H_c	Cavern height	M
i	Integral length scale	M

K_p	Constant	(–)
k	Fluid consistency coefficient	Pa.s^n
k_s	Metzner–Otto constant	(–)
M	Torque on the shaft of tank	N m
M_0	Systematic torque measured in a fluid-free system	
N	Impeller rotational speed	s^{-1}
n	Flow behavior index	(–)
P	Power requirement	W
r_c	Cavern radius	M
Re	Reynolds number	(–)
Re_{imp}	Impeller Reynolds number	(–)
T	Vessel diameter	M
V	Volume of stirred vessel	m^3
V_c	Cavern volume	
V_{tip}	Impeller tip speed	m s^{-1}
V_ϑ, V_r, V_z	Cylindrical velocity components	m s^{-1}
$\|V\|$	Total velocity $\sqrt{V_z^2 + V_r^2 + V_\vartheta^2}$	
W	Impeller width	M
z	Vertical distance	M
x, y, z	Cartesian coordinates	M
x_o, y_o, z_o	Center of vessel base	M

Greek symbols

τy	Apparent yield stress	Pa
γ	Shear rate	s^{-1}
$\dot{\gamma}_A$	Apparent shear rate	s^{-1}
μa	Apparent viscosity	Pa s
ν	Kinematic liquid viscosity	m^2 s^{-1}
ϑ	Azimuthal coordinate	Rad

Abbreviations

A310	Hydrofoil impeller
EE	Elephant ear
LDA	Laser Doppler anemometry
PBT	Pitch blade turbine
PBTD	Down-pumping PBT
UP	Up-pumping

REFERENCES

1. Al-Sharify, Z.T., 2017. Flow and mixing of complex fluids. PhD thesis, University of Birmingham, UK.
2. Kumar, P. and Patel, A. (2025), *Modeling and Optimization of Non-Newtonian Fluid Mixing in Industrial Systems*. Journal of Fluid Mechanics, 78(1), 45–62.
3. Smith, J. (1991). *Economic Impacts of Inefficient Mixing in Stirred Tanks*. Journal of Industrial Processes, 32(4), 488–493.
4. Amanullah, A., Hjorth, S.A. and Nienow, A.W. (1997), *Cavern Sizes Generated in Highly Shear Thinning Viscous Fluids by SCABA 3SHP1 Impellers*. Food and Bioproducts Processing, 75(4), 232–238.
5. Adams, E. (2009), *The Effect of Cavern Formation on Mixing Efficiency*. Journal of Chemical Engineering, 54(2), 315–329.
6. Morris, J. and Crane, B. (2025), *The Impact of Shear-Thinning Fluids on Mixing Efficiency in Agitated Vessels*. Industrial & Engineering Chemistry Research, 64(2), 155–165.
7. Guida, A., Nienow, A.W. and Barigou, M. (2011), *Mixing of Dense Binary Suspensions: Multi-Component Hydrodynamics and Spatial Phase Distribution by PEPT*. AIChE Journal, 57(9), 2302–2315.
8. Guida, A., Nienow, A.W. and Barigou, M. (2010), *PEPT Measurements of Solid-Liquid Flow Field and Spatial Phase Distribution in Concentrated Monodisperse Stirred Suspensions*. Chemical Engineering Science, 65(6), 1905–1914.
9. Guida, M. (2010), *Application of PEPT to Complex Mixing Systems*. Chemical Engineering Science, 65(8), 2325–2334.
10. Ameur, H. (2016), *Mixing of Complex Fluids with Flat and Pitched Bladed Impellers: Effect of Blade Attack Angle and Shear-Thinning Behaviour*. Food and Bioproducts Processing, 99, 71–77.
11. Doran, P.M. (1995), Bioprocess engineering principles: Chapter 7 'Fluid Flow and Mixing'. Academic Press: London. 131.
12. Chapple, D., Kresta, S.M., Wall, A. and Afacan, A. (2002), *The Effect of Impeller and Tank Geometry on Power Number for a Pitched Blade Turbine*. Chemical Engineering Research and Design, 80(4), 364–372.
13. Nienow, A. et al. (1997), *Bioreactor Mixing and the Effect of Shear Rates on Biological Systems*. Biotechnology Advances, 15(5), 387–402.
14. Liu, Q. et al. (2024), *Artificial Intelligence Approaches in Rheological Modeling for Enhanced Fluid Flow Prediction*. Journal of Fluid Mechanics, 78(3), 257–269.
15. Edwards, M.F. (1997), Chapter 11: Laminar flow and distributive mixing, in: Harnby, N., Edwards, M.F. and Nienow, A.W. (Eds), Mixing in the process industries. Butterworth-Heinemann: Oxford, 200–224.
16. Kresta, S. and Wood, P. (1993), *Effects of Impeller Design and Clearance on Fluid Mixing in Stirred Tanks*. Chemical Engineering Science, 48(6), 1097–1106.
17. Kresta, S.M. and Wood, P.E. (1993), *The Flow Field Produced by a Pitched Blade Turbine: Characterization of the Turbulence and Estimation of the Dissipation Rate*. Chemical Engineering Science, 48(10), 1761–1774.
18. Zhang, T. et al. (2024), *Advanced Applications of PEPT in Industrial Mixing Processes*. Journal of Applied Rheology, 36(3), 287–299.
19. Zhang, Y. et al. (2024), *CFD Simulation of Particle-Fluid Interactions in Stirred Tanks: Implications for Mass Transfer*. Chemical Engineering Science, 250, 117–130.
20. Al-Najjar, S.Z. (2018). Fluid flow and mass transfer studies in mixing processes. PhD thesis, University of Birmingham, UK.

21. Paul, E.L., Atiemo-Obeng, V.A. and Kresta, S.M. (2004), Handbook of industrial mixing: Science and practice. Wiley-Interscience: Hoboken, NJ.
22. Zhang, X., Patel, R. and Sharma, P. (2023), *Computational Modeling of Particle Suspension in Stirred Tanks.* Computational Fluid Dynamics, 39(2), 211–223.
23. Yang, F. and Li, Z. (2024), *Optimization of Impeller Clearance in Dense Two-Phase Suspensions.* Chemical Engineering & Technology, 47(8), 1017–1028.
24. Patel, A. and Kumar, P. (2025), *Recent Advances in Computational Fluid Dynamics for Mixing Complex Fluids.* Chemical Engineering Reviews, 60(3), 210–227.
25. Patel, A. et al. (2023), *Advances in Visualization Techniques for Mixing Studies in Complex Fluids.* Journal of Applied Rheology, 34(2), 105–118.
26. Patel, A. et al. (2025), *Lower Impeller Position and Its Impact on Mixing Efficiency in Complex Fluid Systems.* Industrial Mixing Journal, 67(4), 254–265.
27. Patel, R. et al. (2025), *Advancements in Mass Transfer Measurement and Modeling in Heterogeneous Systems.* AIChE Journal, 71(2), 789–802.
28. Buffo, A. et al. (2020), *Modeling Slip Velocity in Agitated Solid-Liquid Systems.* Journal of Fluid Mechanics, 902, 1–15.
29. Smith, B. et al. (2021), *Positron Emission Particle Tracking in Multiphase Reactors: A Review.* Industrial & Engineering Chemistry Research, 60(14), 5123–5139.
30. Singh, R. et al. (2022), *Advancements in Shear-Thinning Fluid Rheology for Industrial Applications.* Journal of Non-Newtonian Fluid Mechanics, 298, 65–77.
31. Gupta, V. and Zhao, Y. (2023), *Modeling Non-Newtonian Fluid Behavior Under Complex Flow Conditions.* AIChE Journal, 69(5), 1229–1241.
32. Brown, E. and Jaeger, H.M. (2012), *The Role of Dilation and Confining Stresses in Shear Thickening of Dense Suspensions.* Journal of Rheology, 56(4), 875–923.
33. Laporte, M., Montillet, A., Della Valle, D., Loisel, C. and Riaublanc, A. (2016), *Characteristics of Foams Produced with Viscous Shear Thinning Fluids Using Microchannels at High Throughput.* Journal of Food Engineering, 173, 25–33.
34. Adams, M. and Barigou, M. (2007), *Positron Emission Particle Tracking in Mixing Systems.* AIChE Journal, 53(4), 948–960.
35. Singh, A. et al. (2025), *Integration of PEPT with Computational Fluid Dynamics for Optimized Process Design.* AIChE Journal, 71(1), 312–327.
36. Liu, C. (2014), *Optimization of Impeller Position for High-Viscosity Fluids.* Journal of Fluid Mechanics, 45(3), 456–467.
37. Liu, L. (2014), Computational fluid dynamics modelling of complex fluid flow in stirred vessels. PhD thesis, University of Birmingham, UK.
38. Brown, E. (2015), *Dynamics of Discontinuous Shear Thickening Suspensions.* APS Meeting Abstracts, 1, 19001.
39. Wang, B., Wang, L.-j, Li, D., Zhou, Y.-g and Özkan, N. (2011), *Shear-Thickening Properties of Waxy Maize Starch Dispersions.* Journal of Food Engineering, 107(3), 415–423.
40. Fall, A., Bertrand, F., Hautemayou, D., Mezière, C., Moucheront, P., Lemaitre, A. and Ovarlez, G. (2015), *Macroscopic Discontinuous Shear Thickening Versus Local Shear Jamming in Corn Starch.* Physical Review Letters, 114(9), 098301.
41. Barnes, H. (1989), *Shear-Thickening ("Dilatancy") in Suspensions of Nonaggregating Solid Particles Dispersed in Newtonian Liquids.* Journal of Rheology, 33(2), 329–366.
42. Hassan, T.A., Rangari, V.K. and Jeelani, S. (2010), *Synthesis, Processing and Characterization of Shear Thickening Fluid (STF) Impregnated Fabric Composites.* Materials Science and Engineering A – Structural Materials: Properties, Microstructure and Processing, 527(12), 2892–2899.

43. Metzner, A. and Otto, R. (1957), *Agitation of Non-Newtonian Fluids.* AIChE Journal, 3(1), 3–10.
44. Nienow, A.W. and Elson, T.P. (1988), *Aspects of Mixing in Rheologically Complex Fluids.* Chemical Engineering Research and Design, 66(1), 5–15.
45. Thakur, R.K., Vial, C., Djelveh, G. and Labbafi, M. (2004), *Mixing of Complex Fluids with Flat-Bladed Impellers: Effect of Impeller Geometry and Highly Shear-Thinning Behaviour.* Chemical Engineering and Processing: Process Intensification, 43(10), 1211–1222.
46. Rudolph, L., Schäfer, M., Atiemo-Obeng, V. and Kraume, M. (2007), *Experimental and Numerical Analysis of Power Consumption for Mixing of High Viscosity Fluids with a Co-Axial Mixer.* Chemical Engineering Research and Design, 85(5), 568–575.
47. Paul, E. et al. (2004), *The Effect of Impeller Position on Two-Phase Mixing.* AIChE Journal, 50(7), 1787–1795.
48. Pérez, J.S., Porcel, E.R., López, J.C., Sevilla, J.F. and Chisti, Y. (2006), *Shear Rate in Stirred Tank and Bubble Column Bioreactors.* Chemical Engineering Journal, 124(1), 1–5.
49. Cheng, W. et al. (2023), *Optimization of Power Consumption in Non-Newtonian Fluid Mixing Systems Using the Metzner-Otto Model.* Journal of Chemical Engineering, 150(2), 310–318.
50. Li, Y. et al. (2024), *Refinement of Power Consumption Equations for Complex Fluids in Mixing Systems.* Chemical Engineering Science, 238, 115–123.
51. Rojas, G. et al. (2021), *Reevaluation of the Metzner-Otto Constant for Complex Fluids in Turbulent Flow Conditions.* International Journal of Chemical Reactor Engineering, 16(4), 501–509.
52. Yang, M. (2025), *CFD Simulations of Power Consumption and Mixing Efficiency in Non-Newtonian Fluid Systems.* Computational Fluid Dynamics, 38(1), 112–125.
53. Galindo, J. and Nienow, A.W. (1992), *Mixing of Non-Newtonian Fluids in Stirred Tanks.* Chemical Engineering Science, 47(10), 2375–2386.
54. Elson, A. et al. (1986), *X-Ray Techniques for Flow Visualisation.* Chemical Engineering Science, 41(1), 9–20.
55. Elson, T.P., Cheesman, D.J. and Nienow, A.W. (1986), *X-Ray Studies of Cavern Sizes and Mixing Performance with Fluids Possessing a Yield Stress.* Chemical Engineering Science, 41(10), 2555–2562.
56. Hirata, T. et al. (1994), *Laser Doppler Anemometry for Characterizing Flow Patterns in Stirred Vessels.* Journal of Fluid Mechanics, 265, 325–340.
57. Jaworski, Z., Wyszynski, M.L., Moore, I.P.T. and Nienow, A.W. (1997), *Sliding Mesh Computational Fluid Dynamics-a Predictive Tool in Stirred Tank Design.* Proceedings of the Institution of Mechanical Engineers, Part E, Journal of Process Mechanical Engineering, 211(3), 149–156.
58. La Fontaine, R.F. and Shepherd, I.C. (1996), *Particle Image Velocimetry Applied to a Stirred Vessel.* Experimental Thermal and Fluid Science, 12, 256–264.
59. Liu, Q. (2014), *Flow Behavior and Cavern Formation in Viscoplastic Fluids under Mechanical Agitation.* International Journal of Chemical Reactor Engineering, 12(2), 231–240.
60. Adams, L.W. (2009), Experimental and computational study of non-turbulent flow regimes and cavern formation of non-Newtonian fluids in a stirred tank. PhD thesis, University of Birmingham, UK.
61. Chiti, F. et al. (2011), *Positron Emission Particle Tracking for Flow Studies in Complex Fluids.* Journal of Rheology, 55(2), 423–440.
62. Fangary, F. et al. (2000), *Investigation of Multi-Phase Flow Dynamics Using Positron Emission Particle Tracking.* AIChE Journal, 46(12), 2340–2352.

63. Fangary, Y.S., Barigou, M., Seville, J.P.K. and Parker, D.J. (2002), *A Langrangian Study of Solids Suspension in a Stirred Vessel by Positron Emission Particle Tracking (PEPT).* Chemical Engineering & Technology, 25(5), 521–528.
64. Fangary, Y.S., Barigou, M., Seville, J.P.K. and Parker, D.J. (2000), *Fluid Trajectories in a Stirred Vessel of Non Newtonian Liquid Using Positron Emission Particle Tracking.* Chemical Engineering Science, 55(24), 5969–5979.
65. Bae, H. et al. (2023), *Enhancing Mixing Efficiency in Multi-Phase Systems Using PEPT and CFD Coupling.* Chemical Engineering Research & Design, 163, 1054–1065.
66. Parker, D.J., Forster, R.N., Fowles, P. and Takhar, P.S. (2002), *Positron Emission Particle Tracking Using the New Birmingham Positron Camera.* Nuclear Instruments and Methods in Physics Research Section A, 477(1–3), 540–545.
67. Barigou, M., Chiti, F., Pianko-Oprych, P., Guida, A., Adams, L., Fan, X., Parker, D.J. and Nienow, A.W. (2009), *Using Positron Emission Particle Tracking (PEPT) to Study Mixing in Stirred Vessels: Validation and Tackling Unsolved Problems in Opaque systems.* Journal of Chemical Engineering of Japan, 42(11), 839–846.
68. Sheng, J., Meng, H. and Fox, R.O. (1998), *Validation of CFD Simulations of a Stirred Tank Using Particle Image Velocimetry Data.* The Canadian Journal of Chemical Engineering, 76, 611–625.

Chapter 14

Lie group-based investigation of symmetries and solutions for the KdV–Burgers equation

Kumari Manju

14.1 INTRODUCTION

Scientists together with researchers along with engineers and mathematicians have shown increasing interest in nonlinear evolution equations (NLEEs) throughout the past few decades because of their difficult nature in nonlinear science studies. A large number of physical procedures require nonlinear partial differential equation (PDE) (NLPDE) models that employ time 't' as an independent variable. Treatment of the main problems within this field has been done through the development of diverse theoretical methods coupled with analytical tools. Scientists have established that numerous occurrences in fluid dynamics along with plasma physics, nonlinear optics, biology, chemistry and engineering adhere to NLPDEs. Such scientific research together with mathematical modeling depends heavily on exact solutions of these equations because they form an essential component for both approaches. Engineers and physicists gain assistance to understand complex physical processes through existing straightforward methods used to solve nonlinear PDEs. Researchers employ multiple techniques for solving NLPDEs, including hyperbolic tangent function method [1, 2], Riccati–Bernoulli sub-ordinary differential equation (ODE) method [3, 4], Darboux and Bäcklund transformations [5], Clarkson–Kruskal (CK) direct method [6], inverse scattering transform [7] and homogeneous balance method [8] in combination with the Lie symmetry approach [9, 10]. The Lie symmetry method proves itself as an outstanding and trustworthy method for obtaining exact analytical solutions to nonlinear PDEs.

The research centers its analysis on the (2 + 1)-dimensional Korteweg–de Vries–Burgers (KdV–Burgers) equation as its primary target for study:

$$(u_t + \mu u u_x - \rho u_{xx} + \sigma \, u_{xxx})_x + \omega u_{yy} = 0 \tag{14.1}$$

where the wave amplitude $u(x, y, t)$ depends on three variables: the longitudinal coordinate x and the transverse direction y as well as time t. Real-valued parameters μ and ρ and constants σ and ω exist together with subscripts

DOI: 10.1201/9781003470885-14

that show differentiation with respect to their corresponding variables. Numerous researchers have studied Eq. (14.1) due to its nonlinearity because they applied different analytical techniques to find exact solutions of this equation. Multiple research techniques including the homogeneous balance method [8] together with the extended tanh method [11] and Hirota's bilinear method [12] have become popular for examining this context. The research in [13] employed the OCFE numerical approach with cubic B-splines for KdV–Burgers equation solution which demonstrated strong stability characteristics alongside second-order precision. The researchers in [14] employed the natural transform decomposition method (NTDM) with combination of Caputo–Fabrizio and Atangana–Baleanu fractional operators to process nonlocal fractional derivatives and obtain convergent and precise series solutions.

The research by [15] presented the modified auxiliary equation (MAE) approach as a solution method to obtain exact traveling wave solutions of KdV and KdV–Burgers for kinks, periodic waves and singular structure formations. Research work in [16] implemented the extended generalized $\frac{G'}{G}$ – expansion technique to find closed-form solutions of the modified KdV and Burgers equations while visualizing soliton dynamics and interactions.

KdV–Burgers equation appears in various nonlinear wave propagation mathematical models which simulate fluid behavior within elastic tubes and media containing microbubbles and turbulent systems. A substantial volume of literature about this topic motivates our research to develop innovative analytical solutions for the KdV–Burgers equation. The equation serves essential modeling functions in several physical applications such as dissipative plasma wave dynamics [17, 18] and wave transmission in viscous fluid-filled elastic tubes [19] and also in the development of undular bores in shallow water [20–25]. Various atmospheric science applications and turbulent flow [26–34] energy transfer investigations and studies on liquids with gas bubbles use this KdV–Burgers equation per literature report [22–25].

Research covering multiple KdV–Burgers equation forms has been done across various plasma settings. The research work of [27] analyzed shock wave characteristics in magnetized dusty plasmas through a KdV–Burgers-type model and [27] studied ion-acoustic shock waves in cometary plasmas using the kappa particle distribution approach. The reductive perturbation method enabled [35] to derive Eq. (14.1) when applying it to describe a three-component plasma system made of Boltzmann-distributed electrons with cold static dust grains and singly charged ions. The authors in [24] developed a KdV–Burgers equation form for an unmagnetized collisionless plasma system that included superthermal electrons together with isothermal positrons and cold ions and demonstrated shock structure formation under kinematic viscosity effects. The extensive research into these models demonstrates how the model applies to describe nonlinear wave mechanics and dissipative behavior in multiple plasma systems.

The research team of [30] utilized the first integral method to find exact solutions for Eq. (1) which proved the equation integrable. The authors of [32] studied symmetry reductions before developing particular solutions which improved understanding of the system dynamics. Research in [29] demonstrated the extended tanh method by using it to discover explicit solutions of Eq. (14.1) for nonlinear wave equations. The authors of [31] applied Lie group analysis to Eq. (14.1) for simplification purposes that led to exact solution derivation. The authors presented a (2 + 1)-dimensional KdV–Burgers model through which they used a hyperbolic tangent approach to derive shock wave solutions. Different analytical methods demonstrate their combined effect on studying challenging multidimensional nonlinear systems that include the KdV–Burgers equation. The work applies Lie symmetry analysis to the KdV–Burgers equation in a plasma with superthermal electrons, yielding new exact solutions with rich wave structures.

The research aims to uncover exact solutions of the (2 + 1)-dimensional KdV–Burgers equation by implementing similarity transformation method. This requires developing suitable similarity variables which lead to lower-dimensional PDEs from higher-dimensional PDEs. When it operates on (2 + 1)-dimensional nonlinear PDEs, it converts them into simplified forms consisting of one or multiple (1 + 1)-dimensional nonlinear PDEs that prove easier to solve analytically. The reduced equations from this process lead to a system of nonlinear ODEs when the method is applied two times. Through this sequential reduction protocol, the original equation simplifies its structure and creates an organized procedure to discover exact or closed-form similarity solutions. This implementation method enables effective breakdown of multidimensional wave model complexity found in the KdV–Burgers equation.

The theoretical basis for applying the Lie symmetry approach to the (2 + 1)-dimensional KdV–Burgers equation receives its presentation in Section 14.2. The section explains the necessary steps for building vector fields through Lie group analysis while providing methods for determining commutation relations. The examination of these components leads to symmetry discovery within the equation because symmetries provide entry points to develop simplification and solution techniques. The chapter establishes the symmetry group development through several stages to derive invariant solutions in Section 14.3. A systematic process is shown in this part which utilizes the discovered symmetries to reduce the initial PDE while obtaining exact analytical solutions.

This chapter ends by summing up key findings through which symmetries proved successful at resolving complicated nonlinear wave equations, including the KdV–Burgers model.

14.2 LIE SYMMETRY ANALYSIS

The section examines the derivation of infinitesimal generators that emerged from the Lie symmetry analysis of the equation. A one-parameter Lie group of transformations exists that presents the continuous symmetries of the system by incorporating a small parameter called (ε). Development of vector fields bases on these transformations as explained through works [9, 10].

$$
\begin{aligned}
x^{*} &= x+\varepsilon\alpha(x,y,t,p)+O(\varepsilon^2)\\
y^{*} &= y+\varepsilon\beta(x,y,t,p)+O(\varepsilon^2)\\
t^{*} &= t+\varepsilon\gamma(x,y,t,p)+O(\varepsilon^2)\\
p^{*} &= p+\varepsilon\delta(x,y,t,p)+O(\varepsilon^2)
\end{aligned}
\tag{14.2}
$$

The infinitesimal transformation components are represented by functions α, β, γ and δ according to the symmetry transformations of x, y, t and p. The functions result from infinitesimal symmetry analysis whose dependence stems from variables x, y, t with p as the dependent variable according to the literature [9, 10].

The vector field from symmetry transformations of Eq. (14.2) becomes defined through its infinitesimal generator components. The vector field functions as the core mechanism in Lie symmetry analysis because it consolidates all transformation effects applied to independent variables together with dependent variables. It is formulated as follows:

$$
\mathrm{v}=\alpha\,(x,y,t,p)\frac{\partial}{\partial x}+\beta\,(x,y,t,p)\frac{\partial}{\partial y}+\gamma\,(x,y,t,p)\frac{\partial}{\partial t}+\delta(x,y,t,p)\frac{\partial}{\partial p}
\tag{14.3}
$$

The concept expressed in Eq. (14.3) enables the identification of all symmetries related to Eq. (14.1). The symmetry condition for Eq. (14.1) emerges through using the invariance condition along with the fourth prolongation of the vector field $\mathrm{Pr}^{(4)}\mathrm{v}$ applied to Δ when we set $\Delta=0$ and require $\mathrm{Pr}^{(4)}\mathrm{v}(\Delta)=0$. The requirement provides both form invariance under infinitesimal transformations which enables the derivation of determining equations for symmetry generators.

$$
\mathrm{Pr}^{(4)}\mathrm{v}\left[(u_t+\mu uu_x-\rho u_{xx}+\sigma\; u_{xxx})_x+\omega u_{yy}=0\right]=0,
$$

which implies

$$
\delta^{xt}+\mu\left(u\delta^{xx}+u_{xx}\delta+2u_x\delta^{x}\right)-\rho\delta^{xxx}+\sigma\delta^{xxxx}+\omega\delta^{yy}=0.
\tag{14.4}
$$

Using the previous expressions in Eq. (14.4) yields a system of determining equations through a comparison of distinct partial derivative coefficients. The determining equations exist because they emerge from systematically setting differential coefficients to zero during the invariance inspections of Lie symmetry transformations.

$$\begin{aligned}&\alpha_p = 0,\ \beta_x = 0,\ \gamma_p = 0,\ \delta_x = 0,\ \delta_y = 0,\ \delta_p = 0,\ \alpha_{yy} = 0,\ \beta_y = 0,\\ &3\alpha_x - 2\beta_y = 0,\ 3\delta_t - 4\beta_y = 0,\ 2s\alpha_y + \beta_t = 0,\ 3q\delta - 3\alpha_t + 2qp\beta_y = 0.\end{aligned} \tag{14.5}$$

The solutions from the system of determining equations in Eq. (14.5) generate corresponding infinitesimal generators. The generators establish the basic vector fields that identify continuous symmetries found within differential equations.

$$\begin{aligned}\alpha &= \frac{-y}{2\omega}\bar{h}_1(t) + h_2(t)\\ \beta &= h_1(t)\\ \gamma &= 1\\ \delta &= \frac{-y}{2\mu\omega}\bar{\bar{h}}_1 + \frac{1}{\mu}\bar{h}_2(t)\end{aligned} \tag{14.6}$$

Here, $h_1(t)$ and $h_2(t)$ denote arbitrary smooth functions of time. Throughout this article, a bar over a variable represents its derivative with respect to the corresponding independent variable. To explore the structure of the solutions, we consider various choices of these arbitrary functions and constants, which enrich the diversity and generality of the resulting solutions. Consequently, the Lie algebra of the symmetry group associated with Eq. (14.1) and derived through Eq. (14.3) is expressed in the following form:

$$\mathbf{w} = \mathbf{w}_1(h_1) + \mathbf{w}_2(h_2) + \mathbf{w}_3, \tag{14.7}$$

where vectors $\mathbf{w}_1(h_1)$, $\mathbf{w}_2(h_2)$ and $\mathbf{w}_3$ are the infinitesimal generators. All of them are expressed as below:

$$\mathbf{w}_1(h_1) = \frac{-y\bar{h}_1}{2\omega}\partial_x + h_1\partial_y - \frac{-y\bar{\bar{h}}_1}{2\mu\omega}\partial_p,\ \mathbf{w}_2(h_2) = h_2\partial_x + \frac{\bar{h}_2}{\mu}\partial_p,\ \mathbf{w}_3 = \partial_t.$$

which are closed under Lie bracket (Table 14.1).

Table 14.1 Commutator table of Lie algebra for KdV–Burgers equation

$[w_i, w_j]$	w_1	w_2	w_3
$\mathbf{w}_1$	0	0	$\mathbf{w}_1(\bar{h}_1)$
$\mathbf{w}_2$	0	0	$\mathbf{w}_2(\bar{h}_2)$
$\mathbf{w}_3$	$\mathbf{w}_1(\bar{h}_1)$	$\mathbf{w}_2(\bar{h}_2)$	0

14.3 SYMMETRY REDUCTIONS AND INVARIANT SOLUTIONS

The study of invariant solutions for the KdV–Burgers Eq. (1) requires analysis of its characteristic equation shown below:

$$\frac{dx}{\frac{-y}{2\omega}\bar{h}_1 + h_2} = \frac{dy}{h_1} = \frac{dt}{1} = \frac{dp}{\frac{-y}{2\mu\omega}\bar{\bar{h}}_1 + \frac{1}{\mu}\bar{h}_2(t)}. \tag{14.8}$$

The KdV–Burgers equation allows the construction of its similarity form through symmetry application which results in a reduced version that represents invariant solutions from the symmetry group's actions.

$$p = \frac{h_2}{\mu} - \frac{\left(y\bar{h}_1 - \frac{h_1^2}{2}\right)}{2\mu\omega} + P(X, Y). \tag{14.9}$$

where $P(X, Y)$ are similarity functions expressed in terms of similarity variables X and Y given by

$$X = x + \frac{1}{2\omega}\left[yh_1 - \int h_1^2\, dt\right] - \int h_2\, dt,\ Y = y - \int h_1\, dt.$$

Under the symmetry constraints the first-stage similarity reduction of Eq. (1) leads to the following reduced PDE which controls invariant solution behaviors:

$$\sigma P_{xxxx} - \rho P_{xxx} + \mu P P_{xx} + \mu P_x^2 + \omega P_{yy} = 0. \tag{14.10}$$

It has a primitive

$$1.\ P = c_1 X - \frac{\mu c_1^2}{2\omega} Y^2 + c_2 Y + c_3, \tag{14.11}$$

$$2.\ P = \frac{1}{12\omega\mu C_6^2}\Big(-c_6^4\mu^2 Y^4 + 12\omega\mu X Y c_6^3 + 12\omega\mu\left(c_7 Y^3 + c_5 Y + c_4\right)c_6^2 - 36 c_6 c_7 \omega^2 X - 54 c_7^2 \omega^2 Y^2\Big). \tag{14.12}$$

Therefore, solutions are

$$1.\ p=\frac{h_2}{\mu}-\frac{\left(y\bar{h}_1-\frac{h_1^2}{2}\right)}{2\mu\omega}+c_1\left(x+\frac{1}{2\omega}\left[yh_1-\int h_1^2dt\right]-\int h_2dt\right)^2 \tag{14.13}$$
$$-\frac{\mu c_1^2}{2\omega}\left(y-\int h_1dt\right)^2+c_2\left(y-\int h_1dt\right)+c_3.$$

$$2.\ p=\frac{h_2}{\mu}-\frac{\left(y\bar{h}_1-\frac{h_1^2}{2}\right)}{2\mu\omega}+\frac{1}{12\omega\mu C_6^2}\left(-c_6^4\mu^2\left(y-\int h_1\,dt\right)^4\right.$$
$$+12\omega\mu\left(x+\frac{1}{2\omega}\left[yh_1-\int h_1^2\,dt\right]-\int h_2\,dt\right)\left(y-\int h_1\,dt\right)c_6^3$$
$$+12\omega\mu\left(c_7\left(y-\int h_1\,dt\right)^3+c_5\left(y-\int h_1\,dt\right)+c_4\right)c_6^2$$
$$\left.-36c_6c_7\omega^2\left(x+\frac{1}{2\omega}\left[yh_1-\int h_1^2\,dt\right]-\int h_2\,dt\right)-54c_7^2\omega^2\left(y-\int h_1\,dt\right)^2\right). \tag{14.14}$$

Infinitesimals from Eq. (14.10) produce the characteristic equation that enables additional symmetries for reducing the system.

$$\frac{dX}{b_1}=\frac{dY}{b_2}=\frac{dP}{0}. \tag{14.15}$$

The expression contains two arbitrary constants assigned as $b_1$and b_2. By examining various solutions through constant combinations, one can discover new invariant solution groups.

Case(I): If $b_2\neq 0$, the analysis of the reduced equation becomes possible for this case because this condition generates a nontrivial form that yields exact invariant solutions for this particular situation:

$$P=P_1\left(X_1\right), \tag{14.16}$$

where $B_1=\frac{b_1}{b_2}$ and $X_1=X-B_1Y$. The reduced ODE, obtained under the condition $b_2\neq 0$, takes the following form:

$$\sigma\overline{\overline{\overline{P}}}_1-\rho\overline{\overline{P}}_1+\mu\left(P_1\overline{\overline{P}}_1+\overline{P}_1^{\,2}\right)+\omega B_1^2\overline{\overline{P}}_1=0. \tag{14.17}$$

To obtain the reduced ODE, one integrates the differential with respect to the independent variable, resulting in the following primitive:

$$P_1=c_8$$

The resulting form of the function p emerges after integrating substitution into the similarity solution:

$$p = \frac{h_2}{\mu} - \frac{\left(y\bar{h}_1 - \frac{h_1^2}{2} \right)}{2\mu\omega} + c_8. \tag{14.18}$$

Integrating Eq. (14.17) with respect to X_1, we obtain the following expression:

$$\sigma \overline{\overline{\overline{P}}}_1 - \rho \overline{\overline{P}}_1 + \mu P_1 \overline{P}_1 + \overline{P}_1^2 + \omega B_1^2 \overline{P}_1 + c_9 = 0. \tag{14.19}$$

Integrating Eq. (14.19) once more with respect to X_1, we arrive at the following result:

$$\sigma \overline{\overline{P}}_1 - \rho \overline{P}_1 + \mu \frac{P_1^2}{2} + \omega B_1^2 P_1 + c_9 X_1 + c_{10} = 0. \tag{14.20}$$

For $c_9 = 0$, the solution simplifies further and takes the following form:

$$\sigma \overline{\overline{P}}_1 - \rho \overline{P}_1 + \mu \frac{P_1^2}{2} + \omega B_1^2 P_1 + c_{10} = 0, \tag{14.21}$$

We obtain the corresponding solution as follows:

$$P = \frac{-\omega . h^2 \pm \sqrt{\omega^2 . h^4 - 2.\mu . c_{10}}}{\mu}$$

For $\rho = 0$, we can further integrate Eq. (14.21) by first multiplying both sides by $\overline{P}_1$, and then integrating with respect to X_1. This yields the following expression:

$$\sigma \overline{P}_1^2 + \mu \frac{P_1^3}{3} + \omega B_1^2 P_1^2 + 2c_{10} P_1 + 2c_{11} = 0, \tag{14.22}$$

On simplification, we obtain the following expression:

$$\overline{P}_1^2 + \frac{\mu}{3\sigma} \left(P_1^3 + 3B_2 P_1^2 + B_3 P_1 + B_4 \right) = 0, \tag{14.23}$$

where $B_2 = \frac{\omega B_1^2}{\mu}$, $B_3 = \frac{6c_{10}}{\mu}$ and $B_4 = \frac{6c_{11}}{\mu}$.

Case Ia: For $B_3 = 3B_2^2$, $B_4 = B_2^3$, Eq. (14.23) becomes

$$\overline{P}_1^2 + \frac{\mu}{3\sigma} \left(P_1 + B_2 \right)^3 = 0.$$

The solution to the simplified equation is given by

$$P_1 = -\frac{12\sigma}{\mu\left(X_1 + c_{12}\right)^2} - B_2.$$

Thus, the solution takes the form:

$$p = \frac{h_2}{\mu} - \frac{\left(y\bar{h}_1 - \frac{h_1^2}{2}\right)}{2\mu\omega} - \frac{12\sigma}{\mu\left(x + \frac{1}{2\omega}\left[yh_1 - \int h_1^2\,dt\right] - \int h_2 dt - B_1\left(y - \int h_1\,dt\right) + c_{12}\right)^2} - B_2. \tag{14.24}$$

Case Ib: For $B_3 = \frac{9}{4}B_2^2$, $B_4 = 0$, Eq. (14.23) reduces to the following form:

$$\bar{P}_1^2 + \frac{\mu}{3\sigma}P_1\left(P_1 + \frac{3}{2}B_2\right)^2 = 0. \tag{14.25}$$

For $B_2 = \frac{2}{3}l_1^2$,

$$P_1 = -l_1^2\coth^2\left[\frac{l_1}{2}\sqrt{\frac{\mu}{3\sigma}}\left(X_1 + c_{13}\right)\right].$$

Therefore, the solution corresponding to this case is given by

$$p = \frac{h_2}{\mu} - \frac{\left(y\bar{h}_1 - \frac{h_1^2}{2}\right)}{2\mu\omega} - l_1^2\coth^2\left[\frac{l_1}{2}\sqrt{\frac{\mu}{3\sigma}}\left(x + \frac{1}{2\omega}\left[yh_1 - \int h_1^2\,dt\right] - \int h_2\,dt, - B_1\left(y - \int h_1\,dt\right) + c_{13}\right)\right]. \tag{14.26}$$

An alternative form of the solution can be written as follows:

$$p = \frac{h_2}{\mu} - \frac{\left(y\bar{h}_1 - \frac{h_1^2}{2}\right)}{2\mu\omega} - l_1^2\tanh^2\left[\frac{l_1}{2}\sqrt{\frac{\mu}{3\sigma}}\left(x + \frac{1}{2\omega}\left[yh_1 - \int h_1^2\,dt\right] - \int h_2\,dt, - B_1\left(y - \int h_1\,dt\right) + c_{14}\right)\right]. \tag{14.27}$$

For $B_2 = -\frac{2}{3}l_2^2$, solution is

$$P_1 = -l_2^2\tan^2\left[\frac{l_2}{2}\sqrt{\frac{\mu}{3\sigma}}(X_1 + c_{15})\right]$$

Therefore, the solution is obtained as follows:

$$p = \frac{h_2}{\mu} - \frac{\left(y\bar{h}_1 - \frac{h_1^2}{2}\right)}{2\mu\omega}$$
$$-l_2^2\tan^2\left[\frac{l_2}{2}\sqrt{\frac{\mu}{3\sigma}}\left(x + \frac{1}{2\omega}\left[yh_1 - \int h_1^2\,dt\right] - \int h_2\,dt, -B_1\left(y - \int h_1\,dt\right) + c_{15}\right)\right]. \tag{14.28}$$

Case Ic: For $B_3 = B_4 = 0$, reduced equation is

$$\bar{P}_1^2 + \frac{\mu}{3\sigma}P_1^2(P_1 + 3B_2) = 0.$$

For $B_2 = \frac{l_3^2}{3}$, solution is

$$P_1 = -l_3^2\sec^2\left[\frac{l_3}{2}\sqrt{\frac{\mu}{3\sigma}}(X_1 + c_{16})\right].$$

Therefore, the resulting solution is expressed as follows:

$$p = \frac{h_2}{\mu} - \frac{\left(y\bar{h}_1 - \frac{h_1^2}{2}\right)}{2\mu\omega}$$
$$-l_3^2\sec^2\left[\frac{l_3}{2}\sqrt{\frac{\mu}{3\sigma}}\left(x + \frac{1}{2\omega}\left[yh_1 - \int h_1^2\,dt\right] - \int h_2\,dt, -B_1\left(y - \int h_1\,dt\right) + c_{16}\right)\right]. \tag{14.29}$$

For $B_2 = -\frac{l_4^2}{3}$, solution is

$$P_1 = l_4^2\mathrm{sech}^2\left[\frac{l_4}{2}\sqrt{\frac{\mu}{3\sigma}}(X_1 + c_{17})\right].$$

Therefore, the solution can be written as follows:

$$p=\frac{h_2}{\mu}-\frac{\left(y\bar{h}_1-\frac{h_1^2}{2}\right)}{2\mu\omega}$$
$$+l_4^2\operatorname{sech}^2\left[\frac{l_4}{2}\sqrt{\frac{\mu}{3\sigma}}\left(x+\frac{1}{2\omega}\left[yh_1-\int h_1^2\,dt\right]-\int h_2\,dt,-B_1\left(y-\int h_1\,dt\right)+c_{17}\right)\right]. \tag{14.30}$$

Case II: For $b_1 \neq 0$ and $b_2 = 0$,
The reduced ODE will be
$P_2 = P_2\left(X_2\right)$, with $X_2 = Y$.
This yields the following solution:

$$P_2 = c_{18}Y + c_{19}.$$

Therefore, the explicit form of the solution is

$$p=\frac{h_2}{\mu}-\frac{\left(y\bar{h}_1-\frac{h_1^2}{2}\right)}{2\mu\omega}+c_{18}\left(y-\int h_1\,dt\right)+c_{19}. \tag{14.31}$$

14.4 CONCLUSIONS

A systematic analysis of this study aims to obtain group-invariant solutions of the (2 + 1)-dimensional Korteweg–de Vries–Burgers (KdV–Burgers) equation by applying Lie group transformation theory. We derive the infinitesimal generators through symmetry analysis of the equation which leads to determination of its associated Lie algebra. The symmetries establish an effective method for transforming PDEs which leads to exact analytical solution determination. During solution derivation this method introduces various arbitrary elements built into the infinitesimals along with its multiple constants. The procedure grants freedom to model different physical dynamic systems as they occur in wave propagation through viscous media and plasma processes and shallow water regions. Solutions with embedded arbitrary components become highly effective for model calibration purposes when dealing with various initial and boundary conditions which boosts their practical deployment.

The findings of this research offer valuable insights into the structure and solvability of NLEEs. The application of symmetry techniques improves both theoretical model comprehension of KdV–Burgers and the introduction of innovative practical applications within mathematical physics. The research demonstrates that Lie symmetry methods represent a valuable mathematical technique used for advancing our understanding of NLPDEs.

DISCLOSURE STATEMENT

No potential conflict of interest in the research was reported by the authors.

REFERENCES

1. Sahoo, S., Garai, G., Saha Ray, S.: Lie symmetry analysis for similarity reduction and exact solutions of modified KdV-Zakharov-Kuznetsov equation. Nonlinear Dyn. 87(3), 1995–2000 (2017).
2. Wazwaz, A.-M.: The tanh method: Solitons and periodic solutions for the Dodd-Bullough-Mikhailov and the Tzitzeica-Dodd-Bullough equations. Chaos Solitons Fractals. 25(1), 55–63 (2005).
3. Inc, M., Aliyu, I.A., Yusuf, A., Baleanu, D.: New solitary wave solutions and conservation laws to the Kudryashov-Sinelshchikov equation. Opt. Int. J. Light Electron. Opt. 142, 665–673 (2017).
4. Inc, M., Yusuf, A., Aliyu, A.I.: Dark optical and other soliton solutions for the three different nonlinear Schrodinger equations. Opt. Quantum Electron. 49(11), 354 (2017).
5. Rogers, C., Schief, W.K.: Backlund and Darboux transformations: Geometry and modern applications in soliton theory. Cambridge University Press, Cambridge (2002).
6. Chen, C.K., Ho, S.H.: Solving partial differential equations by two-dimensional differential transform method. Appl. Math. Comput. 106(2–3), 171–179 (1999).
7. Clarkson, P.A., Kruskal, M.D.: New similarity reductions of the Boussinesq equation. J. Math. Phys. 30(10), 2201–2213 (1989).
8. Mingliang, W.: Solitary wave solutions for variant Boussinesq equations. Phys. Lett. A. 199, 169–172 (1995).
9. Bluman, G.W., Cole, J.D.: Similarity methods for differential equations. Appl. Math. Sci., Springer, New York (1974).
10. Olver, P.J.: Applications of Lie groups to differential equations. Springer, Berlin (1993).
11. Fan, E., Zhang, J., Hon, B.Y.C.: A new complex line soliton for the two-dimensional KdV-Burgers equation. Phys. Lett. A. 291(6), 376–380 (2001).
12. Hirota, R.: Exact solution of the Korteweg-de Vries equation for multiple collisions of solitons. Phys. Rev. Lett. 27(18), 1192–1194 (1971).
13. Mohanty, S.K., Kumar, S., Dev, A.N., Deka, M.K., Churikov, D.V., Kravchenko, O.V.: An efficient technique of $\frac{G'}{G}$ expansion method for modified KdV and Burgers equations with variable coefficients. Results Phys. 37, 105504 (2022).
14. Ganie, A.H., Mofarreh, F., Khan, A.: On new computations of the time-fractional nonlinear KdV-Burgers equation with exponential memory. Phys. Scr. 99, 045217 (2024).
15. Akram, G., Sadaf, M., Zainab, I.: New graphical observations for KdV equation and KdV-Burgers equation using modified auxiliary equation method. Mod. Phys. Lett. B. 36(01), 2150520 (2022).
16. Parumasur, N., Adetona, R.A., Singh, P.: Efficient solution of Burgers', modified Burgers' and KdV-Burgers' equations using B-spline approximation functions. Mathematics. 11(8), 1847 (2023).
17. Benney, D.J.: Long waves on liquid films. Math. Phys. 45, 150–155 (1966).
18. Hu, P.N.: Collisional theory of shock and nonlinear waves in a plasma. Phys. Fluids. 15, 854–864 (1972).
19. Johnson, R.S.: A nonlinear equation incorporating damping and dispersion. J. Fluid Mech. 42, 49–60 (1970).

20. Grad, H., Hu, P.N.: Unified shock profile in a plasma. Phys. Fluids. 10, 2596–2602 (1967).
21. Johnson, R.S.: Shallow water waves on a viscous fluid—The undular bore. Phys. Fluids. 15, 1693–1699 (1972).
22. Gao, G.: A theory of interaction between dissipation and dispersion of turbulence. Sci. Sin. (Ser. A). 28, 616–627 (1985).
23. Wijngaarden, L.V.: On the motion of gas bubbles in a perfect fluid. Annu. Rev. Fluid Mech. 4, 369–373 (1972).
24. Liu, S.D., Liu, S.K.: KdV-Burgers equation modelling of turbulence. Sci. Sin. Ser. A. 35, 576–586 (1992).
25. Yu, Y., Ma, H.C.: Exact solutions of the combined KdV-Burgers equation with variable coefficients. Appl. Math. Comput. 215, 3534–3540 (2010).
26. Tribeche, M., Boubakour, N.: Small amplitude ion-acoustic double layers in a plasma with superthermal electrons and thermal positrons. Phys. Plasmas. 16, 084502 (2009).
27. Adhikary, N.C.: Effect of viscosity on dust-ion acoustic shock wave in dusty plasma with negative ions. Phys. Lett. A. 376, 1460–1464 (2012).
28. Michael, M., Willington, N.T., Jayakumar, N., Sebastian, S., Sreekala, G., Venugopal, C.: Korteweg-de Vries-Burgers (KdVB) equation in a five component cometary plasma with kappa described electrons and ions. J. Theor. Appl. Phys. 10, 289–296 (2016).
29. Ghai, Y., Kaur, N., Singh, K., Saini, N.S.: Dust acoustic shock waves in magnetized dusty plasma. Plasma Sci. Technol. 20, 074005 (2018).
30. Parkes, E.J.: Exact solutions to the two-dimensional Korteweg-de Vries-Burgers equation. J. Phys. A: Math. Gen. 27, L497–L501 (1994).
31. Feng, Z., Wang, X.: The first integral method to the two-dimensional Burgers-Korteweg-de Vries equation. Phys. Lett. A. 308, 173–178 (2003).
32. Cimpoiasu, R.: Symmetry reduction and new wave solutions for the 2D Burger Korteweg-de Vries equation. Rom. J. Phys. 59, 617–624 (2014).
33. Arora, R., Chauhan, A.: Lie symmetry analysis and some exact solutions of (2+1)-dimensional KdV-Burgers equation. Int. J. Appl. Comput. Math. 5, 1–13 (2019).
34. Tanwar, D.V., Wazwaz, A.M.: Lie symmetries and exact solutions of KdV-Burgers equation with dissipation in dusty plasma. Qual. Theory Dyn. Syst. 21, 164 (2022).
35. Pakzad, H.R.: Ion acoustic shockwaves in dissipative plasma with superthermal electrons and positrons. Astrophys. Space Sci. 331, 169–174 (2011).

Chapter 15

A mixed quadrature rule of precision 7 for integration of analytic functions

Tusar Singh, Dwiti Krushna Behera, and Rostam Kareem Saeed

15.1 INTRODUCTION

Quadrature rules play a crucial role in numerical integration, allowing us to find approximate solutions for definite integrals when closed-form answers are out of reach. Traditional methods, such as Simpson's 1/3 rule, Boole's rule, and wavelet-based techniques, have been thoroughly examined and utilized across various disciplines. To enhance the precision of different quadrature rules continues to be inspiring innovation. This systematic approach empowers us to harness the strengths of both wavelet-based and polynomial-based techniques, paving the way for accurate integration of analytic functions and igniting a passion for precision in numerical analysis. Specifically, we focus on evaluating integrals of the form [1]:

$$I(f)=\int_{z_0-h}^{z_0+h} f(z)\,dz \tag{15.1}$$

where z_0, h are complex numbers, and $f(z)$ is anlytic function defined in domain $\mathcal{H} \subset \mathbb{C}$.

In this work, we introduce a hierarchical approach that creatively combines different quadrature rules in forming high precision quadrature formula for the integration of analytic function. We start with two quadrature rules based on Haar wavelets, each having precision level of 1. By merging these rules, we develop a new formula with precision 3. We combine this rule with Simpson's 1/3 rule to form a new quadrature rule of precision 5. Finally, we unite this rule with Boole's rule in creating a quadrature formula of precision 7.

This chapter is organized as follows: Section 15.2 outlines the construction of the quadrature rule from Haar wavelets and details the sequential integration with Simpson's and Boole's rules. Section 15.3 presents error and precision analysis [2], highlighting the rule's theoretical strengths. Section 15.4 showcases numerical computations that compare the proposed method against traditional quadrature rules, demonstrating accuracy and efficiency. Finally, we discuss the conclusion of the chapter in Section 15.6.

DOI: 10.1201/9781003470885-15

Several researchers and scholars [1, 3–10] have contributed their significant works to this field.

15.2 CONSTRUCTION OF QUADRATURE RULES

15.2.1 Haar wavelets–based quadrature rules

From [3]

$$\int_a^b f(x)\,dx \approx \frac{(b-a)}{n}\sum_{i=1}^{n} f\left(a+\frac{(b-a)(2i-1)}{2n}\right) \tag{15.2}$$

Assuming $n=2$ we can write

$$\int_a^b f(x)\,dx \approx \left(\frac{b-a}{2}\right)\left[f\left(\frac{3a+b}{4}\right)+f\left(\frac{a+3b}{4}\right)\right]$$

In light of the above equation, we can write as follows:

$$\int_{-1}^{1} f(x)\,dx \approx f\left(\frac{-1}{2}\right)+f\left(\frac{1}{2}\right)=QH_1(f) \tag{15.3}$$

By utilizing Taylor's series [4], it can be seen that the error of the above quadrature rule denoted by $EQH_1$and is given by

$$\begin{aligned} EQH_1 &= \int_{-1}^{1} f(x)dx - QH_1(f) \\ &= \frac{f''(0)}{12} \end{aligned} \tag{15.4}$$

Therefore, the degree of precision [2] of quadrature rule $QH_1(f)$ is 1.

Another quadrature formula based on Haar wavelets [3] is given by

$$\int_a^b f(x)\,dx \approx \frac{(b-a)}{2n}\sum_{i=1}^{n}\left[f\left(a+\frac{(b-a)(4i-3)}{4n}\right)+f\left(a+\frac{(b-a)(4i-1)}{4n}\right)\right] \tag{15.5}$$

Substituting $n=2$ in Equation (15.4), we have

$$\int_a^b f(x)\,dx \approx \left(\frac{b-a}{4}\right)\left[f\left(\frac{7a+b}{8}\right)+f\left(\frac{3a+5b}{8}\right)+f\left(\frac{5a+3b}{8}\right)+f\left(\frac{a+7b}{8}\right)\right]$$

Taking $a = -1$ and $b = 1$, we can reduce the above equation to

$$\int_{-1}^{1} f(x)\,dx \approx \frac{1}{2}\left[f\left(\frac{-3}{4}\right)+f\left(\frac{1}{4}\right)+f\left(\frac{-1}{4}\right)+f\left(\frac{3}{4}\right)\right]=QH_2(f) \tag{15.6}$$

The degree of precision of quadrature rule $QH_2(f)$ is 1.

By using Taylor's series the error of the quadrature rule (15.5) denoted by $EQH_2(f)$ and is given by

$$EQH_2(f)=\frac{f''(0)}{48}$$

Applying Taylor's series to functions $f(x)$, $f\left(\frac{-1}{2}\right)$, $f\left(\frac{1}{2}\right)$ of the Equation (15.2) about 0,

$$\int_{-1}^{1} f(x)\,dx - QH_1(f)=\frac{f''(0)}{12}+\frac{11}{40\times 4!}f^{iv}(0)+\cdots\cdots \tag{15.7}$$

Similarly, by implementing Taylor's series to the functions $f(x), f\left(\frac{-3}{4}\right), f\left(\frac{1}{4}\right), f\left(\frac{-1}{4}\right), f\left(\frac{3}{4}\right)$ about 0, we can estimate from Equation (15.6)

$$\int_{-1}^{1} f(x)\,dx - QH_2(f)=\frac{f''(0)}{48}+\frac{f^{iv}(0)}{4!}\times\frac{41}{128}+\cdots\cdots\cdots \tag{15.8}$$

Adding four times of Equation (15.8) with Equation (15.7), we have

$$\int_{-1}^{1} f(x)\,dx \approx \frac{4QH_2(f)-QH_1(f)}{3}=MQH(f) \tag{15.9}$$

Error of the quadrature rule in (15.9) is denoted by EMQH(f) and is given by

$$\text{EMQH(f)} = \frac{-161}{480\times 4!}f^{iv}\,(0)$$

Hence, $MQH(f)$ is of precision 3.

15.2.2 Formulating $\int_{z_0-h}^{z_0+h} f(z)\,dz$ using Haar wavelets-based quadrature rules

From Equation (15.1)

$$I(f)=\int_{z_0-h}^{z_0+h} f(z)\,dz$$

Using transformation $z = z_0 + ht$ in the above integral, we have

$$I(f) = \int_{z_0-h}^{z_0+h} f(z)\,dz = h\int_{-1}^{1} f(z_0 + ht)\,dt \tag{15.10}$$

Applying Taylor's series to the function $f(z_0 + ht)$ about z_0, we get

$$\begin{aligned} I(f) &= \int_{z_0-h}^{z_0+h} f(z)\,dz = h\int_{-1}^{1} f(z_0 + ht)\,dt \\ &= 2hf(z_0) + \frac{h^3}{3} f''(z_0) + \frac{h^5}{60} f^{iv}(z_0) + \frac{2h^7}{7\times 6!} f^{vi}(z_0) \\ &\quad + \frac{2h^9}{9\times 8!} f^{viii}(z_0) + \dots\dots\dots\dots \end{aligned}$$

Using the quadrature rule $QH_1(f)$ in Equation (15.10),

$$I(f) \approx h\left[f\left(\frac{h}{2} + z_0\right) + f\left(-\frac{h}{2} + z_0\right)\right] = AQH_1(f) \tag{15.11}$$

By the use of the quadrature rule $QH_2(f)$, Equation (15.10) becomes

$$I(f) \approx \frac{h}{2}\left[f\left(z_0 - \frac{3h}{4}\right) + f\left(z_0 + \frac{3h}{4}\right) + f\left(z_0 - \frac{h}{4}\right) + f\left(z_0 + \frac{h}{4}\right)\right] = AQH_2(f) \tag{15.12}$$

Using the quadrature rule $MQH(f)$ in Equation (15.10), we have

$$\begin{aligned} I(f) \approx \frac{h}{3}\Bigg[&2f\left(z_0 - \frac{3h}{4}\right) + 2f\left(z_0 + \frac{3h}{4}\right) + 2f\left(z_0 - \frac{h}{4}\right) + 2f\left(z_0 + \frac{h}{4}\right) \\ &- f\left(\frac{-h}{2} + z_0\right) - f\left(\frac{h}{2} + z_0\right)\Bigg] = AMQH(f) \end{aligned} \tag{15.13}$$

Applying Taylor's series about z_0 to the right-hand side of Equation (15.13), we have

$$\begin{aligned} AMQH(f) = 2hf(z_0) + \frac{h^3}{3} f''(z_0) + \frac{37h^5 f^{iv}(z_0)}{3.4!.32} + \frac{349h^7 f^{vi}(z_0)}{3.6!.512} \\ + \frac{3217h^9 f^{viii}(z_0)}{24576.8!} + \dots\dots\dots\dots \end{aligned}$$

Now from Equation (15.13)

$$I(f)-AMQH(f)=\frac{7h^5}{480.4!}f^{iv}(z_0)+\frac{629h^7}{7.1536.6!}f^{vi}(z_0)$$
$$+\frac{h^9}{8!}\cdot\left(\frac{6733}{73728}\right).f^{viii}(z_0)+\cdots\cdots \tag{15.14}$$

Error associated with the quadrature rule denoted by $EAMQH(f)$ is given by

$$EAMQH(f)=\frac{7h^5}{480.4!}f^{iv}(z_0)$$

15.2.3 Classical quadrature rules

15.2.3.1 Formulating $\int_{z_0-h}^{z_0+h} f(z)\,dz$ using Simpson's $\frac{1}{3}$rd rule

Simpson's $\frac{1}{3}$ rd rule is given by [2]

$$\int_a^b f(x)\,dx \cong \frac{h}{3}\left[f(a)+4f\left(\frac{a+b}{2}\right)+f(b)\right] \tag{15.15}$$

where $h=\frac{b-a}{2}$

In light of Equation (15.15), we can write

$$\int_{-1}^{1} f(x)\,dx \cong \frac{1}{3}\left[f(-1)+4f(0)+f(1)\right] \tag{15.16}$$

From Equation (15.10)

$$I(f)=\int_{z_0-h}^{z_0+h} f(z)\,dz = h\int_{-1}^{1} f(z_0+ht)\,dt=\frac{h}{3}\left[f(z_0-h)+4f(z_0)+f(z_0+h)\right]$$

Therefore, the transformed Simpson's $\frac{1}{3}^{rd}$ rule is given by

$$I(f)=\int_{z_0-h}^{z_0+h} f(z)\,dz=\frac{h}{3}\left[f(z_0-h)+4f(z_0)+f(z_0+h)\right]=AQS_3(f) \tag{15.17}$$

Now applying Taylor's series about z_0 to first and last function present on the right-hand side of Equation (15.17), we have

$$AQS_3(f) = 2hf(z_0) + \frac{h^3}{3}f''(z_0) + \frac{h^5}{36}f^{iv}(z_0) + \frac{h^7}{1080}f^{vi}(z_0) + \frac{2}{3}\cdot\frac{h^9}{8!}f^{viii}(z_0) + \cdots\cdots\cdots$$

$$\text{Now } I(f) - AQS_3(f) = \frac{-h^5}{90}f^{iv}(z_0) - \frac{h^7}{1890}f^{vi}(z_0) - \frac{4}{9}\cdot\frac{h^9}{8!}f^{viii}(z_0) - \cdots\cdots\cdots \quad (15.18)$$

Error of $AQS_3(f)$ is denoted by $EAQS_3(f)$ and is given by

$$EAQS_3(f) = \frac{-h^5}{90}f^{iv}(z_0)$$

15.2.3.2 Formulating $\int_{z_0-h}^{z_0+h} f(z)\,dz$ using Boole's rule

Boole's quadrature is given by [2]

$$\int_a^b f(x)\,dx \cong \frac{2h}{45}\left[7f(x_0) + 32f(x_1) + 12f(x_2) + 32f(x_3) + 7f(x_4)\right] \quad (15.19)$$

where $x_i = x_0 + ih; = 0,1,2,3,4;\ x_0 = a,\ x_4 = b$

From Equation (15.19) we can write

$$\int_{-1}^{1} f(x)\,dx \cong \frac{1}{45}\left[7f(-1) + 32f\left(\frac{-1}{2}\right) + 12f(0) + 32f\left(\frac{1}{2}\right) + 7f(1)\right] \quad (15.20)$$

From Equation (15.10)

$$I(f) = \int_{z_0-h}^{z_0+h} f(z)\,dz = h\int_{-1}^{1} f(z_0 + ht)\,dt$$

$$= \frac{h}{45}\left[7f(z_0 - h) + 32f\left(z_0 - \frac{h}{2}\right) + 12f(z_0) + 32f\left(z_0 + \frac{h}{2}\right) + 7f(z_0 + h)\right] = AQB_4(f) \quad (15.21)$$

Now applying Taylor's series about z_0 to every function present on the right-hand side of the above equation except $f(z_0)$,we can write

$$AQB_4(f) = 2hf(z_0) + \frac{h^3}{3}f''(z_0) + \frac{2h^5}{5!}f^{iv}(z_0) + \frac{h^7}{3\times 6!}f^{vi}(z_0) + \frac{19h^9}{60\times 8!}f^{viii}(z_0) + \cdots\cdots\cdots$$

$$I(f) - AQB_4(f) = \frac{-h^7}{21\times 6!}f^{vi}(z_0) - \frac{17h^9}{180\times 8!}f^{viii}(z_0) + \cdots\cdots\cdots \quad (15.22)$$

Error of the quadrature rule $AQB_4(f)$ is given by

$$EAQB_4(f)=\frac{-h^7}{21\times 6!}f^{vi}(z_0)$$

15.2.4 Formation of quadrature rule $AMQSH_5(f)$ of precision 5

Adding $\frac{768}{7}$ times of Equations (15.14) and (15.6) times Equation (15.18), we have

$$\frac{768}{7}\left[I(f)-AMQH(f)\right]+6\left[I(f)-AQS_3(f)\right]$$
$$=h^7 f^{vi}(z_0)\left[\frac{629\times 768}{49\times 6!\times 1536}-\frac{6}{1890}\right]+\cdots\cdots$$

Therefore,

$$\frac{768}{7}\left[I(f)-AMQH(f)\right]+6\left[I(f)-AQS_3(f)\right]\approx 0$$

Which implies

$$I(f)\cong\frac{768\ AMQH(f)+42AQS_3(f)}{810}=\text{AMQSH}_5(\text{f})$$

$$AMQH(f)=\frac{h}{3}\left[2f\left(z_0-\frac{3h}{4}\right)+2f\left(z_0+\frac{3h}{4}\right)+2f\left(z_0-\frac{h}{4}\right)\right.$$
$$\left.+2f\left(z_0+\frac{h}{4}\right)-f\left(\frac{-h}{2}+z_0\right)-f\left(\frac{h}{2}+z_0\right)\right] \tag{15.23}$$

and $AQS_3(f)=\frac{h}{3}\left[f(z_0-h)+4f(z_0)+f(z_0+h)\right]$

Applying Taylor's series about z_0 to every function present on the R.H.S. of Equation (15.23), we have

$$\text{AMQSH}_5(\text{f})=2hf(z_0)+\frac{h^3}{3}f''(z_0)+\frac{h^5}{60}f^{iv}(z_0)+\frac{h^7}{2{,}880}f^{vi}(z_0)$$
$$+\frac{457}{2880}\frac{h^9}{8!}f^{viii}(z_0)+\cdots\cdots \tag{15.24}$$

Now $I(f)-\text{AMQSH}_5(\text{f})=\frac{h^7}{20160}f^{vi}(z_0)+\frac{61}{960}\cdot\frac{h^9}{8!}f^{viii}(z_0)+\cdots\cdots$

Error of the quadrature rule $\mathrm{AMQSH}_5(\mathrm{f})$ denoted by $\mathrm{EAMQSH}_5(\mathrm{f})$ and is given by

$$\mathrm{EAMQSH}_5(\mathrm{f}) = I(f) - \mathrm{AMQSH}_5(\mathrm{f}) = \frac{h^7}{20160} f^{vi}(z_0)$$

15.2.5 Formation of quadrature rule $AMQBSH_6(f)$ of precision 7

Adding 10080 times Equation (15.24) and 7560 times of (22), we get

$$10080\left[I(f) - \mathrm{AMQSH}_5(\mathrm{f})\right] + 7560\left[I(f) - AQB_4(f)\right] = \frac{-5163}{8}\cdot\frac{h^9}{8!}.f^{viii}(z_0) + \cdots\cdots$$

$$\Rightarrow 17640\ I(f) - 10080\ \mathrm{AMQSH}_5(\mathrm{f}) + 7560\ I(f) - 7560\ AQB_4(f)$$

$$= \frac{-5163}{8}\cdot\frac{h^9}{8!}.f^{viii}(z_0) + \cdots\cdots\cdot$$

$$\Rightarrow I(f) - \left[\frac{10080\ \mathrm{AMQSH}_5(\mathrm{f}) + 7560\ AQB_4(f)}{17640}\right] = \frac{-5163}{17640\times 8}\cdot\frac{h^9}{8!}.f^{viii}(z_0) + \cdots\cdots\cdot$$

$$\text{Thus } I(f) - \left[\frac{10080\ \mathrm{AMQSH}_5(\mathrm{f}) + 7560\ AQB_4(f)}{17640}\right] \approx 0$$

$$\Rightarrow I(f) \approx \left[\frac{10080\ \mathrm{AMQSH}_5(\mathrm{f}) + 7560\ AQB_4(f)}{17640}\right] = AMQSHB_4(f) \tag{15.25}$$

Error of the quadrature rule $AMQSHB_4(f)$ is presented by $EAMQSHB_4(f)$ and is given by

$$EAMQSHB_4(f) = \frac{-5163}{17640\times 8}\cdot\frac{h^9}{8!}.f^{viii}(z_0)$$

15.3 ERROR ANALYSIS AND PRECISION ANALYSIS

In this section we presented the error analysis and precision analysis of different quadrature rules that are involved in this chapter.

Theorem 1

Let $f(z)$ be a sufficiently smooth analytic and continuous function on the interval $\left[z_0 - h,\ z_0 + h\ \right]$ with derivatives up to order eight exists. Then the quadrature rule $AMQSHB_4(f)$ is given by

$$I(f) = \int_{z_0-h}^{z_0+h} f(z)\ dz \cong \left[\frac{10080\ \mathrm{AMQSH}_5(\mathrm{f}) + 7560\ AQB_4(f)}{17640}\right] = AMQSHB_4(f)$$

where the component quadrature rules are given by

$$\mathrm{AMQSH}_5(\mathrm{f}) = \frac{768\ AMQH(f) + 42AQS_3(f)}{810}$$

$$AMQH(f) = \frac{h}{3}\left[2f\left(z_0 - \frac{3h}{4}\right) + 2f\left(z_0 + \frac{3h}{4}\right) + 2f\left(z_0 - \frac{h}{4}\right) + 2f\left(z_0 + \frac{h}{4}\right) - f\left(\frac{-h}{2} + z_0\right) - f\left(\frac{h}{2} + z_0\right)\right]$$

$$AQS_3(f) = \frac{h}{3}\left[f(z_0 - h) + 4f(z_0) + f(z_0 + h)\right]$$

$$AQB_4(f) = \frac{h}{45}\left[7f(z_0 - h) + 32f\left(z_0 - \frac{h}{2}\right) + 12f(z_0) + 32f\left(z_0 + \frac{h}{2}\right) + 7f(z_0 + h)\right]$$

Then, the error for the quadrature rule $AMQSHB_4(f)$ is given by

$$EAMQSHB_4(f) = \frac{-5163}{17640 \times 8} \cdot \frac{h^9}{8!} \cdot f^{viii}(z_0)$$

Theorem 2

Let $M = max_{z \in [z_0 - h,\ z_0 - h\ ]} \left|f^{viii}(z_0)\right|$ which represents the maximum of the absolute value of the eighth derivative of $f(z)$ on the interval. Then we can bound the error as follows:

$$\left|EAMQSHB_4(f)\right| \le \frac{5163\ h^9}{5{,}68{,}99{,}58{,}400} M$$

Proof
Proof follows from theorem 1.

Theorem 3

Let $f(z)$ be a sufficiently smooth analytic and continuous function on the interval $\left[z_0 - h,\ z_0 - h\ \right]$ with derivatives up to order eight exists. Then the quadrature rule $AMQSHB_4(f)$ has degree of precision 7. Where degree of precision of $\mathrm{AMQSH}_5(\mathrm{f})$, $AQB_4(f)$, $AQS_3(f)$, $AMQH(f)$ are 5, 5, 3, and 3, respectively.

Proof
Proof follows from theorem 1.

15.4 NUMERICAL COMPUTATION

In this section, we assess the performance of the newly constructed quadrature rule by applying it to test functions. The goal is to evaluate the accuracy of the proposed method $AMQSHB_4(f)$ compared to quadrature methods $\mathrm{AMQSH}_5(\mathrm{f})$, $AQB_4(f)$, $AQS_3(f)$, $AMQH(f)$. We have taken the following test functions to ensure the computational efficiency of new quadrature rule $AMQSHB_4(f)$.

1. $I_1 = \int_{-i}^{i} e^z \, dz$, *Exact Value* $= 1.682941969615793\mathrm{i}$
2. $I_2 = \int_{-\frac{i}{2}}^{\frac{i}{2}} \cos z \, dz$, *Exact value* $= 1.0421906109874948\mathrm{i}$

The numerical results for each test function are summarized in Tables 15.1 and 15.2. The tables provide the error, approximate value, error value, execution time and degree of precision for each quadrature method $AMQSHB_4(f)$, $\mathrm{AMQSH}_5(\mathrm{f})$, $AQB_4(f)$, $AQS_3(f)$ and $AMQH(f)$.

The following figure (Figure 15.1) represents a comparison between the execution time and degree of precision of different quadrature rules discussed in this chapter for problem I_1.

$$Rule1 = AMQSHB_4(f),\ Rule2 = \mathrm{AMQSH}_5(\mathrm{f}),\ Rule3 = AMQH(f),$$
$$Rule4 = AQS_3(f),\ Rule5 = AQB_4(f)$$

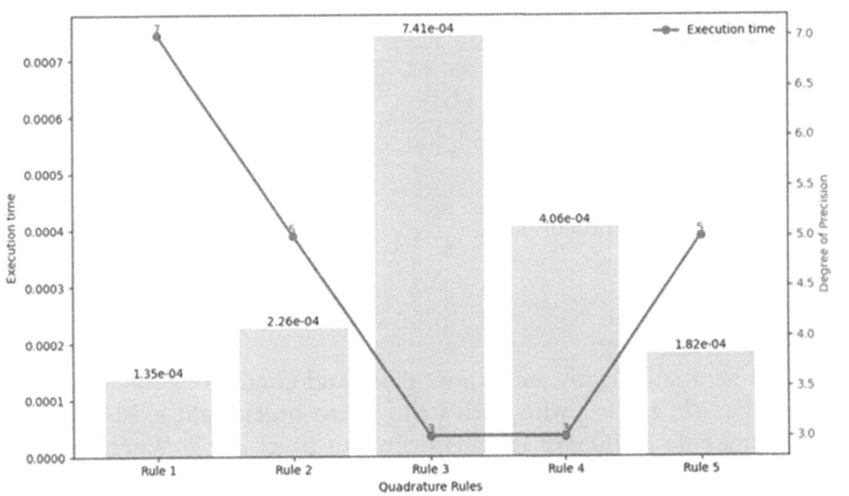

Figure 15.1 Comparison between the execution time and degree of precision of different quadrature rules discussed in this chapter for problem I_1.

Table 15.1 Computation of approximate values, error values, execution time, degree of precision for the integral $\boldsymbol{I_1} = \int_{-i}^{i} \mathbf{e}^{z}\,\mathbf{dz}$

Quadrature rules	*Exact value*	*Approximate value*	*Absolute error*	*Execution time (t)*	*Degree of precision*
$AMQSHB_4(f)$	1.682941969615793i	1.6829420696418276i	$1.0002603456626957 \times 10^{-7}$	0.00013494491577148438	7
$AMQSH_5(f)$	1.682941969615793i	1.6829900178209011i	$4.804820510795338 \times 10^{-5}$	0.0002264977	5
$AMQH(f)$	1.682941969615793i	1.6824133461857056i	0.000528623430087416	0.0007414817810058594	3
$AQS_3(f)$	1.682941969615793i	1.69353487057876i	0.010592900962967056	0.0004062652587890625	3
$AQB_4(f)$	1.682941969615793i	1.682878138736396i	$6.383087939698662 \times 10^{-5}$	0.00018215179443359375	5

Table 15.2 Computation of approximate values, error values, execution time, degree of precision for the integral $\boldsymbol{I_2} = \int_{-\frac{i}{2}}^{\frac{i}{2}} \mathbf{\cos z\, dz}$

Quadrature rules	*Exact value*	*Approximate value*	*Error value*	*Execution time (t)*	*Degree of precision*
$AMQSHB_4(f)$	1.04219061 09874948*i*	1.042190611190977i	$2.0348212004250854 \times 10^{-10}$	0.0000624657	7
$AMQSH_5(f)$	1.04219061 09874948*i*	1.0421902203743592i	$3.9061313561461475 \times 10^{-7}$	0.0002095699	5
$AMQH(f)$	1.04219061 09874948*i*	1.0421709830603405i	$1.962792715426431 3 \times 10^{-5}$	0.0017962456	3
$AQS_3(f)$	1.04219061 09874948*i*	1.0425419884021268i	0.00035137741463198147	0.0000970364	3
$AQB_4(f)$	1.04219061098 74948*i*	1.0421911322798003i	$5.212923055442076 \times 10^{-7}$	0.0001859665	5

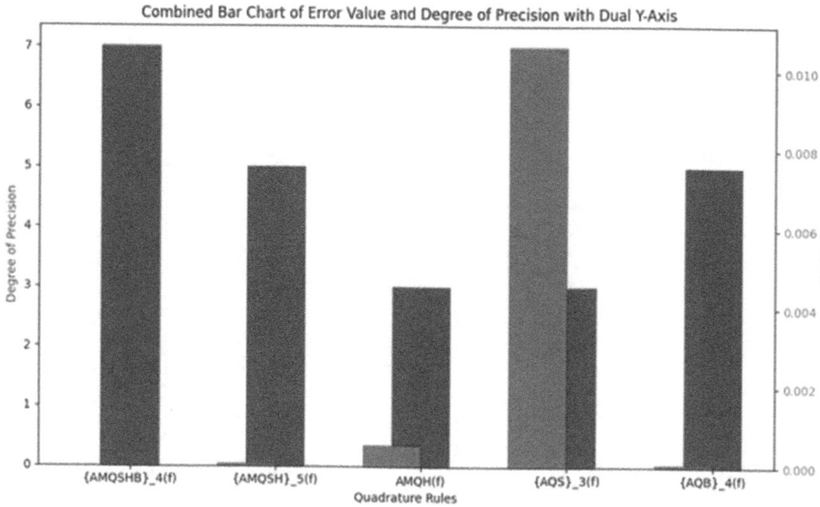

Figure 15.2 Comparison between error value and degree of precision of different quadrature rules for problem I_1.

The following graph (Figure 15.2) presents the comparison between error value and degree of precision of different quadrature rules for problem I_1.

The graph (Figure 15.3) represents a study of comparison between execution time and degree of precision of different quadrature rules for the problem I_2.

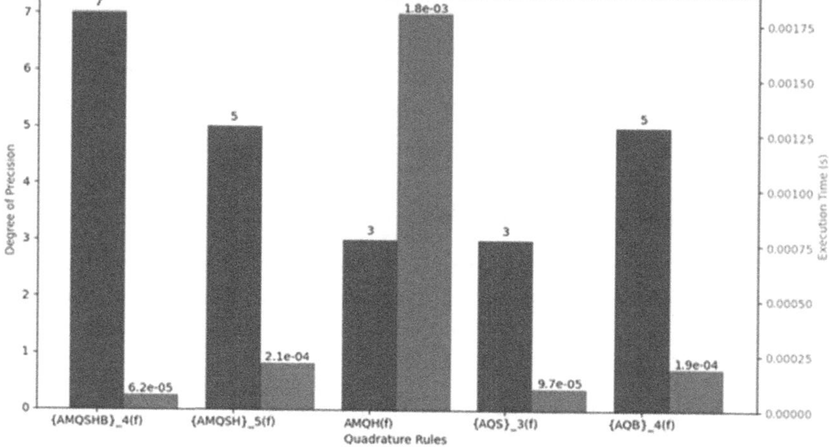

Figure 15.3 Comparison between execution time and degree of precision of different quadrature rules for the problem I_2.

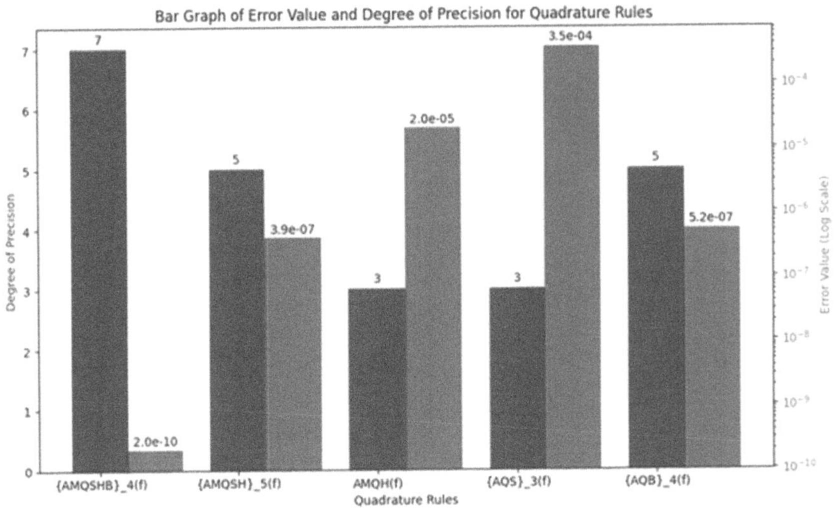

Figure 15.4 Comparison between error value and degree of precision of different quadrature rules for the problem I_2.

The graph (Figure 15.4) represents a study of comparison between error value and degree of precision of different quadrature rules for the problem I_2.

15.5 APPLICATIONS OF QUADRATURE RULES IN LAYERED STRUCTURE ANALYSIS

15.5.1 Finite Element Analysis (FEA)

Quadrature rules are used in solving weak form of governing equations of FEA, where the equation of motion is linked to vibrational analysis. Quadrature rules are used to evaluate the approximate value of these integrals over elements of structure.

15.5.1.1 Element Stiffness and Mass Matrices

The stiffness matrix and mass matrix of each finite element are determined by integration of material properties and shape functions. Quadrature rules are used in solving integrals which deal with higher order shape functions.

15.5.1.2 Time Integration in Transient Analysis

In time integration, equations of motions are integrated using quadrature rules. The Newmark and Runge–Kutta methods use the quadrature rule to solve integrations.

15.5.1.3 Integration of Material Properties

In structural analysis material properties vary from layer to layer. Therefore, quadrature rules are less important in integrating the properties about thickness of layer in composite materials. Computation of stiffness matrix of each layer by integrating across its thickness is very important in case of multi-layered plates. In such situations, quadrature rules are used to get more accurate results for these integrals.

15.5.1.4 Adaptive Quadrature

Variation of material properties across layers can be seen in complex problems. In such cases quadrature rules are suitable. Adaptive quadrature methods are performed in regions where material properties are changeable.

15.5.2 Boundary Element Method (BEM)

Displacement and stress at boundary of a given domain can be evaluated by quadrature rules in BEM. In determining dynamic behaviour of layered structures in fluid–structure interaction situations, BEM is used. It is valid when integration of boundary conditions at the interface between materials is required.

15.5.2.1 Interaction Between Solid and Fluid Layers

In fluid structure interaction, quadrature methods are used in calculating boundary integrals when fluid layer is in contact with solid structure. For better accuracy and efficiency of resultant solution, suitable quadrature rules should be chosen.

15.5.3 Wave Propagation Analysis

In the context of wave propagation analysis within layered structures, quadrature techniques are commonly employed to address wave equations.

15.5.3.1 Dispersion Relations

Wave propagation in layered media solves integrals emerging from dispersion relations. To navigate the complexities of these scenarios, numerical quadrature methods are utilized to provide approximations for these integrals, when closed-form analytical solutions may not be feasible.

15.5.3.2 Spectral Methods

Spectral methods can be advantageous for analysing wave propagation in both periodic and layered media. In these cases, quadrature rules are implemented to enhance the efficiency of computing spectral integrals.

15.5.4 Vibration and Modal Analysis

Quadrature methods are used in computing modal frequencies and shape using integrals over structure's domain.

15.6 CONCLUSION

The mixed quadrature rule of precision 7 has been constructed in this paper. Four quadrature rules have been combined by using Taylor's series in producing it. We employed two test examples in proving the efficiency of newly developed mixed quadrature rule. From numerical computation section it is concluded that new method requires less execution time as compared to other rules. Also the absolute error is very negligible as compared to other quadrature formulas. Therefore, newly formed quadrature rule is more efficient than other quadrature rules.

REFERENCES

1. B.P. Acharya, Compound Birkhoff-Young rule for numerical integration of analytic functions, *International Journal of Mathematical Education in Science and Technology*, 1983, vol. 14, no. 1, 101–103.
2. R.L. Burden, J.D Faires, *Numerical Analysis*, Cengage Learning, Ninth Edition, 2011.
3. I. Aziza, Siraj-ul-Islamb, W. Khanb, Quadrature rules for numerical integration based on Haar wavelets and hybrid functions, *Computers and Mathematics with Applications*, 2011, vol. 61, 2770–2781.
4. L. Ma, The Taylor's theorem and its application, *Highlights in Science, Engineering and Technology*, 2023, vol. 72, 1368–1372.
5. J. Zhoua, P.J. Barendrecht, M. Barton, J. Kosinka, Numerical quadrature for Gregory quads, *Applied Mathematics and Computation*, 2023, vol. 453, 1–14.
6. H. Xiaoa, Z. Gimbutas, Numerical algorithm for the construction of efficient quadrature rules in two and higher dimensions, *Computers and Mathematics with Applications*, 2010, vol. 59, 663–676.
7. P.A.A. Magalhaes, P.A.A. Magalhaes Junior, C.A Magalhaes, A.L.M.A. Magalhaes, New formulas of numerical quadrature using spline interpolation, *Archives of Computational Methods in Engineering*, 2021, vol. 28, 553–576.
8. S. Mahesar, M.M. Shaikh, M.S. Chandio, A.W. Shaikh, Some new time and cost efficient quadrature formulas to compute integrals using derivatives with error analysis, *Symmetry*, 2022, 14, no. 12, 2611.
9. E. Denich, P. Novati, Numerical quadrature for integrals involving oscillating functions, *Journal of Approximation Software*, 2024, vol. 1, 1–14.
10. L. Fermo, L. Reichel, G. Rodriguez, M.M. Spalevi, Averaged Nyström interpolants for the solution of Fredholm integral equations of the second kind, *Applied Mathematics and Computation*, 2024, vol. 467.

Index

For Product Safety Concerns and Information please contact our EU representative GPSR@taylorandfrancis.com Taylor & Francis Verlag GmbH, Kaufingerstraße 24, 80331 München, Germany